IMAGE AND VIDEO COMPRESSION STANDARDS
Algorithms and Architectures

Second Edition

Books are to be returned on or before
the last date below.

IMAGE AND VIDEO COMPRESSION STANDARDS
Algorithms and Architectures

Second Edition

by

Vasudev Bhaskaran
Konstantinos Konstantinides
Hewlett-Packard Laboratories

KLUWER ACADEMIC PUBLISHERS
Boston / Dordrecht / London

Distributors for North, Central and South America:
Kluwer Academic Publishers
101 Philip Drive
Assinippi Park
Norwell, Massachusetts 02061 USA
Telephone (781) 871-6600
Fax (781) 871-6528
E-Mail <kluwer@wkap.com>

Distributors for all other countries:
Kluwer Academic Publishers Group
Distribution Centre
Post Office Box 322
3300 AH Dordrecht, THE NETHERLANDS
Telephone 31 78 6576 000
Fax 31 78 6576 254
E-Mail <orderdept@wkap.nl>

 Electronic Services <http://www.wkap.nl>

Library of Congress Cataloging-in-Publication Data

Bhaskaran, Vasudev, 1956-
   Image and video compression standards : algorithms and
architectures / by Vasudev Bhaskaran, Konstantinos Konstantinides. -
- 2nd ed.
    p. cm. -- (The Kluwer international series in engineering and
computer science ; SECS 408)
  Includes bibliographical references and index.
  ISBN 0-7923-9952-8 (alk. paper)
  1. Image compression. 2. Video compression. I. Konstantinides,
Konstantinos, 1956-   . II. Title. III. Series.
TA1632.B49   1997
621.36'7--dc21
                                                  97-17141
                                                    CIP

Printed on acid-free paper.

Printed in the United Kingdom

This printing is a digital duplication of the original edition.

to Achan, Chechi,
Balu, Kishore, Shanti, Paula,
Anastasia, Anneta, and Niki

CONTENTS

PREFACE

This book presents an introduction to the algorithms and architectures that form the underpinnings of the image and video compression standards, including JPEG (compression of still-images), H.261 and H.263 (video teleconferencing), and MPEG-1 and MPEG-2 (video storage and broadcasting). The next generation of audiovisual coding standards, such as MPEG-4 and MPEG-7, are also briefly described. In addition, the book covers the MPEG and Dolby AC-3 audio coding standards and emerging techniques for image and video compression, such as those based on wavelets and vector quantization.

The emphasis of the book is on the foundations of these standards, namely, techniques such as predictive coding, transform-based coding, motion estimation, motion compensation, and entropy coding, as well as how they are applied in the standards. We avoid the implementation details of each standard. However, we do provide all the material necessary to understand the workings of each of the compression standards, including information that can be used by the reader to evaluate the efficiency of various software and hardware implementations conforming to these standards. We place particular emphasis on those algorithms and architectures that have been found to be useful in practical software or hardware implementations.

No prior knowledge of image and video compression theory and architectures is assumed; however, some background in high-level hardware design is expected for a better understanding of the chapters covering the hardware implementation of the standards.

Chapters 1 to 4 review the fundamentals of image and video compression. In Chapter 1 we cover the key principles of image compression and explain the basic terminology. We present a compression taxonomy and discuss design issues and tradeoffs in selecting a compression algorithm. In Chapter 2 we review methods and standards for lossless compression, including differential, Huffman, and arithmetic coding. We present several practical algorithms for the design of Huffman codes, and we describe the current lossless JPEG algorithm, work in progress on a new lossless and near-lossless JPEG standard (JPEG-LS), and the standards for lossless facsimile transmission (Group 3, Group 4, and JBIG).

In Chapter 3 we examine techniques for lossy compression. We review the source-coding theorem and compute the rate-distortion function for several image models relevant to image compression. We place particular emphasis on DCT-based coding, and describe several schemes for the efficient implementation of the DCT. In Chapter 4 we present the fundamentals of video coding, including motion estimation and motion compensation.

In Chapter 5 we describe lossy JPEG and its application to the compression of still-images. We cover recent extensions to JPEG and the JPEG-based standard for color facsimile. In Chapter 6 we describe the coding algorithms and syntax for the MPEG-1 and MPEG-2 video compression standards, the basic video coding aspects of MPEG-4, and objectives and target application areas for MPEG-7. In Chapter 7 we describe the H.320 and H.324 video teleconferencing standards, with emphasis on their video coding parts, namely, H.261 and H.263. Similarities and differences among the video compression standards are also presented and reviewed.

Next, we move to the hardware implementation of the standards. In Chapter 8 we define a measure of computational complexity, and we use it to measure the complexity of various algorithms used in the standards. Bearing in mind that most hardware implementations of the standards are based on either RISC or DSP cores, in Chapter 9 we review the principles of RISC and DSP designs and describe their main differences and similarities. Even though we present the latest designs in image and video processors, we place more emphasis on the designs of the core algorithms; that is, the discrete cosine transform (Chapter 10), motion estimation (Chapter 11), and entropy coding (Chapter 12). In Chapter 13 we describe architectures for the JPEG standard, and in Chapter 14 we discuss programmable and dedicated video processors for MPEG and videoconferencing standards. Key features of recent VLIW media processors are also described.

In Chapter 15 we review recent developments in the design of multimedia-enhanced general-purpose RISC processors, such as the PA-8000 from Hewlett-Packard, the UltraSPARC from Sun Microsystems, and Intel's MMX-based Pentium processors. In Chapter 16 we present the key standards on audio coding, and we describe in more detail the MPEG and Dolby AC-3 coding schemes. In Chapter 17 we provide an overview of techniques based on vector quantization and wavelets. Such techniques are used in many proprietary image and video compression schemes and may play a significant role in future standards.

Given the wide use of color imagery as input to image and video coders, Appendix A provides a quick reference to basic color transformations. Appendix B provides a summary of common abbreviations used in the text and can be used as a quick reference for the various standard image formats (SIF, CIF, and CCIR 601). Information about sites with public software implementations of the standards is included in Appendix C.

The material in this book is partitioned into the following categories: fundamentals, video standards, architectures, audio standards, and emerging techniques. However, most of the chapters are self contained and allow the reader to easily switch from one category to another.

This book is aimed at the professional who wants to have a basic understanding of the latest developments and applications of image compression standards. It is based on tutorials given by the authors at several forums, including the IEEE International Conference on Acoustics Speech and Signal Processing, the ACM Multimedia Conference, and Hewlett-Packard. It provides a reference for any engineer planning to work in this field, either in basic implementation or in research and development. It is intended for self-study; however, it can also be used as a companion textbook in any course on data compression, video coding, computer architecture, signal or image processing, or the design of signal and video processors.

New topics in the second edition: Since the publication of the first edition, we have seen a number of new developments in standards-related activities. The original JPEG and MPEG standards have been refined, and several new compression standards have emerged. In lossless compression, we introduce JPEG-LS, a new standard for lossless and near-lossless compression of still-images. In video coding, we describe two new MPEG-2 profiles: 4:2:2 and multiview. Activities in two new MPEG standards, MPEG-4 and MPEG-7, are also discussed here. In videoconferencing, we describe the new H.263 standard and compare it with H.261. New developments in fast motion-estimation algorithms, a major component of all MPEG and videoconferencing standards, are also presented. In audio coding, new standards, such as MPEG-2 AAC and G.723.1, are also discussed. Since the first edition, there have been significant advances in the hardware support for multimedia and compression standards. We describe enhancements to several general purpose processors, including the Intel MMX and the newly emerging class of media processors. The information on commercially available audio and video coders has also been updated.

Acknowledgments: This book would not be possible without the support from the management at Hewlett-Packard Laboratories, especially Fred L. Kitson, Daniel T. Lee, and Ho John Lee. We also thank our colleagues at Hewlett-Packard for many helpful discussions. Balas K. Natarajan has been particularly helpful from the beginning of this project and we thank him for many valuable discussions, comments, and for reviewing parts of this manuscript. The review of our sections on Golomb-Rice coding and JPEG-LS by Guillermo Sapiro and Gadiel Seroussi is gratefully acknowledged. We thank Bob Rau for his comments on our tutorials and Ruby Lee for providing us with a different perspective on the hardware implementation of the video standards. We thank Nariman Farvardin and David Daut for their comments on the computation of the rate-distortion function. We thank Akio Yamamoto for his help in the example on wavelet coding, José Fridman for many valuable comments and suggestions, Masoud Khansari for his comments on the video teleconferencing standards, and Elias Manolakos. Feedback from the readers of the first edition of the book has been extremely valuable in helping us to improve the clarity of the presentation and to select new material. We are grateful to all of them, and especially to Thanos Skordas and Aggelos Katsaggelos, for their suggestions and corrections.

We also thank Robert Holland, Jr. and his staff from Kluwer Academic Press for their support, and Suzanne M. Rumsey for helping with many typesetting questions. Finally, we dedicate this book to our teachers and to our families, and especially to Achan, Chechi, Balu, Kishore, Shanti, Paula, Anastasia, Anneta, and Niki.

1

COMPRESSION FUNDAMENTALS

1.1 INTRODUCTION

In recent years, there have been significant advancements in algorithms and architectures for the processing of image, video, and audio signals. These advancements have proceeded along several directions. On the algorithm front, new techniques have led to the development of robust methods to reduce the size of the image, video, or audio data. Such methods are extremely vital in many applications that manipulate and store digital data. Informally, we refer to the process of size reduction as a compression process. We will define this process in a more formal way later. On the architecture front, it is now feasible to put sophisticated compression processes on a relatively low-cost single chip; this has spurred a great deal of activity in developing multimedia systems for the large consumer market.

One of the exciting prospects of such advancements is that multimedia information comprising image, video, and audio has the potential to become just another data type. This usually implies that multimedia information will be digitally encoded so that it can be manipulated, stored, and transmitted along with other digital data types. For such data usage to be pervasive, it is essential that the data encoding be standard across different platforms and applications. This will foster widespread development of applications and will also promote interoperability among systems from different vendors. Furthermore, standardization can lead to the development of cost-effective implementations, which in turn will promote the widespread use of multimedia information. This is the primary motivation behind the emergence of image and video compression standards.

1.2 BACKGROUND

Compression is a process intended to yield a compact *digital representation* of a signal. In the literature, the terms *source coding, data compression, bandwidth compression,* and *signal compression* are all used to refer to the process of compression. In the cases where the signal is defined as an image, a video stream, or an audio signal, the generic problem of compression is to minimize the bit rate of their digital representation. There are many applications that benefit when image, video, and audio signals are available in compressed form. Without compression, most of these applications would not be feasible.

Example 1: Let us consider facsimile image transmission. In most facsimile machines, the document is scanned and digitized. Typically, an 8.5 × 11 inch page is scanned at 200 dpi, thus resulting in 3.74 Mbits. Transmitting this data over a low-cost 14.4 kbits/s modem would require 5.62 minutes. With compression, the transmission time can be reduced to 17 seconds. This results in substantial savings in transmission costs.

Example 2: Let us consider a video-based CD-ROM application. Full-motion video, at 30 frames per second (fps) and a 720 × 480 resolution, generates data at 20.736 Mbytes/s. At this rate, only 31 seconds of video can be stored on a 650 MByte CD-ROM. Compression technology can increase the storage capacity to 74 minutes, for VHS-grade video quality.

There are many other applications that benefit from data compression technology. Table 1.1 lists a representative set of such applications for image, video, and audio data, as well as typical data rates of the corresponding compressed bit streams. Typical data rates for the uncompressed bit streams are also shown.

Image, video, and audio signals are amenable to compression due to the following factors:

Application	Data Rate	
	Uncompressed	**Compressed**
Voice *8 ksamples/s, 8 bits/sample*	64 kbps	2 - 4 kbps
Slow-motion video (10 fps) *framesize 176 × 120, 24 bits/pixel*	5.07 Mbps	8 - 16 kbps
Audio conference *8 ksamples/s, 16 bits/sample*	128 kbps	6 - 64 kbps
Video conference (15 fps) *framesize 352 × 240, 24 bits/pixel*	30.41 Mbps	64 - 768 kbps
Digital audio (stereo) *44.1 ksamples/s, 16 bits/sample*	1.5 Mbps	128 - 768 kbps
Video file transfer (15 fps) *framesize 352 × 240, 24 bits/pixel*	30.41 Mbps	384 kbps
Digital video on CD-ROM (30 fps) *framesize 352 × 240, 24 bits/pixel*	60.83 Mbps	1.5 - 4 Mbps
Broadcast video (30 fps) *framesize 720 × 480, 24 bits/pixel*	248.83 Mbps	3 - 8 Mbps
HDTV (59.94 fps) *framesize 1280 × 720, 24 bits/pixel*	1.33 Gbps	20 Mbps

Table 1.1   Applications for image, video, and audio compression.

■   There is considerable statistical redundancy in the signal.

1. Within a single image or a single video frame, there exists significant correlation among neighbor samples. This correlation is referred to as *spatial correlation*.

2. For data acquired from multiple sensors (such as satellite images), there exists significant correlation among samples from these sensors. This correlation is referred to as *spectral correlation*.

3. For temporal data (such as video), there is significant correlation among samples in different segments of time. This is referred to as *temporal correlation*.

■   There is considerable information in the signal that is irrelevant from a perceptual point of view.

■ Some data tend to have high-level features that are redundant across space and time, that is, the data is of a fractal nature.

For a given application, compression schemes may exploit any one or all of the above factors to achieve the desired compression data rate.

A systems view of the compression process is depicted in Figure 1.1. The core

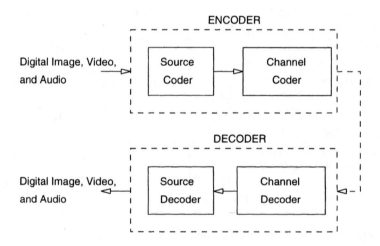

Figure 1.1 Generic compression system.

of the encoder is the source coder. The source coder performs the compression process by reducing the input data rate to a level that can be supported by the storage or transmission medium. The bit rate output of the encoder is measured in bits per sample or bits per second. For image or video data, a pixel is the basic element; thus, bits per sample is also referred to as bits per pixel or bits per pel. In the literature, the term *compression ratio*, denoted as c_r, is also used instead of *bit rate* to characterize the capability of the compression system. An intuitive definition of c_r is

$$c_r = \frac{source\ coder\ input\ size}{source\ coder\ output\ size}. \tag{1.1}$$

This definition is somewhat ambiguous and depends on the data type and the specific compression method that is employed. For a still-image, size could refer to the bits needed to represent the entire image. For video, size could refer to the bits needed to represent one frame of video. Many compression methods for video do not process each frame of video; hence, a more commonly used notion for size is the bits needed to represent one second of video.

In a practical system, the source coder is usually followed by a second level of coding: the channel coder (Figure 1.1). The channel coder translates the compressed bit stream into a signal suitable for either storage or transmission. In most systems, source coding and channel coding are distinct processes. In recent years, methods to perform combined source and channel coding have also been developed. Note that, in order to reconstruct the image, video, or audio signal, one needs to reverse the processes of channel coding and source coding. This is usually performed at the decoder.

From a systems design viewpoint, one can restate the compression problem as a bit rate minimization problem, where several constraints may have to be met, including the following:

- Specified levels of signal quality. This constraint is usually applied at the decoder.

- Implementation complexity. This constraint is often applied at the decoder, and in some instances at both the encoder and the decoder.

- Communication delay. This constraint refers to the end-to-end delay, and is measured from the start of encoding a sample to the complete decoding of that sample.

Note that these constraints have different importance in different applications. For example, in a two-way teleconferencing system, the communication delay might be the major constraint, whereas in a television broadcasting system, signal quality and decoder complexity might be the main constraints.

1.3 COMPRESSION TAXONOMY

A typical classification of compression methods is shown in Figure 1.2.

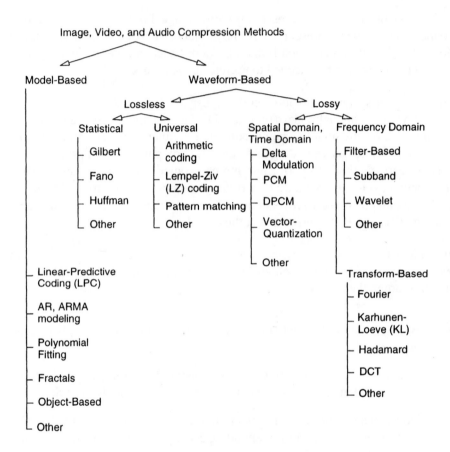

Figure 1.2   A taxonomy of image, video, and audio compression methods.

1.3.1   Lossless versus Lossy Compression

Lossless compression

In many applications, the decoder has to reconstruct the original data without any loss. For a lossless compression process, the reconstructed data and the original data must be identical in value for each and every data sample. This is also referred to as a reversible process. In lossless compression, for a specific application, the choice of a compression method involves a tradeoff along the three dimensions depicted in Figure 1.3, that is, coding efficiency, coding complexity, and coding delay.

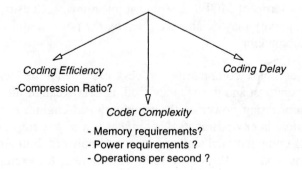

Figure 1.3   Tradeoffs in lossless compression.

Coding Efficiency: This is usually measured in bits per sample or bits per second (bps). Coding efficiency is usually limited by the information content or *entropy* of the source. In intuitive terms, the entropy of a source X provides a measure for the "randomness" of X. From a compression theory point of view, sources with large entropy are more difficult to compress (for example, random noise is very hard to compress).

In literature on lossless compression, the coding efficiency is sometimes characterized by the entropy at the output of a preprocessor within the source coder; the implication here is that the source coder does not increase this entropy figure significantly. Note that, in practice, the source is composed from a finite-alphabet; thus, the entropy is finite. For an analog source, the entropy would be infinite; however, in most compression schemes, digitization of the analog source leads to a finite-alphabet representation, which limits the source entropy.

Coding Complexity: The complexity of a compression process is analogous to the computational effort needed to implement the encoder and decoder functions. The computational effort is usually measured in terms of memory requirements and number of arithmetic operations. The operations count is characterized by the term millions of operations per second and is often referred to as MOPS. Here, by operation, we imply a basic arithmetic operation that is supported by the computational engine. In the compression literature, the term MIPS (millions of instructions per second) is sometimes used. This is specific to a computational engine's architecture; thus, in this text we refer to coding complexity in terms of MOPS. In some applications, such as portable devices, coding complexity may be characterized by the power requirements of a hardware implementation.

Coding Delay: A complex compression process often leads to increased coding delays at the encoder and the decoder. Coding delays can be alleviated by increasing the processing power of the computational engine; however, this may be impractical in environments where there is a power constraint or when the underlying computational engine cannot be improved. Furthermore, in many applications, coding delays have to be constrained; for example, in interactive communications. The need to constrain the coding delay often forces the compression system designer to use a less sophisticated algorithm for the compression processes.

From this discussion, it can be concluded that these tradeoffs in coding complexity, delay, and efficiency are usually limited to a small set of choices along these axes. In a subsequent section, we will briefly describe the tradeoffs within the context of specific lossless compression methods.

Lossy compression

The majority of the applications in image or video data processing do not require that the reconstructed data and the original data be identical in value. Thus, some amount of loss is permitted in the reconstructed data. A compression process that results in an imperfect reconstruction is referred to as a lossy compression process. This compression process is irreversible. In practice, most irreversible compression processes rapidly degrade the signal quality when they are repeatedly applied on previously decompressed data.

The choice of a specific lossy compression method involves tradeoffs along the four dimensions shown in Figure 1.4. Due to the additional degree of freedom, namely, in the signal quality, a lossy compression process can yield higher compression ratios than a lossless compression scheme.

Signal Quality: This term is often used to characterize the signal at the output of the decoder. There is no universally accepted measure for signal quality.

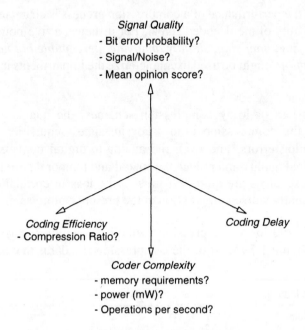

Figure 1.4   Tradeoffs in lossy compression.

One measure that is often cited is the signal-to-noise ratio (SNR), which can be expressed as

$$SNR = 10\,log_{10}\,\frac{encoder\ input\ signal\ energy}{noise\ signal\ energy}. \qquad (1.2)$$

The noise signal energy is defined as the energy measured for a hypothetical signal that is the difference between the encoder input signal and the decoder output signal. Note that SNR as defined here is given in decibels (dB). In the case of images or video, PSNR (peak signal-to-noise ratio) is used instead of SNR. The calculations are essentially the same as in the case of SNR; however, in the nominator, instead of using the encoder input signal, one uses a hypothetical signal with a signal strength of 255 (the maximum decimal value of an unsigned 8-bit number, such as in a pixel).

High SNR or PSNR values do not always correspond to signals with perceptually high quality. Another measure of signal quality is the mean opinion

score, where the performance of a compression process is characterized by the
subjective quality of the decoded signal. For instance, a five-point scale such
as *very annoying, annoying, slightly annoying, perceptible but not annoying,*
and *imperceptible* might be used to characterize the impairments in the decoder
output.

In either lossless or lossy compression schemes, the quality of the input
data affects the compression ratio. For instance, acquisition noise, data
sampling timing errors, and even the analog-to-digital conversion process
affect the signal quality and reduce the spatial and temporal correlation. Some
compression schemes are quite sensitive to the loss in correlation and may
yield significantly worse compression in the presence of noise.

Both lossless and lossy compression methods fit within the general model
depicted in Figure 1.5. Most of the compression standards that are described

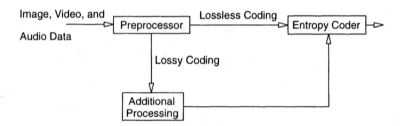

Figure 1.5   Lossless and lossy compression coding framework.

in this book fit within this framework and employ both lossy and lossless
compression schemes to achieve high coding efficiency. These compression
methods will be described in detail in later chapters.

1.3.2   Variable Bit Rate versus Constant Bit Rate

From Figure 1.3, the tradeoffs employed in the selection of a lossless compression method can lead to several types of coder designs. In one approach,
one can envision a coder where the coding delay is fixed. This could lead to a
lossless coder where the coding efficiency fluctuates from sample to sample,
but the output symbols are delivered at a constant rate. On the other hand,

if coding delay is not critical, one can process multiple samples at a time to obtain a higher coding efficiency. In this case, the encoder output datasize is fixed, but output symbols are delivered at irregular time intervals. Thus, depending on the application environment, one can obtain either fixed length symbols at irregular time intervals or variable length symbols at constant time intervals. Note that, using the generic coder structure shown in Figure 1.1, it is possible to convert a variable length *source-coder* output bit stream to a fixed length *channel-coder* output bit stream.

In the case of lossy compression, the tradeoffs shown in Figure 1.4 can lead to several classes of coders. In one approach, the coding efficiency could be constrained to a fixed value. For such coders, the signal quality would tend to fluctuate. Such a coder might be required in a communication system where the transmission rate cannot exceed a specified value. Another application of such coders is in storage applications with fixed storage capacity. These coders are referred to as *constant bit rate* coders. If the transmission or storage requirements are not fixed and if the user desires specific quality at the output of the decoder, then the coding efficiency can be allowed to fluctuate. These coding schemes are referred to as *constant quality* or *variable bit rate* coders.

1.3.3   Single- or Multiple-Sample-Based Compression

Within the compression taxonomy shown in Figure 1.2, one can classify the coding schemes as single- or multiple-sampled-based approaches. Single-sample-based compression schemes process one input sample at a time and generate a compressed representation for each input sample. In the case of image or video data, such schemes are referred to as *pixel-based* or *pel-based* schemes. Multiple-sample-based schemes process several input samples at a time and generate one or several compressed representations for the entire set of input samples. For image or video data, such compression methods are referred to as *block-based* schemes. The benefit of block-based compression over pel-based compression is well founded in Shannon's coding theorem, where it is shown that if the the blocklength increases, then the coding efficiency can reach its theoretical limit. However, block-based schemes tend to have higher complexity than pel-based schemes.

1.3.4   Comparison of Compression Domains

Compression processes can exploit redundancies in the data across space, time, or frequency. Single-sample-based compression schemes are often spatial-domain or time-domain based; in this case, there are no inherent bit savings if frequency domain approaches are adopted. Block-based compression schemes fit into three main categories: spatial-domain-based, time-domain-based, and frequency-domain-based. If a group of samples is highly correlated in the spatial domain, then it tends to also have a very compact frequency-domain representation; thus, a frequency-domain-based compression process is preferred for such data sets. For multidimensional data, such as video, that have both spatial and temporal components, a hybrid spatial- and frequency-domain approach is adopted. Such hybrid techniques are the basis of all the image and video compression standards which will be discussed in subsequent chapters of this text.

1.4   ISSUES IN COMPRESSION METHOD SELECTION

In this chapter, we have introduced some fundamental concepts related to image, video, and audio compression. When choosing a specific compression method, one should consider the following issues:

- Lossless or lossy. This is usually dictated by the coding efficiency requirements.

- Coding efficiency. Even in a lossy compression process, the desirable coding efficiency might not be achievable. This is especially the case when there are specific constraints on output signal quality.

- Variability in coding efficiency. In some applications, large variations in coding efficiency among different data sets may not be acceptable.

- Resilience to transmission errors. Some compression methods are more robust to transmission errors than others. If retransmissions are not permitted, then this requirement may impact on the overall encoder-decoder design.

- Complexity tradeoffs. In most implementations, it is important to keep the overall encoder-decoder complexity low. However, certain applications may require only a low decoding complexity.

- Nature of degradations in decoder output. Lossy compression methods introduce artifacts in the decoded signal. The nature of artifacts depends on the compression method that is employed. The degree to which these artifacts are judged objectionable also varies from application to application. In communication systems, there is often an interplay between the transmission errors and the coding artifacts introduced by the coder. Thus, it is important to consider all types of errors in a system design.

- Data representation. In many applications, there is a need to support two decoding phases. In the first phase, decoding is performed to derive an intelligible signal; this is the case in data browsing. In the second phase, decoding is performed to derive a higher quality signal. One can generalize this notion to suggest that some applications require a hierarchical representation of the data. In the compression context, we refer to such compression schemes as *scalable compression methods*. The notion of scalability has been adopted in the compression standards that are described later in the text.

- Multiple usage of the encoding-decoding tandem. In many applications, such as video editing, there is a need to perform multiple encode-decode operations using results from a previous encode-decode operation. This is not an issue for lossless compression; however, for lossy schemes, resilience to multiple encoding-decoding cycles is essential.

- Interplay with other data modalities, such as audio and video. In a system where several data modalities have to be supported, the compression methods for each modality should have some common elements. For instance, in an interactive videophone system, the audio compression method should have a frame structure that is consistent with the video frame structure. Otherwise, there will be unnecessary requirements on buffers at the decoder and a reduced tolerance to timing errors.

- Interworking with other systems. In a mass-market environment, there will be multiple data modalities and multiple compression systems. In such an environment, transcoding from one compression method to another may be needed. For instance, video editing might be done on a frame-by-frame basis; hence, a compression method that does not exploit temporal redundancies might be used here. After video editing, there might be a need to broadcast this video. In that case, temporal redundancies could be exploited to achieve a higher coding efficiency. In such a scenario, it is important to select compression methods that support transcoding from one compressed stream format to another. Interworking is important in many communications environments as well.

1.5   TO PROBE FURTHER

In this chapter, we have briefly reviewed some of the basic concepts associated with the compression of image, video, and audio signals. Additional details, including descriptions of specific compression methods, can be found in Jayant [120]. Some specific applications that are based on the concepts described in this chapter can be found in Chapter 2 of [188]. These basic concepts will be refined in subsequent chapters as we describe specific compression methods for image, video, and audio.

2

METHODS AND STANDARDS FOR LOSSLESS COMPRESSION

2.1 INTRODUCTION

Lossless compression refers to compression methods for which the original uncompressed data set can be recovered exactly from the compressed stream. The need for lossless compression arises from the fact that many applications, such as the compression of digitized medical data, require that no loss be introduced from the compression method. Bitonal image transmission via a facsimile device also imposes such requirements. In recent years, several compression standards have been developed for the lossless compression of such images. We discuss these standards later in this chapter. In general, even when lossy compression is allowed, the overall compression scheme may be a combination of a lossy compression process followed by a lossless compression process as depicted in Figure 1.5. Various image, video, and audio compression standards follow this model, and several of the lossless compression schemes used in these standards are described in this chapter.

2.2 PRELIMINARIES

Consider a source S that generates random symbols s_1, s_2, $\cdots$, s_N. For example, S may be a digital image, and s_i may represent one of N possible pixel values. If p_i denotes the probability of occurrence for symbol s_i, then

$$I(s_i) = log \frac{1}{p_i} = -log\, p_i \qquad (2.1)$$

is defined as the *self-information* for symbol s_i, that is, the information we get from receiving s_i. If the base of the logarithm is two, then the self-information is measured in *bits*. If the base is ten, then self-information is measured in *nats* (natural digits). For the remainder of this book, we always assume that information is measured in bits per symbol.

According to Shannon, the average information or entropy of a source S is defined as

$$H(S) = \sum_i p_i \, log_2 \frac{1}{p_i} \, . \qquad (2.2)$$

From information theory, if the symbols are distinct, then the average number of bits needed to encode them is always bounded by their entropy. In practice, the entropy of a source is in general unknown, and estimates for the entropy depend on the *probability model* we adopt for the structure of the symbols.

For example, consider the following sequence of symbols

$$4\,5\,8\,6\,4\,7\,8\,9\,4\,8\,.$$

In one probability model, we can assume that our source generates six distinct symbols ($s_1 = 4$, $s_2 = 5$, $s_3 = 6$, $s_4 = 7$, $s_5 = 8$, and $s_6 = 9$) and that each symbol is generated with equal probability, that is, $p_i = \frac{1}{6}$, for $i = 1, 2, \cdots, 6$. From the definition of entropy (2.2), under this model

$$H(S) = 6 \times \frac{1}{6} log_2 6 = \frac{log_{10} 6}{log_{10} 2} = 2.585. \qquad (2.3)$$

In other words, using this model, we cannot find a coding scheme for this sequence that can code better than 2.585 bits per sample.

In a different model, we can assume that the probability of each symbol corresponds to the number of occurrences of that symbol in our sequence. Then, for the symbols defined before, $p_1 = p_3 = \frac{3}{10}$, and $p_2 = p_4 = p_5 = p_6 = \frac{1}{10}$. Using this model, $H(S) = 2.371$.

The general model of a lossless compression scheme is as depicted in Figure 2.1. Given an input set of symbols, a modeler generates an estimate of the probability distribution of the input symbols. This probability model is then used to map symbols into codewords. The combination of the modeling and the symbol-to-codeword mapping functions is usually referred to as *entropy*

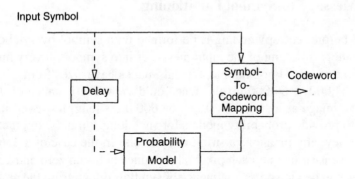

Figure 2.1   A generic model for lossless compression.

coding. The key idea of entropy coding is to use short codewords for symbols that occur with high probability and long codewords for symbols that occur with low probability.

The probability model can be derived either from the input data or from a priori assumptions about the data. Note that, for decodability, the same model must also be generated by the decoder. Thus, if the model is dynamically estimated from the input data, causality constraints require a delay function between the input and the modeler. If the model is derived from a priori assumptions, then the delay block is not required; furthermore, the model function need not have access to the input symbols. The probability model does not have to be very accurate, but the more accurate it is, the better the compression will be. Note that, compression is not always guaranteed. If the probability model is wildly inaccurate, then the output size may even expand. However, even then the original input can be recovered without any loss.

Decompression is performed by reversing the flow of operations shown in Figure 2.1. This decompression process is usually referred to as *entropy decoding*.

2.2.1   Message-to-Symbol Partitioning

As noted before, entropy coding is performed on a symbol-by-symbol basis. Appropriate partitioning of the input messages into symbols is very important for efficient coding. For example, typical images have sizes from 256 × 256 pixels to 64,000 × 64,000 pixels. One could view one instance of, say, the 256 × 256 image as a single message, 64,000 units long; however, it is very difficult to provide probability models for such long symbols. In practice, we typically view any image as a string of symbols. In the case of a 256 × 256 image, if we assume that each pixel takes values between zero and 255, then this image can be viewed as a sequence of symbols drawn from the alphabet 0, 1, 2,.., 255. The modeling problem now reduces to finding a good probability model for the 256 symbols in this alphabet.

For some images, one might partition the data set even further. For instance, if we have an image with 12 bits per pixel, then this image can be viewed as a sequence of symbols drawn from the alphabet 0, 1,..., 4,095. Hardware and/or software implementations of the lossless compression methods may require that data be processed in 8-, 16-, 32-, or 64-bit units. Thus, one approach might be to take the stream of 12-bit pixels and artificially view it as a sequence of 8-bit symbols. In this case, we have reduced the alphabet size. This reduction compromises the achievable compression ratio; however, the data are matched to the processing capabilities of the computing element.

Other data partitions are also possible; for instance, one may view the data as a stream of 24-bit symbols. This approach may result in higher compression since we are combining two pixels into one symbol. In general, the partitioning of the data into blocks, where a block is composed of several input units, may result in higher compression ratios, but also increases the coding complexity.

2.2.2   Differential Coding

Another preprocessing technique that improves the compression ratio is differential coding. Differential coding skews the symbol statistics so that the resulting distribution is more amenable to compression. Image data tend to have strong interpixel correlation. If, say, the pixels in the image are in the order $x_1, x_2, x_3, ..., x_N$, then instead of compressing these pixels, one might process the sequence of differentials $y_i = x_i - x_{i-1}$, where $i = 1, 2, ..., N$, and $x_0 = 0$. In compression terminology, y_i is referred to as the *prediction residual* of x_i. The notion of compressing the prediction residual instead of x_i is used in all the image and video compression standards. For images, a typical probability distribution for x_i and the resulting distribution for y_i are shown in Figure 2.2.

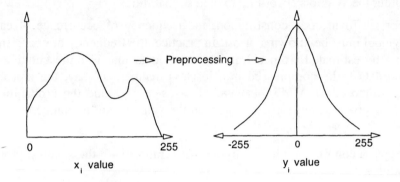

Figure 2.2 Typical distribution of pixel values *x* and residuals *y*. The corresponding probability of occurrence is shown on the vertical axis.

Let symbol s_i have a probability of occurrence p_i. From coding theory, the ideal symbol-to-codeword mapping function will produce a codeword requiring $I(s_i) = log_2 \frac{1}{p_i}$ bits. A uniform distribution for p_i ($p_i \approx \frac{1}{255}$), such as the one shown in the left plot of Figure 2.2, will result in codewords that on average require eight bits; thus, no compression is achieved. On the other hand, for a skewed probability distribution, such as the one shown in the right plot of Figure 2.2, the mapping function can on average yield codewords requiring less than eight bits per symbol and thereby achieve compression.

2.3 HUFFMAN ENCODING

In 1952, D. A. Huffman developed a code construction method that can be used to perform lossless compression. In Huffman coding, the modeling and the symbol-to-codeword mapping functions of Figure 2.1 are combined into a single process. As discussed earlier, the input data are partitioned into a sequence of symbols so as to facilitate the modeling process. In most image and video compression applications, the size of the alphabet composing these symbols is restricted to at most 64,000 symbols. The Huffman code construction procedure evolves along the following steps:

1. Order the symbols according to their probabilities.

   For Huffman code construction, the frequency of occurrence of each symbol must be known a priori. In practice, the frequency of occurrence can be estimated from a training set of data that is representative of the data to be compressed in a lossless manner. If, say, the alphabet is composed of N *distinct* symbols $s_1, s_2, ..., s_N$ and the probabilities of occurrence are $p_1, p_2, ..., p_N$, then the symbols are rearranged so that $p_1 \geq p_2 \geq p_3 ... \geq p_N$.

2. Apply a contraction process to the two symbols with the smallest probabilities.

   Suppose the two symbols are s_{N-1} and s_N. We replace these two symbols by a hypothetical symbol, say, H_{N-1}, that has a probability of occurrence $p_{N-1} + p_N$. Thus, the new set of symbols has $N - 1$ members: $s_1, s_2, ..., s_{N-2}, H_{N-1}$.

3. We repeat the previous step until the final set has only one member.

The recursive procedure in Step 2 can be viewed as the construction of a binary tree, since at each step we are merging two symbols. At the end of the recursion process, all the symbols $s_1, s_2, .., s_N$ will be leaf nodes of this tree. The codeword for each symbol s_i is obtained by traversing the binary tree from its root to the leaf node corresponding to s_i.

We illustrate the code construction process with the following example depicted in Figure 2.3. The input data to be compressed are composed of symbols in the alphabet *k, l, u, w, e, r, ?*. In Step 1, we sort the probabilities. In Step 2, we merge the two symbols *k* and *w* to form the new symbol (*k, w*). The probability of occurrence for the new symbol is the sum of the probabilities of occurrence for *k* and *w*. We sort the probabilities again and perform the merger on the pair of least frequently occurring symbols, as shown in the boxed region in Step 3 of Figure 2.3. We repeat this process through Step 6. By traversing this process from right to left and visualizing this as a binary tree, as shown in

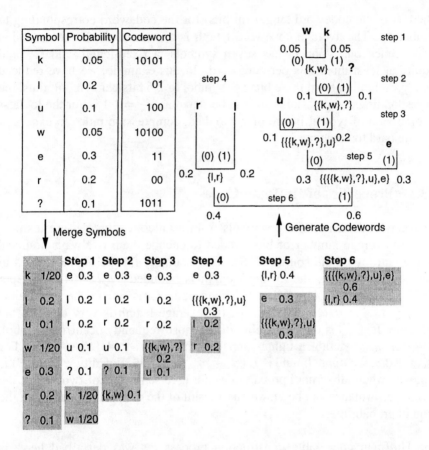

Figure 2.3 An example of Huffman codeword construction.

this figure, one can determine the codewords for each symbol. For example, to reach the symbol *u* from the root of the tree, one traverses nodes that were assigned the bits 1, 0, and 0. Thus, the codeword for *u* is 100.

In this example, the average codeword length is 2.6 bits per symbol. In general, the average codeword length is defined as

$$l_{avg} = \sum l_i p_i, \tag{2.4}$$

where l_i is the codeword length (in bits) for the codeword corresponding to symbol s_i. The average codeword length is a measure of the compression ratio. Since our alphabet has seven symbols, a fixed-length coder would require at least three bits per codeword. In this example, we have reduced the representation from three bits per symbol to 2.6 bits per symbol; thus, the corresponding compression ratio can be stated as $\frac{3}{2.6} = 1.15$. For the lossless compression of typical image or video data, compression ratios in excess of two are hard to come by.

2.3.1   Properties of Huffman Codes

As mentioned earlier, if the symbols from a random source S are distinct, then the average number of bits needed to encode them is always bounded by the entropy of the source $H(S)$. For example, for the alphabet used in the previous section, the average length is bounded by 2.546439 bits per symbol. It can be shown that Huffman codewords satisfy the constraints $H(S) \leq l_{avg} < H(S) + 1$; that is, the average length is very close to the optimum. If p_{max} is the probability of the most frequently occurring symbol, then, for $p_{max} < 0.5$, a tighter upper bound is $l_{avg} < H(S) + p_{max}$. For $p_{max} \geq 0.5$, the upper bound is $H(S) + p_{max} + 0.086$. Equality with $H(S)$ is achieved when all symbol probabilities are inverse powers of two. Thus, the loss in performance can be viewed as a result of the quantization effect on the symbol probabilities.

The Huffman code table construction process, as was described here, is referred to as a *bottom-up* method, since we perform the contraction process on the two least frequently occurring symbols. In recent years, *top-down* construction methods have also been published in the literature.

The code construction process has a complexity of $O(N \log_2 N)$. With presorting of the input symbol probabilities, code construction methods with complexity $O(N)$ are presently known.

In the example, one can observe that no codeword is a prefix for another codeword. Such a code is referred to as a *prefix-condition* code. Huffman codes always satisfy the prefix-condition. If codelengths l_i satisfy the constraint $\sum 2^{-l_i} \leq 1$, then it can be shown that the resulting codewords satisfy the prefix-condition. Furthermore, the corresponding codewords can be constructed as the first l_i bits in the fractional representation of a_i,

$$a_i = \sum_{j=1}^{i-1} 2^{-l_j}, i = 1, 2, ..., N. \tag{2.5}$$

Due to the prefix-condition property, Huffman codes are *uniquely decodable*. Not every uniquely decodable code satisfies the prefix-condition. A code such as $0, 01, 011, 0111$ does not satisfy the prefix-condition, since zero is a prefix for all of the codewords; however, every codeword is uniquely decodable, since a zero signifies the start of a new codeword.

If we have a binary representation for the codewords, the complement of this representation is also a valid set of Huffman codewords. The choice of using the codeword set or the corresponding complement set depends on the application. For instance, if the Huffman codewords are to be transmitted over a noisy channel where the probability of error of a one being received as a zero is higher than the probability of error of a zero being received as a one, then one would choose the codeword set for which the bit zero has a higher probability of occurrence. This will improve the performance of the Huffman coder in this noisy channel.

In Huffman coding, fixed-length input symbols are mapped into variable-length codewords. Since there are no fixed-size boundaries between codewords, if some of the bits in the compressed stream are received incorrectly or if they are not received at all due to dropouts, all the data are lost. This potential loss can be prevented by using special markers within the compressed bit stream to designate the start or end of a compressed stream packet.

2.4 HUFFMAN DECODING

The Huffman encoding process is relatively straightforward. The symbol-to-codeword mapping table provided by the modeler is used to generate the codewords for each input symbol. On the other hand, the Huffman decoding process is somewhat more complex.

2.4.1   Bit-Serial Decoding

For the example shown in Figure 2.3, let us assume that the binary coding tree is also available to the decoder. In practice, this tree can be reconstructed from the symbol-to-codeword mapping table that is known to both the encoder and the decoder. The decoding process consists of the following steps:

1. Read the input compressed stream bit by bit and traverse the tree until a leaf node is reached.

2. As each bit in the input stream is used, it is discarded. When the leaf node is reached, the Huffman decoder outputs the symbol at the leaf node. This completes the decoding for this symbol.

We repeat these steps until all of the input is consumed. For the example discussed in the previous section, since the longest codeword is five bits and the shortest codeword is two bits, the decoding bit rate is not the same for all symbols. Hence, this scheme has a fixed input bit rate but a variable output symbol rate.

2.4.2   Lookup-Table-Based Decoding

Lookup-table-based methods yield a constant decoding symbol rate. The lookup table is constructed at the decoder from the symbol-to-codeword mapping table. If the longest codeword in this table is L bits, then a 2^L entry lookup table is needed. In our example, $L = 5$. Specifically, the lookup table construction for each symbol s_i is as follows:

- Let c_i be the codeword that corresponds to symbol s_i. Assume that c_i has l_i bits. We form an L-bit address in which the first l_i bits are c_i and the remaining $L - l_i$ bits take on all possible combinations of zero and one. Thus, for the symbol s_i, there will be 2^{L-l_i} addresses.

- At each entry we form the two-tuple (s_i, l_i).

Decoding using the lookup table approach is relatively easy:

1. From the compressed input bit stream, we read in L bits into a buffer.

2. We use the L-bit word in the buffer as an address into the lookup table and obtain the corresponding symbol, say s_k. Let the codeword length be l_k. We have now decoded one symbol.

3. We discard the first l_k bits from the buffer, and we append to the buffer the next l_k bits from the input, so that the buffer has again L bits.

4. We repeat Steps 2 and 3 until all of the symbols have been decoded.

The primary advantages of lookup-table-based decoding are that it is fast and that the decoding rate is constant for all symbols, regardless of the corresponding codeword length. However, the input bit rate is now variable. For image or video data, the longest codeword could be around 16 to 20 bits. Thus, in some applications, the lookup table approach may be impractical due to space constraints.

Variants on the basic theme of lookup-table-based decoding include using hierarchical lookup tables and combinations of lookup table and bit-by-bit decoding. In the next section, we describe codeword construction methods that facilitate lookup-table-based decoding by constraining the maximum codeword length to a fixed-size L.

2.5 HUFFMAN CODES WITH CONSTRAINED LENGTH

In the previous section, we described several decoding procedures for Huffman decoding. For fast decoding, lookup-table-based approaches are preferable. In many applications, depending on the symbol probabilities, the Huffman codebook design may yield codewords that may require a large number of bits. This outcome is possible when some of the symbol probabilities are extremely small. Thus, lookup-table-based decoding may not be feasible if the memory requirements are prohibitive. To alleviate this problem, a shortened Huffman code can be used.

The basic idea of the shortened Huffman code is to view the Huffman code representation as a hierarchical representation. Suppose, due to lookup table size constraints, we require that no codeword length exceed L bits. Let S be the set of N symbols $s_1, s_1, ..., s_N$ for which a Huffman code is to be designed. Let p_i be the occurrence frequency for symbol s_i. We further assume that the symbols are ordered so that $p_1 \geq p_2 \geq p_3 \cdots \geq p_N$. The design procedure for a shortened Huffman code is as follows:

1. Let s_t be the first symbol for which $p_t \leq \frac{1}{2^L}$. Partition S into two sets S_1 and S_2 as

$$S_1 = \{s_i | p_i > \frac{1}{2^L}\} = \{s_1, s_2, \cdots, s_{t-1}\}, \qquad (2.6)$$

$$S_2 = \{s_i | p_i \leq \frac{1}{2^L}\} = \{s_t, s_{t+1}, \cdots, s_N\}. \qquad (2.7)$$

2. Create a special symbol Q such that its frequency of occurrence is

$$q = \sum_{i=t}^{N} p_i. \qquad (2.8)$$

3. Augment S_1 by Q to form a new set W. The new set has the occurrence frequencies corresponding to symbols in S_1 and the special symbol Q. We then construct an optimal prefix code for W using the design procedure for unconstrained-length Huffman codewords, as explained in a previous section. This design procedure will yield codewords c_{s1} for symbols in the set S_1 and a codeword c_q for the symbol Q. c_q is the shortened prefix-code for symbols in S_2.

   If l_i is the length of the i-th codeword of S_1, then

$$\max_{s_i \in S_1} l_i = \max_{s_i \in S_1} \left\lceil log_2 \frac{1}{p_i} \right\rceil \leq L. \qquad (2.9)$$

   Thus, all symbols in S_1 yield codewords that are at most L bits. Codeword c_q will also not exceed L bits. ($\lceil x \rceil$ denotes the smallest integer larger or equal to x.)

Huffman encoding of a message string $m_1 m_2 m_3 ... m_k$ is relatively straightforward.

- For all $m_i \in S_1$, output the corresponding codeword from c_{s1}.

- For all $m_i \in S_2$, output the codeword c_q followed by an L-bit fixed-length binary representation for m_i. (Actually, one can use fewer than L bits, since, if there are N_{s2} symbols in S_2 and $N_{s2} \leq N$, then the fixed-length binary representation for each m_i is $\lceil log_2 N_{s2} \rceil$ bits).

When the symbol occurrence frequencies are known a priori, one can estimate the worst-case performance degradation for the shortened Huffman codes as compared with the unconstrained Huffman code table design.

Let l_{sh} be the average codeword length for the shortened codeword design approach. Let l_W be the average codelengths for W, and let $H(W)$ be the entropy for W. We have

$$l_{sh} = l_W + qL, \tag{2.10}$$

$$H(W) \leq l_{sh} \leq H(W) + 1. \tag{2.11}$$

Furthermore, l_{sh} is bounded by

$$H(S) \leq l_{sh} \leq H(W) + qL + 1, \tag{2.12}$$

where $H(S)$ denotes the entropy of S. However,

$$H(W) = \sum_{i=1}^{t-1} p_i \, log_2 \frac{1}{p_i} + q \, log_2 \frac{1}{q}, \tag{2.13}$$

$$qL = \sum_{i=t}^{N} p_i L = \sum_{i=t}^{N} p_i \, log_2 \frac{1}{2^{-L}} \leq \sum_{i=t}^{N} p_i \, log_2 \frac{1}{p_i}, \tag{2.14}$$

and

$$H(S) = \sum_{i=1}^{t-1} p_i \, log_2 \frac{1}{p_i} + \sum_{i=t}^{N} p_i \, log_2 \frac{1}{p_i}. \tag{2.15}$$

Hence,

$$H(S) \leq l_{sh} \leq H(S) + 1 + q \, log_2 \frac{1}{q}. \tag{2.16}$$

For an optimal unconstrained Huffman code, the average codelength is bounded by $H(S) \leq l_{avg} \leq H(S) + 1$. Hence, the worst-case increase in the average codeword length for the shortened code is $q \, log_2 \frac{1}{q}$ bits per symbol (this function attains a maximum value of $\frac{1}{2}$).

Huffman decoding for shortened prefix codes using a lookup-table-based approach is similar to the procedure described in the previous section. In this case, a two-level decoding is done when the codeword c_q is decoded. After c_q is decoded, the next L bits are read from the input bit stream, and the corresponding symbol m_i is then computed. We illustrate the code construction process with an example.

Example 1: Let $S = s_0, s_1, ..., s_{15}$. The occurrence frequencies and the resulting unconstrained Huffman code are as shown in Table 2.1. Here, the Huffman code table design is based on the approach described in the previous section. The average codeword length is $l_{avg} = \sum_{i=0}^{15} l_i p_i = 2.694$ bits per

Symbol s_i	p_i	l_i	Codeword
0	0.28200	2	11
1	0.27860	2	10
2	0.14190	3	011
3	0.13890	3	010
4	0.05140	4	0011
5	0.05130	4	0010
6	0.01530	5	00011
7	0.01530	5	00010
8	0.00720	6	000011
9	0.00680	6	000010
10	0.00380	7	0000011
11	0.00320	7	0000010
12	0.00190	7	0000001
13	0.00130	8	00000001
14	0.00070	9	000000001
15	0.00040	9	000000000

Table 2.1   Unconstrained length Huffman codewords.

symbol. Note that, the longest codeword is nine bits. Thus, a 512-entry table is needed for lookup-table-based decoding. Let us suppose that only a 128-entry lookup table can be permitted. Thus, we need to design a seven-bit shortened Huffman code; that is, all codeword lengths must be less than or equal to seven bits.

The code table design procedure is as follows:

1. We form the two sets S_1 and S_2. Symbols s_8 to s_{15} belong to S_2, since $p_i \leq \frac{1}{128}$ for $i = 8$ to 15. The occurrence frequency of the special symbol Q is $\sum_{i=8}^{15} p_i = 0.0253$.

2. The codeword design process for the shortened Huffman codes is illustrated in Figure 2.4. The corresponding Huffman code table for the

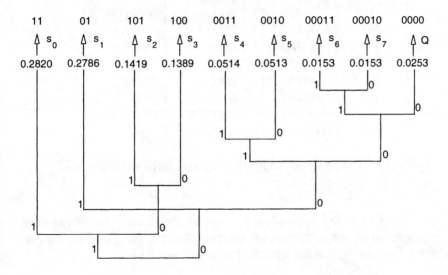

Figure 2.4   Shortened Huffman code.

shortened Huffman codes is shown in Table 2.2. Note that, for all symbols in S_2, we need a prefix code of four bits followed by a seven-bit representation for the specific symbol in S_2. The average codeword length is now $l_{sh} = \sum_{i=0}^{7} l_i p_i + \sum_{i=8}^{15} 11 p_i = 2.8057$ bits per symbol.

During encoding, if the input symbol is in set S_1, then the encoder outputs the corresponding Huffman codeword. If the input symbol is in set S_2, then the encoder outputs a codeword with the prefix 0000 followed by a codeword which is the seven-bit representation of the symbol in the set S_2. For instance,

Symbol i	p_i	l_i	Codeword	Additional
0	0.28200	2	11	
1	0.27860	2	01	
2	0.14190	3	101	
3	0.13890	3	100	
4	0.05140	4	0011	
5	0.05130	4	0010	
6	0.01530	5	00011	
7	0.01530	5	00010	
8	0.00720	11	0000	0001000
9	0.00680	11	0000	0001001
10	0.00380	11	0000	0001010
11	0.00320	11	0000	0001011
12	0.00190	11	0000	0001100
13	0.00130	11	0000	0001101
14	0.00070	11	0000	0001110
15	0.00040	11	0000	0001111

Table 2.2   Constrained-length (L = 7 bits) Huffman codewords.

if, say, symbol s_9 is to be encoded, then the Huffman code for this symbol will be made up of two codewords 0000 and 0001001. Note that the longest codeword does not exceed the specified length of seven bits.

Decoding is performed as follows:

1. We first construct a lookup table as explained in the previous section. Huffman decoding for each output symbol is as follows:

2. From the input bit stream we fetch bits into a buffer until the buffer has seven bits. We access the lookup table location, using the seven bits as an address. This lookup table location contains the symbol for the Huffman code and the codelengths for the Huffman code. Let m_k be the symbol and l_k the codelengths.

3. The first l_k bits in the buffer are discarded by shifting the buffer contents to the left by l_k bit positions.

- If m_k is not the special symbol Q, m_k is one of the symbols $s_0, s_1, ..., s_7$, and thus we have correctly decoded this symbol.

- If m_k is the special symbol Q, additional bits from the input bit stream are needed for decoding. We fetch l_k bits from the input bit stream to fill up the buffer. The buffer now contains the binary representation for one of the symbols $s_8, s_9, ..., s_{15}$, and thus we have correctly decoded a symbol from S_2.

- We repeat Steps 2 and 3 until the complete message has been decoded.

As illustrated by this example, a shortened prefix code can be decoded with modest lookup table requirements, and decoding is quite efficient. However, the average codelengths are worse, compared to the unconstrained Huffman code table design. Table 2.3 shows the maximum limits for the additional bits per symbol for a shortened Huffman code versus an unconstrained Huffman code. The data in this table are computed for the example cited in this section and use the bounds derived in this section.

Lookup Table Size (entries)	Worst case $l_{sh} - l_{avg}$ (bits/symbol)
16	0.4213
32	0.2326
64	0.2326
128	0.1342
256	0.0731
512	0.0338

Table 2.3 Average codeword length l_{sh} degradations compared to an unconstrained Huffman code for various values of L. Code table size is 2^L entries.

The key disadvantage of two-level decoding is that we cannot guarantee a constant symbol rate at the Huffman decoder output. This may be a requirement for many applications. In the next section, we describe other approaches to constrained-length Huffman codeword design. These approaches avoid the two-level decoding that is needed for the approach developed in this section. Furthermore, the approaches described in the next section are capable of yielding an average codeword length which is close to the optimum.

2.6   CONSTRAINED-LENGTH HUFFMAN CODES: AD HOC DESIGN

In the previous section, we developed a Huffman code table design in which the codeword lengths were constrained to be at most L bits. The design entailed using a prefix code followed by an L-bit representation for infrequently occurring message symbols. Here, we develop an ad hoc design methodology that does not use a prefix code. This methodology has the advantage that with a lookup-table-based decoder, the output decoding rate will be constant.

The codebook design process is as follows:

Let the N members of the symbol alphabet be $s_1, s_2, ..., s_N$. We denote this as the set S. Messages that are to be Huffman encoded are composed of symbols in S. We assume that the occurrence frequencies (probabilities) for each symbol in S is known; the i-th symbol s_i has the occurrence probability denoted as p_i.

- For a maximum codeword length of L bits, we define a threshold $T = 2^{-L}$.

- Sort $s_i, i = 1, 2, ..., N$ so that $p_k \geq p_{k+1}$.

- For each p_i, if $p_i \leq T$, set $p_i = T$.

- Design the codebook using the modified p_i values and the unconstrained-length Huffman code table design approach.

Since p_i is restricted to at most 2^{-L}, no codeword length will exceed L bits. However, this ad hoc design does not guarantee that there will be at least one codeword that has L bits. Hence, this design approach may be overly aggressive. In Table 2.4, we show the codewords that result from this ad hoc design method. (Ignore for the time being the column labeled *Reordered*.) The symbol frequencies are the same as those used in the previous section.

The resulting codewords, shown in the fourth column of this table, have a maximum codeword length of seven bits and an average codeword length l_{avg} of 2.7308 bits per symbol. Note that the unconstrained-length Huffman code table would have required a maximum codelength of nine bits and would yield an average codelength of 2.6904 bits per symbol. As expected, constraining

Symbol i	p_i	l	Codeword	Reordered l	Reordered Codeword
0	0.28200	2	11	2	11
1	0.27860	2	01	2	01
2	0.14190	3	101	3	101
3	0.13890	3	100	3	100
4	0.05140	4	0010	4	0010
5	0.05130	4	0001	4	0001
6	0.01530	6	001100	6	001100
7	0.01530	6	001101	6	001101
8	0.00720	7	0011110	6	000010
9	0.00680	7	0011111	6	000011
10	0.00380	7	0011100	6	000000
11	0.00320	7	0011101	6	000001
12	0.00190	6	000010	7	0011110
13	0.00130	6	000011	7	0011111
14	0.00070	6	000000	7	0011100
15	0.00040	6	000001	7	0011101
l_{avg}			2.7308		2.7141

Table 2.4 Constrained-length ($L = 7$ bits) Huffman codewords: Ad hoc design.

the codeword lengths results in a slightly worse average codelength. Also observe that symbols s_8 to s_{11} have codewords with more bits than symbols s_{12} to s_{15}, even though they have a higher probability of occurrence. This is due to the fact that some of the probabilities were set to the threshold T, and hence the ordering among the probabilities is obscured.

Intuitively, we expect symbols with a low probability of occurrence to require longer codewords. Thus, in our ad hoc design approach, we can apply this notion to rearrange the final codewords. Rearranging is done by simply sorting the codeword lengths in ascending order of magnitude and associating this sorted list to the corresponding list of codewords. The use of post sorting on the symbol lengths yields the codeword lengths shown in the fifth column. The average codeword length has now been reduced to 2.7141 bits per symbol. This ad hoc design has a lower average codeword length than the prefix-code design approach developed in the previous section. The average codeword length can be lowered further using a method that we describe in the next section.

2.7   CONSTRAINED-LENGTH HUFFMAN CODES: THE VOORHIS METHOD

This method yields codewords with an average length that is close to the optimum, subject to the constraint that no codeword exceed L bits. The design process is in three steps.

1. If the message is composed of symbols drawn from a set of symbols $s_1, s_2, ..., s_N$ and if the symbol occurrence frequencies are known a priori, sort the symbols so that $p_1 \geq p_2... \geq p_N$.

2. Determine codeword lengths $l_1, l_2, ..., l_N$ that minimize $\sum_{i=1}^{N} l_i p_i$ subject to the constraint $1 \leq l_i \leq L$.

   For uniquely decodable codes, we also require that $\sum_{i=1}^{N} 2^{-l_i} \leq 1$. The resulting codeword lengths will be such that $1 \leq l_1 \leq l_2... \leq l_N \leq L$.

3. The i-th codeword is the first l_i bits of the fraction computed by $\sum_{k=1}^{i-1} 2^{-l_k}$.

The major difficulty in this process is in the second step, namely, in the calculation of the codeword lengths. The algorithm due to Voorhis proposes a recursive solution. The recursion assumes that codeword lengths $l_1, l_2, .., l_{r-1}$ have been found so far. Hence, the binary representation for the corresponding codewords, denoted as $c_1, c_2, ..., c_{r-1}$, is known. Define

$$k \quad = \quad l_{r-1}, \tag{2.17}$$

$$s \quad = \quad 2^k \left(1 - \sum_{i=1}^{r-1} 2^{-l_i}\right). \tag{2.18}$$

Then s k-bit prefixes are available for the codewords $c_r, ..., c_N$. This is referred to as the [k,r,s] problem, and the goal is to find an optimal way to partition s k-bit prefixes of codespace among the codewords $c_r, ..., c_N$, subject to the condition $l_i \in [k, L] \forall i \in [r, N]$. The computational steps for determining the codelengths $l_1, l_2, ..., l_N$ by solving the [k,r,s] problem are as follows.

1.   **for** $k \in [1, L]$
        for $r \in [1, min(N - 1, 2^k)]$
            for $s \in [N - r + 1, min(2N - 2r, 2^k - r + 1)]$
                $\psi_{k,r,s} = \sum_{i=1}^{N} k p_i$

2.   **for** $k = L - 1, ..., 1$
        for $r = min(N - 1, 2^k), ..., 1$
            for $s = min(N - r, 2^k - r + 1), ..., \lceil (N - r + 1)2^{k-L} \rceil$
                if $s < \lceil (N - r)2^{k-L} \rceil + 1$
                    $\psi_{k,r,s} = \psi_{k+1,r,2s}$
            else
                    $\psi_{k,r,s} = min(k p_r + \psi_{k,r+1,s-1}, \psi_{k+1,r,2s})$

3. $k = 1, r = 1, s = 2$

4.   **if** $k = L$ **or** $N - r + 1 \leq s$ **go to** step 8

5.   **if** $(N - r)2^{-L} > (s - 1)2^{-k}$ **go to** step 7

6.   **if** $k p_i + \psi_{k,r+1,s-1} < \psi_{k+1,r,2s}$ **go to** step 8

7. $k = k + 1, s = 2s$, **go to** step 4

8. $l_r = k, r = r + 1, s = s - 1$

9.   **if** $r \leq N$ **go to** step 4

At this point, we have computed the codelengths $l_1, l_2, ..., l_N$. These codelengths minimize $\sum_{i=1}^{N} l_i p_i$ and satisfy the constraint $1 \leq l_1 \leq l_2... \leq l_N \leq L$. The codewords corresponding to these codelengths can be computed as

$$l_0 = 1, \tag{2.19}$$

$$a_1 = 2^{-l_0}, \tag{2.20}$$

$$a_i = \sum_{k=1}^{i-1} 2^{-l_k}, i = 2, 3, .., N. \tag{2.21}$$

The i-th codeword is the binary representation of a_i truncated to the first l_i bits.

Example 2: We apply the algorithm described here to construct a Huffman code table using the symbol probabilities of Example 1. For a Huffman

codeword length constraint of $L = 7$ bits, the resulting codewords are as shown
in Table 2.5. For comparison, we show the codewords for the unconstrained
Huffman code table and the ad hoc method described earlier. While both the

Symbol i	p_i	Unconstrained	Ad hoc	Voorhis
0	0.28200	11	11	11
1	0.27860	10	01	10
2	0.14190	011	101	011
3	0.13890	010	100	010
4	0.05140	0011	0010	0011
5	0.05130	0010	0001	0010
6	0.01530	00011	001100	00011
7	0.01530	00010	001101	00010
8	0.00720	000011	000010	0000111
9	0.00680	000010	000011	0000110
10	0.00380	0000011	000000	0000101
11	0.00320	0000010	000001	0000100
12	0.00190	0000001	0011110	0000011
13	0.00130	00000001	0011111	0000010
14	0.00070	000000001	0011100	0000001
15	0.00040	000000000	0011101	0000000
l_{avg}		2.6940	2.7141	2.7045

Table 2.5 Voorhis method. Constrained-length (L = 7 bits)
Huffman codewords.

ad hoc design method and the Voorhis method satisfy the maximum codeword
length constraint of $L = 7$ bits, the Voorhis method yields a lower average
codeword length l_{avg}. We further illustrate the performance of these two design
methods in Figure 2.5. Here, we have used the symbol probabilities from
Example 1, and we show the average codeword length for various maximum
codeword length settings. In all instances, the Voorhis method outperforms
the ad hoc method, and the performance gap will widen as the maximum
codeword length is increased.

For a message composed of symbols from an N-symbol alphabet, if $L = log_2 N + d$, then the Voorhis method has a complexity of $O(dN^2)$. The ad
hoc design has a lower complexity of $O(N\,log_2 N)$; however, the average

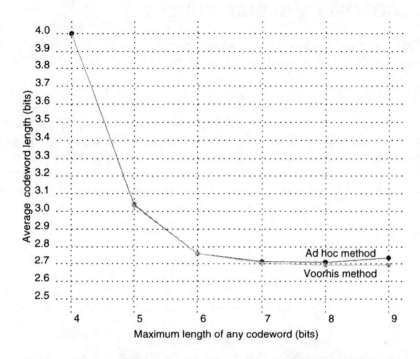

Figure 2.5 Huffman codeword design. Comparison of average code lengths between the ad hoc and the Voorhis methods.

codeword length is not necessarily minimized among all possible codeword choices. In image and video compression standards, N is usually large, say, 256 symbols, and the ad hoc design may be preferred in these instances. Note that, regardless of the design approach, various implementations may impose additional constraints. For instance, in the Huffman tables for the JPEG image compression standard, there should be no Huffman codeword that is all ones, that is, $11111...1$. If the codeword design method yields such a sequence, one can simply take the complement of such a code table. The complement of a Huffman code table is also a valid Huffman code table and will satisfy the unique prefix condition that is characteristic of all Huffman codewords.

2.8 GOLOMB AND RICE CODING

In 1965, Golomb developed a very simple coding scheme for sources with a
geometric probability distribution, that is, of the form

$$P(n) = (1 - p_0)p_0^n, \quad n \geq 0, \quad 0 < p_0 < 1, \tag{2.22}$$

where p_0 is constant. For example, if p_0 denotes the probability of zero, then
$P(n)$ is the probability of a run-length of n zeros. Given a positive integer
m, the Golomb code G_m encodes a nonnegative integer n as follows. Divide
n by m. Let q be the quotient and r be the remainder of this division (that
is, $n = mq + r$). Then, the G_m code for n is the concatenation of the *unary*
representation of q (that is, q zeros followed by a single one) with the *modified
binary* representation of r (that is, using codewords of length $\lfloor log_2 m \rfloor$ bits for
$r < 2^{\lceil log_2 m \rceil} - m$ and $\lceil log_2 m \rceil$ bits otherwise).

For example, let $m = 3$ and $n = 14$. Dividing 14 by 3 yields $q = 4$ and $r = 2$.
The unary representation of 4 is 00001 and the binary representation of 2 is
10. Thus, the G_3 code of 14 is given by concatenating 00001 and 10, that is,
0000110.

For the geometric probability distribution of (2.22), a G_m code is optimal if
m is given by

$$m = \left\lceil -\frac{log(1 + p_0)}{log(p_0)} \right\rceil. \tag{2.23}$$

Later, Rice addressed the problem of coding both negative and nonnegative
values and rediscovered a special case of Golomb codes when m is a power of
two, that is, $m = 2^k$, with very simple encoding and decoding procedures. For
$m = 2^k$, the quotient of $\frac{n}{m}$ is given by shifting to the right by k bits the binary
representation of n, and the remainder of $\frac{n}{m}$ is given by the last k bits of n.
Thus, the Rice code for n is given by concatenating the unary representation
of q with the k least significant bits of n.

For example, let $k = 2$ ($m = 4$) and $n = 22$. The binary representation of 22 is
10110. 10110 shifted to the right by two bits (which corresponds to dividing
22 by 4) yields $101 = 5$, which has a 000001 unary representation. The two
least significant bits of 10110 are 10. Thus, for $k = 2$, the Rice code of 22 is
00000110.

Since both Golomb and Rice made the observation about the simplicity of generating codes when $m = 2^k$, these special codes are also referred to as Golomb-Rice codes.

To decode a Golomb-Rice code, start at the beginning (left end) of the codeword and count the number of 0's preceding the first 1. Let this number be q and discard this first 1. Let the next k bits be the number r. The value of the decoded symbol is given by $n = 2^k q + r$. In a binary representation, n is given by simply concatenating the binary values of q and r. For example, if k = 3, codeword 000001010 is decoded as (5)(010) or 101010 = 42. If there are no leading zeros, then $q = 0$. We discard the first 1, and the decoded value is given simply by the next k bits. For example, for $k = 4$, codeword 10010 is decoded as 2.

The Golomb-Rice codes were designed for coding nonnegative values. If a source has negative values in a two-sided geometric distribution, then one can remap all values to nonnegative values. This also requires a good estimate of the center of this two-sided distribution.

Golomb-Rice codes have been used lately in a number of image compression schemes. This is due to the observation that prediction errors in image coding tend to behave as two-sided geometric distributions centered at zero.

2.9 ARITHMETIC CODING

The Huffman coder generates a new codeword for each input symbol. This implies that the lower limit on compression for a Huffman coder is one bit per input symbol. As observed earlier, higher compression ratios can be achieved by combining several symbols into a single unit. However, in the Huffman coding context, there is a corresponding increase in complexity in codeword construction.

Another problem with Huffman coding is that the coding and the modeling steps are combined into a single process, and thus adaptive coding is difficult. If the probabilities of occurrence of the input symbols change, then one has to redesign the Huffman tables from scratch. An alternative to Huffman coding is arithmetic coding.

Arithmetic coding is a lossless compression technique that benefits from treating multiple symbols as a single data unit but at the same time retains the incremental symbol-by-symbol coding approach of Huffman coding. Arithmetic coding separates the coding from the modeling. This process allows for the dynamic adaptation of the probability model without affecting the design of the coder. Provisions for substituting Huffman coding for arithmetic coding are contained in many of the image compression standards. In this section, we briefly describe the basic concepts in arithmetic coding.

2.9.1   The Encoding Process

In the theory of arithmetic coding, a single codeword is assigned to each possible dataset. Each codeword can be considered a half-open subinterval in the interval [0,1). By assigning enough precision bits to each of the codewords, one can distinguish one subinterval from any other subinterval, and thus uniquely decode the corresponding dataset. Like Huffman codewords, the more probable datasets correspond to larger subintervals and thus require fewer bits of precision.

We describe the encoding process in arithmetic coding via an example. For a comparison with Huffman coding, we use the same alphabet $(k, l, u, w, e, r, ?)$ and the same symbol probabilities as in the example shown in Figure 2.3. The message to be compressed is the string $lluure?$. We use ? as an end-of-message marker to terminate the decoding process. Using the Huffman coder of Figure 2.3, the corresponding codeword would be the 18-bit string 010110010000111011. For the arithmetic coder, the encoding process is shown in Figure 2.6.

1. At the start of the process, the message is assumed to be in the half-open interval [0,1). This interval is split into several subintervals, one subinterval for each symbol in our alphabet. The upper limit of each subinterval is the cumulative probability up to and including the corresponding symbol. The lower limit is the cumulative probability up to but not including the symbol. For our example, the subinterval ranges are given in Table 2.6.

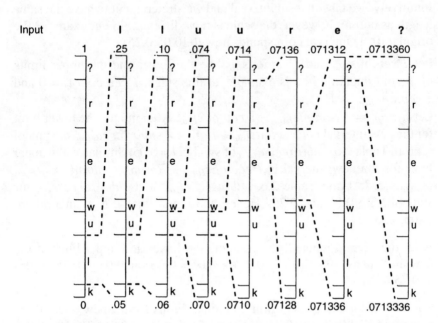

Figure 2.6 Arithmetic coding: example of the encoding process.

s_i	p_i	Subinterval
k	0.05	[0.00,0.05)
l	0.20	[0.05,0.25)
u	0.10	[0.25,0.35)
w	0.05	[0.35,0.40)
e	0.30	[0.40,0.70)
r	0.20	[0.70,0.90)
?	0.10	[0.90,1.00)

Table 2.6 Symbol subinterval ranges.

2. When the first symbol in our message, *l*, appears, we select the corresponding subinterval and make it the new current interval. The intervals of the remaining symbols in our alphabet are then scaled accordingly.

Intuitively, we take that subinterval and we stretch it out to have the same length as before; however, the scale is now different. In our example, the original [0, 1) interval corresponds now to [0.05,0.25).

Let $Previous_{low}$ and $Previous_{high}$ be the lower and the upper limits of the old interval. In this example, at this stage, $Previous_{low} = 0$ and $Previous_{high} = 1$.

Let $Range = Previous_{high} - Previous_{low}$. After input l, the lower limit for the new interval is $Previous_{low} + Range \times subinterval_{low}$ of symbol l. From Table 2.6, $subinterval_{low}$ of symbol $l = 0.05$. Similarly, the upper limit for the new interval is $Previous_{low} + Range \times subinterval_{high}$ of symbol l. From Table 2.6, $subinterval_{low}$ and $subinterval_{high}$ for symbol l are 0.05 and 0.25. Hence, the limits for the new interval are [0.05,0.25).

3. After the second input, l, we compute the lower and upper limit of the new interval as $0.05 + (0.25 - 0.05) \times 0.05 = 0.06$ and $0.05 + (0.25 - 0.05) \times 0.25 = 0.10$.

4. For the third input, u, the lower and upper limits for the new interval are computed as $0.06 + (0.10 - 0.06) \times 0.25 = 0.07$ and $0.06 + (0.10 - 0.06) \times 0.35 = 0.074$.

5. The calculations described in previous steps are repeated using the limits of the previous interval and the subinterval ranges of the current symbol. This yields the limits for the new interval. After the last symbol, ?, the final range is computed. In our example, the final range is [0.0713336, 0.0713360).

6. There is no need to transmit both values of the bounds in the last interval. Instead, we transmit a value that is within the final range. In our example, any number such as 0.0713336, 0.071334, ..., 0.0713355 could be used. We use 0.0713348389. The reader can verify that this number can be represented as $2^{-4} + 2^{-7} + 2^{-10} + 2^{-15} + 2^{-16}$, and hence it can be represented with 16 bits.

In conclusion, the codeword length for the message $lluure?$ is 16 bits with an arithmetic coder, 18 bits with a Huffman coder, and $7 \times 3 = 21$ bits with a fixed-length coder. Arithmetic coding yields better compression because it encodes a message as a whole new symbol instead of as separate symbols.

s_i	i	$cumprob_i$
k	7	0.00
l	6	0.05
u	5	0.25
w	4	0.35
e	3	0.40
r	2	0.70
?	1	0.90
	0	1.00

Table 2.7   Modified decoder table.

2.9.2   The Decoding Process

The decoding process can be described using the same example. Let us assume that the message *lluure?* was encoded with 16 bits and that the corresponding fractional value is 0.0713348389. Given the symbol probabilities, shown in Table 2.6, to each symbol in our alphabet we assign a unique number (i), and we associate a cumulative probability value $cumprob_i$. For our example, the modified decoder table is shown in Table 2.7.

Given the fractional representation of the input codeword *value*, the following algorithm outputs the corresponding decoded message. In our example, *value* = 0.0713348389.

> *DecodeSymbol* (*value*):
> **Begin**
>> $Previous_{low} = 0$.
>> $Previous_{high} = 1$.
>> $Range = Previous_{high} - Previous_{low}$.
>> **repeat**
>>> Find i such that
>>> $$cumprob_i \le \frac{value - Previous_{low}}{Range} < cumprob_{i-1}.$$
>>> **Output symbol** corresponding to i from the decoding table.
>>> Update:
>>> $Previous_{high} = Previous_{low} + Range * cumprob_{i-1}$.
>>> $Previous_{low} = Previous_{low} + Range * cumprob_i$.

$$Range = Previous_{high} - Previous_{low}.$$
    until symbol decoded is ?.
End

We show a few steps of the decoding process in our example.

1. First, $Previous_{low} = 0$, $Previous_{high} = 1$, and $Range = 1$. For i = 6, $cumprob_6 \leq \frac{value - Previous_{low}}{Range} < cumprob_5$. Thus, the first decoded symbol is l. As per the algorithm shown above, we update $Previous_{high}, Previous_{low}$, and $Range$ to 0.25, 0.05, and 0.2, respectively.

2. We repeat the decoding process, and we find that $i = 6$ again satisfies the limits $cumprob_6 \leq \frac{value - 0.05}{0.20} < cumprob_5$. Thus, the second decoded symbol is l. The updated $Previous_{high}, Previous_{low}$, and $Range$ values are 0.10, 0.06, and 0.04, respectively.

3. Repeating the decoding process yields that $i = 5$ satisfies the limits $cumprob_5 \leq \frac{value - 0.06}{0.04} < cumprob_4$. Thus, the third decoded symbol is u. As before, we update $Previous_{high}, Previous_{low}$, and $Range$ to 0.074, 0.070, and 0.004, respectively.

4. We repeat the decoding process as in previous steps. Finally, we decode ?, which signals the end of the message and terminates the decoding algorithm.

2.10   IMPLEMENTATION ISSUES

The encoding and decoding processes described in the previous section have to be modified for a practical implementation of an arithmetic coder. Some of these design issues are discussed next.

■ **Incremental Output**

    The encoding method we have described generates a compressed bit stream only after reading the entire message. However, in most implementations, and particularly in image compression, we desire an incremental transmission scheme.

From Figure 2.6, we observe that after encoding u, the subinterval range is [0.07,0.074). In fact, it will start with the value 0.07; hence, we can transmit the first two digits 07. After the encoding of the next symbol, the final representation will begin with 0.071 since both the upper and the lower limits of this range contain 0.071. Thus, we can transmit the digit 1. We repeat this process for the remaining symbols. Thus, incremental encoding is achieved by transmitting to the decoder each digit in the final representation as soon as it is known. The decoder can perform incremental calculations too.

- **High-precision arithmetic**

As we have illustrated in the example, most of the computations in arithmetic coding use floating-point arithmetic; however, most low-cost hardware implementations support only fixed-point arithmetic. Furthermore, division (used by the decoder) is undesirable in most implementations.

Consider the problem of arithmetic precision in the encoder. During encoding, the subinterval range is narrowed as each new symbol is processed. Depending on the symbol probabilities, the precision required to represent this range may grow; thus, there is a potential for overflow or underflow. For instance, in an integer implementation, if the fractional values are not scaled appropriately, different symbols may yield the same limits for $Previous_{high}$ and $Previous_{low}$; that is, no subdivision of the previous subinterval takes place. At this point, encoding would have to be abnormally terminated.

Let subinterval limits $Previous_{low}$ and $Previous_{high}$ be represented as integers with c bits of precision. The length of a subinterval is equal to the product of the probabilities of the individual events. If we represent this probability with f bits of precision, to avoid overflow or underflow, we require $f \leq c + 2$ and $f + c \leq p$, where p is the arithmetic precision for the computations. For arithmetic coding on a 16-bit computer, $p = 16$. If $c = 9$, then $f = 7$. If the message is composed of symbols from a k-symbol alphabet, then $2^f \geq k$. Thus, a 256-symbol alphabet cannot be correctly encoded using 16-bit arithmetic.

In recent years there have been many developments in arithmetic coding that allow fixed-precision arithmetic with simple operations, such as shifts, additions, and subtractions.

■ **Probability modeling**

Thus far, we have assumed a priori knowledge of the symbol probabilities p_i. In many practical implementations that use the arithmetic coder, symbol probabilities are estimated as the pixels are processed. This allows the coder to adapt better to changes in the input stream. A typical example is a document that includes both text and images. Text and images have quite different probability models. In this case, an adaptive arithmetic coder is expected to perform better than a nonadaptive entropy coder.

Huffman coding and arithmetic coding techniques form the basis of various lossless image compression standards. In the next section, we give a brief overview of the techniques used in the compression standards.

2.11 STANDARDS FOR LOSSLESS COMPRESSION

Standards related to the coding and transmission of signals over public telecommunication channels are developed under the auspices of the telecommunication standardization sector of the International Telecommunication Union (ITU-T). This sector was formerly known as the CCITT. The first standards for lossless compression were developed for facsimile applications. Scanned images used in such applications are bitonal; that is, the pixels take on one of two values, *black* or *white*, and these values are represented with one bit per pixel.

2.11.1 Facsimile Compression Standards

In every bitonal image, there are large regions that are either all white or all black. For instance, in Figure 2.7, we show a few pixels of a line in a bitonal image. Note that, the six contiguous pixels of the same color can be described as a *run* of six pixels with with *value* = 0. Thus, if each pixel of the image is remapped from say, its (*position, value*) to a *run* and *value*, then a more compact description can be obtained. In our example, no more than four bits are needed to describe the six-pixel run. In general, for many document type

Figure 2.7   Sample scanline of a bitonal image.

images, significant compression can be achieved using such preprocessing. Such a mapping scheme is referred to as a run-length coding scheme.

The combination of a run-length coding scheme followed by a Huffman coder forms the basis of the image coding standards for facsimile applications. These standards include the following:

- ITU-T Rec.   T.4 (also known as Group 3).   There are two coding approaches within this standard.

    1. Modified Huffman (MH) code. The image is treated as a sequence of scanlines, and a run-length description is first obtained for each line. A Huffman code is then applied to the (*run, value*) description. A separate Huffman code is used to distinguish between black and white runs, since the characteristics of these runs are quite different. The Huffman code table is static; that is, it does not change from image to image. For error-detection purposes, after each line is coded, an EOL (end of line) codeword is inserted.

    2. Modified Read (MR) code. Here, pixel values in a previous line are used as predictors for the current line. This is followed by a run-length description and a static Huffman code as in the MH code. An EOL codeword is also used. To prevent error propagation, MR coding is mixed with MH coding periodically.

- ITU-T Rec. T.6 (also known as Group 4). The coding technique used here is referred to as a Modified Modified Read (MMR) code. This code is a simplification of the MR code, wherein the error-protection mechanisms in the MR code are removed so as to improve the overall compression ratio.

These compression standards yield good compression (20:1 to 50:1) for business-type scanned documents. For images composed of natural scenes and rendered as bitonal images using a halftoning technique, the compression ratio is severely degraded. In Figure 2.8, the image on the left possesses characteristics representative of business document images, whereas the image on the right is a typical halftone image. The former is characterized by long runs

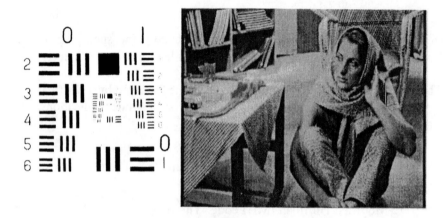

Figure 2.8   Typical bitonal images.

of black or white, and the static Huffman code in the facsimile compression standards is matched to these run-lengths. In the latter image, the run-lengths are relatively short, spanning only one to two pixels, and the static Huffman code is not matched to such runs. An adaptive arithmetic coder is better suited for such images.

2.11.2   The JBIG Compression Standard

Recently, a compression standard was developed to efficiently compress halftone as well as business-document-type images. This is the JBIG (Joint Binary Image Experts Group) compression standard. Its standards nomenclature is ISO/IEC IS 11544, ITU-T Rec. T.82. The JBIG compression standard consists of a modeler and an arithmetic coder. The modeler is used to estimate the symbol probabilities that are then used by the arithmetic coder, as we have described in a previous section. Previously encoded pixels are used as a context for the model, and a probability estimate is derived from this context; thus, with causal modeling, the decoder can mimic the encoder operations

	JBIG - Baselayer	**ITU-T Rec. T.6**
Complexity Parameters	Three-line template, AT-max = 16	2-D runlength, Huffman code
Memory	1,589 bytes	1,024 bytes
Buffer	Three scanlines	Two scanlines
Operations	Add, shift	Add, shift, compare
Compression		
Halftone Image	5.2:1	1.5:1
Letter Image	48:1	33.3:1

Table 2.8 Comparative analysis between JBIG and the ITU-T Rec. T.6 facsimile compression standards.

without any additional information for the modeler. Additional features in the JBIG standard support various image display and browsing modes. In Table 2.8, we provide a simple complexity analysis between the JBIG coding scheme and the ITU-T Rec. T.6 facsimile compression standard. We also provide compression ratios for two typical images: a 202 Kbyte halftone image and a 1 Mbyte image, primarily comprised of text. The latter image is referred to as *letter* in the table. For business-type documents, JBIG yields 20 percent to 50 percent more compression than the facsimile compression standards ITU-T Rec. T.4 and Rec. T.6. For halftone images, compression ratios with JBIG are two to five times more than those obtained with the facsimile compression standards. However, software implementations of JBIG compression on a general purpose computer are two to three times slower than implementations of the ITU-T Rec. T.4 and T.6 standards.

The JBIG standard can also handle grayscale images by processing each plane of a grayscale image as separate bitonal images.

2.11.3   The Lossless JPEG Standard

Most people know JPEG as a transform-based lossy compression standard. JPEG (Joint Photographic Experts Group), like JBIG, has been developed jointly by both the ITU-T and the ISO. We will describe this standard in greater detail in a subsequent chapter; however, here, we describe briefly the lossless mode of compression supported within this standard. The lossless compression method within JPEG is fully independent of transform-based coding. Instead, it uses differential coding to form prediction residuals that are then coded with either a Huffman coder or an arithmetic coder. As explained earlier, the prediction residuals usually have a lower entropy; thus, they are more amenable to compression than the original image pixels.

In lossless JPEG, one forms a prediction residual using previously encoded pixels in the current line and/or the previous line. The prediction residual for pixel X in Figure 2.9 is defined as $r = y - X$, where y can be any of the following functions:

$$y = 0 \tag{2.24a}$$
$$y = a \tag{2.24b}$$
$$y = b \tag{2.24c}$$
$$y = c \tag{2.24d}$$
$$y = a + b - c \tag{2.24e}$$
$$y = a + \frac{b - c}{2} \tag{2.24f}$$
$$y = b + \frac{a - c}{2} \tag{2.24g}$$
$$y = \frac{a + b}{2} \tag{2.24h}$$

Note that, pixel values at pixel positions a, b, and c are available to both the

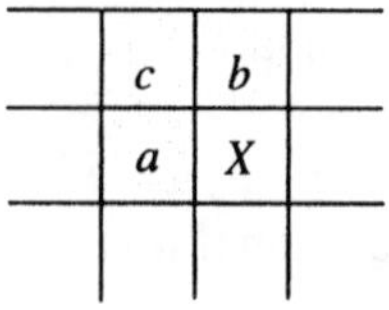

c	b	
a	X	

Figure 2.9   Lossless JPEG prediction kernel.

encoder and the decoder prior to processing X. The particular choice for the y function is defined in the scan header of the compressed stream so that both the encoder and the decoder use identical functions. Divisions by two are computed by performing a one-bit right shift.

Category	Prediction Residual
0	0
1	-1, 1
2	-3, -2, 2, 3
3	-7, .., -4, 4, .., 7
4	-15, .., -8, 8, .., 15
5	-31, .., -16, 16, .., 31
6	-63, .., -32, 32, .., 63
7	-127, .., -64, 64, .., 127
8	-255, .., -128, 128, .., 255
9	-511, .., -256, 256, .., 511
10	-1023, .., -512, 512, .., 1023
11	-2047, .., -1024, 1024, .., 2047
12	-4095, .., -2048, 2048, .., 4095
13	-8191, .., -4096, 4096, .., 8191
14	-16383, .., -8192, 8192, .., 16383
15	-32767, .., -16384, 16384, .., 32767
16	32768

Table 2.9 Prediction residual categories for lossless JPEG compression.

The prediction residual is computed modulo 2^{16}. This residual is not directly Huffman coded. Instead, it is expressed as a pair of symbols: the *category* and the *magnitude*. The first symbol represents the number of bits needed to encode the magnitude. Only this value is Huffman coded. The magnitude categories for all possible values of the prediction residual are shown in Table 2.9. If, say, the prediction residual for X is 42, then from Table 2.9 we determine that this value belongs to category 6; that is, we need an additional six bits to uniquely determine the value 42. The prediction residual is then mapped into the two-tuple (6, 6-bit code for 42). Category 6 is Huffman coded, and the compressed representation for the prediction residual consists of this Huffman codeword followed by the 6-bit representation for the magnitude. In general, if the value of the residual is positive, then the code for the magnitude is its direct binary representation. If the residual is negative, then the code for the magnitude is the one's complement of its absolute value. Therefore, codewords for negative residuals always start with a zero bit.

Example 3: Consider Figure 2.9 with pixel values $a = 100$, $b = 191$, $c = 100$, and $X = 180$. Let $y = \frac{a+b}{2}$; then $y = 145$, and the prediction residual is $r = 145\text{-}180 = \text{-}35$. From Table 2.9, -35 belongs to category 6. The binary number for 35 is 100011, and its one's complement is 011100. Thus, -35 is represented as (6, 011100). If the Huffman code for six is 1110, then -35 is coded by the 10-bit codeword 1110011100. Without entropy coding, -35 would require 16 bits.

In the decoder, the category (that is, 6) is extracted first. Thus, the next six bits, 011100, correspond to the magnitude of the residual. Since the most significant bit is zero, the residual is negative. After taking the one's complement of 011100, the decoded value of the residual r is -35. The a and b bits have already been decoded; thus, $y = 145$ as before, and $X = y + 35 = 180$.

This notion of using a category table is a form of context modeling and simplifies the Huffman coder. Without categorization of the prediction residuals, we would require a Huffman table for an alphabet of 2^{16} symbols. Such a large codeword table would complicate both the codeword construction process and the decoding process.

Lossless JPEG outperforms JBIG for typical grayscale images with more than six bits per pixel. At six bits per pixel or below, JBIG yields better compression ratios than JPEG. For typical images, such as the grayscale version of the halftone image of Figure 2.8, compression ratios in excess of 1.5 to 1 are quite difficult to achieve. The standards committee is currently working on developing new lossless compression techniques that can outperform the simple single-prediction, single-Huffman table coding method currently used in the lossless JPEG compression standard.

2.12   NEW WORK IN JPEG: JPEG-LS

In this section we describe recent work by the JPEG committee for a new standard for lossless and near-lossless compression, referred to as JPEG-LS. Near-lossless compression is defined as the coding process where each pixel at the output of the decoder differs by the corresponding input pixel by no more than a prespecified value.

In addition to supporting a near-lossless mode, JPEG-LS offers improved compression performance over the existing JPEG lossless coder (using both Huffman and arithmetic coding), with only moderate increase in computational complexity over the Huffman-based coder. Figure 2.10 shows a block diagram of the JPEG-LS encoder. It includes: a context modeler, a run-length coder,

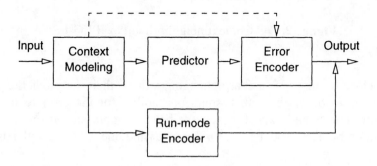

Figure 2.10   Block diagram of the JPEG-LS encoder.

a nonlinear predictor, and an Golomb-Rice entropy coder. These blocks are examined in further detail next.

2.12.1   Context Modeling

A new feature in JPEG-LS is context modeling without using arithmetic coding. In context modeling, coding of a new pixel x depends on values of its previously encoded neighbors. These values are used as a conditioning state or *context*. For each context, one can collect statistics on the pixel to be encoded and then use this information to drive the entropy coder. For example, as shown in Figure 2.11, JPEG-LS uses four neighbor pixels for context modeling.

The potential benefits of context modeling can be shown by revisiting the JPEG lossless coder we described in the previous section. For a pixel x, that coder uses a single probability model for coding the prediction error $y - x$, regardless of the values of the surrounding pixels a, b, and c. If x is in a "flat" area, then one would expect the error distribution of $y - x$ to be concentrated near zero, with a high probability for error to be zero. On the other hand,

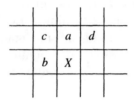

Figure 2.11   Context neighborhood in JPEG-LS.

if x is close to an edge (a sharp changing value), then the prediction error is expected to be large, with a small probability for the error to be zero. Therefore, one would expect to achieve better compression if one had two different probability models: one for flat areas and one for areas with edges.

Context modeling has been used rather effectively for coding bitonal images, for example, in JBIG, however, it is impractical to store and manipulate information about all possible contexts when coding grayscale images. For example, a neighborhood of four 8-bit pixels yields a total of 256^4 different contexts. JPEG-LS uses only 364 contexts, defined as follows.

1. Using the template shown in Figure 2.11, compute the gradients

$$\begin{aligned}
d1 &= \hat{d} - \hat{a}, \\
d2 &= \hat{a} - \hat{c}, \\
d3 &= \hat{c} - \hat{b},
\end{aligned} \qquad (2.25)$$

   where $\hat{a}$ to $\hat{d}$ denote the reconstructed values for pixels a to d. In lossless mode, these are identical to the values of a to d. $d1$ to $d3$ are referred to as local gradients and capture the level of activity (smooth, edges, etc.) surrounding the sample at x.

2. Using three nonnegative thresholds $T1$, $T2$, and $T3$, quantize $d1$, $d2$, and $d3$ to $Q1$, $Q2$, and $Q3$, respectively. The input-output function of the gradient quantizers is shown in Figure 2.12. Each of the gradients is mapped to nine possible values, for example, between -4 and 4. The three thresholds depend on the bits per pixel in the image. For example, for 8-bits per pixel, the default values are $T1 = 3$, $T2 = 7$, and $T3 = 21$. The value of ε in Figure 2.12 refers to the tolerance for error in the

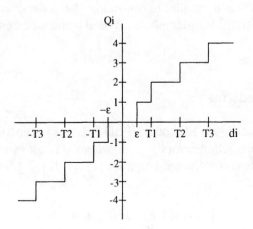

Figure 2.12   Gradient quantization in JPEG-LS.

near-lossless mode. Since each region can take nine possible values, at this point we have $9^3 - 1 = 728$ context values, defined by the vector $(Q1, Q2, Q3)$. (We subtract the case where all gradients are zero, because in that case the coder enters a *run-mode*, which is explained later.)

3. Due to symmetry, contexts of opposite signs can be merged. That is, if the first nonzero element of $(Q1, Q2, Q3)$ is negative, then all signs of $(Q1, Q2, Q3)$ are reversed to obtain $(-Q1, -Q2, -Q3)$. This final step halves the number of possible contexts to 364.

2.12.2   Run-mode Coder

If all the gradients in (2.25) are equal to zero or if in the near-lossless mode all of their absolute values are smaller than ε, then the coder enters into a run-mode. The encoder reads subsequent samples as long as $\hat{x} = \hat{b}$ (or in near-lossless mode, as long as $|\hat{x} - \hat{b}| \le \varepsilon$) or until the end of the current scan line. At the end of the run, the run-mode encoder encodes the length of the run. If the run was not interrupted by an end of row, then the encoder also encodes the last scanned sample. In run-mode, the coder does not use context modeling or prediction. Run-lengths are coded using an extension of Golomb coding.

2.12.3   The Predictor

Like the original lossless JPEG coder, JPEG-LS does not encode the x samples themselves, but prediction errors $y - x$, where y is an estimate of x based on the values of its reconstructed neighbor pixels. JPEG-LS uses a nonlinear predictor defined as

$$y = \begin{cases} min(\hat{a}, \hat{b}) & \text{if } \hat{c} \geq max(\hat{a}, \hat{b}) \\ max(\hat{a}, \hat{b}) & \text{if } \hat{c} \leq min(\hat{a}, \hat{b}) \\ \hat{a} + \hat{b} - \hat{c} & \text{otherwise} \end{cases} \quad . \tag{2.26}$$

This predictor performs an edge-detection test. If no edge is detected, then $y = \hat{a} + \hat{b} - \hat{c}$. Otherwise, for vertical or horizontal edges, $\hat{a}$ or $\hat{b}$ are predicted, respectively.

JPEG-LS uses an entropy coder that assumes that prediction errors follow a two-sided geometric and symmetric distribution, centered between -1 and 0. To alleviate the effects of systematic biases, prediction is followed by a context-dependent bias cancelation step that is aimed at centering the distribution of prediction errors between -1 and 0. The bias term is computed based on the sum of errors incurred so far for samples of the same context. After the bias cancelation step, the prediction error is computed. If the encoder is operating in a near-lossless mode, the prediction error is quantized and an estimate of $\hat{x}$, which will be used to encode further samples, is computed.

2.12.4   Error Coding

For error coding, JPEG-LS uses a Golomb-Rice coder, as described in Section 2.8. In effect, JPEG-LS combines the compression potential of context models with the simplicity of Golomb-Rice coding. In a pure Golomb-Rice (or other Huffman-like) coder, at least one bit per sample is needed. JPEG-LS alleviates this restriction by using the special run-mode in flat areas of the image. Overall, compression ratios from JPEG-LS have been shown to be very close to those obtained by arithmetic coders that are ten times more complex than JPEG-LS.

One of the important parameters in Golomb-Rice coding is the value of k that yields the code with the shortest possible average codelength. In JPEG-LS, the value of k is context-dependent and adaptive. If $N[i]$ denotes the number of prediction residuals seen so far in context i, and $A[i]$ denotes the accumulated sum of magnitudes of prediction errors in context i, then k is computed as

$$k = \left\lceil log_2 \frac{A[i]}{N[i]} \right\rceil . \tag{2.27}$$

In practice, k can be computed by repeatedly multiplying $N[i]$ by 2, until $2^k \times N[i] \geq A[i]$. In C code, this can be computed as

```
k = 0, Z = N[i];
while (Z < A[i]) { Z = 2 * Z; k = k + 1 };.
```

Since prediction errors, say e, may take both negative and positive values, negative and positive values are interleaved into a new nonnegative sequence e'. For $k \neq 0$, e' is defined as

$$e' = \begin{cases} 2e & \text{if } e \geq 0 \\ -2e - 1 & \text{if } e < 0 \end{cases} . \tag{2.28}$$

After this mapping, the mapped errors are encoded using a Golomb-Rice code and context-dependent parameters are updated. For each context i, JPEG-LS stores four parameters: (1) $N[i]$, the number of prediction residuals, (2) $A[i]$, the accumulated sum of the magnitudes of the prediction errors, (3) $B[i]$, the sum of prediction errors after correction, and (4) $C[i]$, a correction value used to offset prediction bias. A and N are used to update the k parameter of the Golomb-Rice coder. N and B are used to update the correction values in C.

Since all of the context-dependent parameters are computed using reconstructed pixel values, none of them has to be passed to the decoder. Instead, they are computed by the decoder the same way as in the encoder. The JPEG-LS decoder is very similar to the encoder, except that it replaces the run-mode and error encoders in Figure 2.10 with run-mode and error decoders. The JPEG-LS standard also defines various modes for color (true-color or palettized) images.

Table 2.10 shows compression ratios for four grayscale (8 bits per pixel) images using lossless JPEG (with both Huffman (H) coding and arithmetic coding (A)), JPEG-LS, and the portable network graphics format (PNG). Images *cmpnd1, cmpnd2* (text with natural images), and *finger* (a fingerprint) are from the JPEG continuous-tone test image set. The rest of the images are commonly used by the image processing research community.

Test Image	JPEG (H)	JPEG (A)	JPEG-LS	PNG
Natural images				
Barbara	1.48	1.53	1.70	1.53
Lena	1.59	1.59	1.74	1.63
Zelda	1.86	1.90	2.05	1.92
Compound documents				
cmpnd1	3.22	5.62	6.38	5.98
cmpnd2	3.05	5.27	5.86	5.34
Line art				
Einstein	1.74	2.13	2.29	2.07
finger	1.37	1.35	1.41	1.38

Table 2.10   Compression ratios for images with 8-bits per pixel using the JPEG, JPEG-LS, and PNG lossless compression algorithms.

PNG is a portable network graphics format for the lossless storage of raster images. It has been developed as a royalty-free replacement for GIF. It supports indexed-color, grayscale, and true color images. For compression, PNG currently supports only a variant of the LZ77 Lempel-Ziv compression scheme, which is commonly used by many text-oriented compression programs, such as *compress, zip,* and *gzip*. However, in the future, PNG may also support other compression schemes. PNG is supported by the worldwide Web consortium (W3C) as the format of choice for storing and displaying images in the internet.

In Table 2.10, for lossless JPEG with Huffman coding, the results are for the best predictor among all JPEG predictors and with custom Huffman tables for each image. For lossless JPEG with arithmetic coding, the predictor $y = \frac{a+b}{2}$ was used. From Table 2.10, JPEG-LS performs the best among all compression schemes. PNG performs better than lossless JPEG, but worse than JPEG-LS.

For lossless JPEG, arithmetic coding provides a significant improvement only in compound documents.

JPEG-LS is now an ISO/IEC committee draft document (ISO/IEC CD 14495). It is expected to become a draft international standard by the summer of 1997 and an international standard in 1998.

2.13 TO PROBE FURTHER

We have reviewed some of the algorithms and standards for lossless compression. Huffman coding, in particular, is the most widely used entropy coder in the standards. The textbook by Sayood [225] provides additional tutorial material and examples on lossless coding. A general discussion on entropy coders can be found in any textbook on information theory (e.g. [70]). The original description of the Huffman coding scheme can be found in [101]. Variations and performance bounds for Huffman coding are developed in detail in [71]. Most Huffman coders have a a complexity of order $O(N \log_2 N)$. In [161], Lu and Chen present a Huffman code generator with complexity of order $O(N)$. In this text, we described the bottom-up approach to Huffman codeword construction. A top-down approach is developed in [137].

In many practical implementations, the maximum codeword length of a Huffman code needs to be constrained. In this chapter, we described several approaches for constructing such a code. The Voorhis method [254] provides a code-construction algorithm that attains the maximum codeword length constraint and yields an optimum average codeword length. In recent years, this problem has been revisited, and additional code generation methods are described in [177], [138], and [160]. Another fast algorithm for optimal length-limited Huffman codes is described in [139]. The Golomb and Rice codes are described in [79] and and [221]. The optimality of the Golomb codes for geometric distributions is shown in [69].

Huffman coding methods are amenable to simpler software and hardware implementations; however, techniques based on arithmetic coding tend to yield a higher compression ratio. A detailed description of arithmetic coding can be found in [20]. An introduction to arithmetic coding that also includes source code is provided in [261]. There are several variants of the arithmetic coder, such as the Q-coder developed by IBM [170]. The Q-coder is an extension of the arithmetic coder described in the text. It includes a probability estimation module for the modeler and replaces time-consuming multiplications with additions and shifts. In [98], Howard and Vitter present another reduced-precision binary arithmetic coder in which most of the operations are performed with table lookups.

The ITU-T Rec. T.4 and T.6 standards can be found in [105] and [106]. A survey paper by Hunter and Robinson [102] provides a detailed description of the T.4 standard. The JBIG and the JPEG standards are described in the ISO/ITU-T documents [110] and [108]. In [12], Arps and Truong provide a detailed comparison of all the standard algorithms for still-image lossless compression, including Group 3, Group 4, JBIG, and lossless JPEG. Some early work on parameter reduction and context selection for the compression of grayscale images was presented in [244]. The JPEG-LS coder is based in great part on the LOCO-I coder developed by Weinberger et al. [257] and on work by Ono et al. [190]. PNG has been developed by the PNG Development group. Details about the PNG format can be found in [2].

3

FUNDAMENTALS OF LOSSY IMAGE COMPRESSION

3.1 INTRODUCTION

Lossy compression of images deals with compression processes where decompression yields an imperfect reconstruction of the original image data. A wide range of lossy compression methods have been developed for compressing still-image data. These methods fall into one of the categories shown in Figure 1.2. In this chapter, we describe some of the basic concepts of lossy compression that have been adopted in practice and that form the basis of the image and video compression standards.

As discussed in Chapter 1, the selection of a particular compression method involves many tradeoffs. However, regardless of the compression method that is being used, given the level of image loss (or distortion), there is always a bound on the minimum bit rate of the compressed bit stream. We begin this chapter by briefly describing such a bound using results from rate-distortion theory. Next, we describe some practical lossy compression schemes, and we compare their performance against the theoretical performance bound. Special emphasis is given to DCT-based compression schemes, since they form the basis of all the image and video compression standards. We conclude with a discussion on the fast implementation of the DCT.

3.2   PRELIMINARIES

3.2.1   A Model for Spatial Correlation in Images

Image data tend to have a high degree of spatial redundancy. Furthermore, image data are eventually presented to a human viewer. Thus, if one views a lossy still-image compression system from end to end, that is, from the creation of the visual information to the eventual display of the information after decompression, then it is prudent to exploit characteristics within each component of this system so as to generate a compressed stream with the least number of bits.   Within such a system, compression is achieved by exploiting both the spatial redundancies within the image and the perceptual characteristics of the human visual system so that loss due to compression may not be discernible to the viewer.

It is also possible to exploit some higher-level image characteristics or, for color images, to exploit correlation of data among different color components. For example, compression schemes based on fractals provide an image representation based on patterns that are repeated at different scales within the same image. Since the compression standards are not based on the existence of such features, we do not discuss any of these compression schemes in this book.

By spatial redundancy, we imply that pixels are correlated across space. For any image X, the spatial redundancy can be modeled by a two-dimensional (2-D) covariance function.   A typical isotropic (nonseparable) covariance function for image data is given by

$$Cov_X(i,j) = \sigma^2 e^{-\alpha\sqrt{i^2+j^2}}, \tag{3.1}$$

where σ^2 is the variance of the image and i and j refer to the distance from the reference pixel about which the covariance function is defined. If we denote the 2-D image samples as $X(i,j)$, let

$$\rho_1 = \frac{E[X(i,j)X(i-1,j)]}{E[X^2(i,j)]}, \tag{3.2}$$

$$\rho_2 = \frac{E[X(i,j)X(i,j-1)]}{E[X^2(i,j)]} \tag{3.3}$$

denote the correlation between two pixels in the vertical and horizontal directions, where $E[\ ]$ is the expectation (or averaging) operator. If we assume $\rho_1 = \rho_2$, that is, there is no difference between the vertical and horizontal correlation of two neighbor pixels, then $e^{-\alpha} = \rho_1$. For image data, typical values for ρ_1 and ρ_2 are around 0.95. A simple calculation of the covariance function for various values of i and j indicates that the covariance function decays rapidly and is quite small beyond $i, j > 8$. This finding suggests that compression schemes that exploit spatial redundancy need not consider blocks of pixels larger than eight pixels in either the vertical or horizontal dimension.

Under the compression context, this mathematical model for the spatial redundancy within an image serves two purposes:

1. It can be used to design components of the image compression system.

2. Given that the distortion between the original and the decompressed image is D, it can be used to derive a theoretical bound on the lowest bit rate $R(D)$ that is achievable with any compression scheme. A common measure for D is the mean square error between the encoded and decoded images, normalized by the variance of the input signal $X(i, j)$.

In the next section, we discuss in more detail the D versus $R(D)$ relationship.

3.2.2   Rate-Distortion Function for Images

Due to the loss of information, lossy compression yields higher compression ratios than lossless compression. However, in any practical lossy compression system, there is a tradeoff between loss or distortion (say D) and the bit rate of the compressed bit stream, (say R). Here, R is expressed in bits per encoder output symbol, and D is normalized by the variance of the encoder input. If the encoder input data are 8-bit pixels, the compression ratio can be expressed as $\frac{8}{R}$. Rate-distortion theory establishes the theoretical minimum bit rate R_{min} so that the compressed input can be reconstructed within the allowed distortion D. For every compression scheme, there is a D versus R relationship. For a given D, the rate-distortion function $R(D)$ is defined as the minimum possible rate R necessary to achieve average distortion D or less. $R(D)$ is independent of the particular compression method and depends only on the underlying stochastic model for the input images and the distortion measure.

As observed in the previous chapter, in lossless compression, the efficacy of a compression method is determined by the entropy of the source at the input of the entropy encoder. Furthermore, the entropy of the source depends only on the underlying statistical model for the source. Similar observations can be made for lossy compression schemes.

In lossy compression, the source-coding theorem states that it is possible to design a coding-decoding scheme of rate $R > R(D)$ so that the average distortion is D or less. The converse of the source-coding theorem states that if a coding-decoding system has rate $R < R(D)$, then it is impossible to achieve average distortion D or less with this system. These two theorems imply that $R(D)$ defines the lowest possible bit rate among all lossy compression methods. When $D = 0$, $R(D)$ is the source entropy.

Since $R(D)$ defines the lower limit for achievable compression ratio for a specified distortion D, it would be informative to compute $R(D)$ for an image. The resulting $R(D)$ can then be used to determine if a desired compression ratio is achievable in practice.

For the isotropic covariance function defined in (3.1), and $\sigma^2 = 1$, the 2-D power spectral density of an image X has the form

$$S_X(f_x, f_y) = \frac{f_0}{2\pi(f^2 + f_0^2)^{\frac{3}{2}}} = S_0(f), \qquad (3.4)$$

where $f = \sqrt{f_x^2 + f_y^2}$. We will assume that the power spectrum is bandlimited such that $S_X(f_x, f_y) = 0$ for $f > \frac{1}{\sqrt{\pi}}$. Let $\rho_1 = \rho_2 = 0.95$, assuming Nyquist sampling and approximating $e^{-x} = 1 - x$; then $f_0 = \frac{0.05}{\pi\sqrt{\pi}}$.

It has been shown by Berger that the optimal encoder-decoder tandem yields distortion $D(\theta)$ and a minimum rate $R(\theta)$ given by

$$D(\theta) = \int_{-\infty}^{\infty}\int_{-\infty}^{\infty} min(\theta, S_X(f_x, f_y))df_x df_y, \qquad (3.5)$$

$$R(\theta) = \int_{-\infty}^{\infty}\int_{-\infty}^{\infty} max\left(0, \frac{1}{2}log_2\frac{S_X(f_x, f_y)}{\theta}\right) df_x df_y. \qquad (3.6)$$

Parameter θ takes on all positive real values and generates the function $R(D)$, which we refer to as the rate-distortion function for this source.

For the bandlimited $S_X(f_x, f_y)$ of (3.4), by varying θ, the function $R(D)$ can be computed using (3.5) and (3.6). Closed form expressions for $D(\theta)$ and $R(\theta)$ can be obtained as follows:

If $f(\theta) \leq \frac{1}{\sqrt{\pi}}$,

$$D(\theta) = \pi\theta f^2(\theta) + f_0\left(\frac{1}{\sqrt{f^2(\theta) + f_0^2}} - \frac{1}{\sqrt{\frac{1}{\pi} + f_0^2}}\right), \tag{3.7}$$

$$R(\theta) = \frac{3\pi}{4\ln 2}\left(f_0^2 \ln \frac{f_0^2}{f^2(\theta) + f_0^2} + f^2(\theta)\right), \tag{3.8}$$

and if $f(\theta) > \frac{1}{\sqrt{\pi}}$,

$$D(\theta) = \theta, \tag{3.9}$$

$$R(\theta) = \frac{3\pi}{4\ln 2}\left(f_0^2 \ln \frac{f_0^2}{\frac{1}{\pi} + f_0^2} + \frac{1}{\pi}\right) + \frac{1}{2}\log_2 \frac{\Phi}{\theta}, \tag{3.10}$$

where θ and Φ are defined as

$$\theta = \frac{f_0}{2\pi(f^2(\theta) + f_0^2)^{\frac{3}{2}}}, \tag{3.11}$$

$$\Phi = \frac{f_0}{2\pi(\frac{1}{\pi} + f_0^2)^{\frac{3}{2}}}. \tag{3.12}$$

Figure 3.1 shows the rate-distortion function for various image models. In this figure, SNR is defined as $10\log_{10}\frac{1}{D}$. For the 2-D isotropic model, we show the $R(D)$ plot computed using (3.7) to (3.10). For comparison, $R(D)$ is also plotted for an image modeled using a one-dimensional (1-D) covariance function and for an image with uncorrelated data. If the image is uncorrelated and the samples have a Gaussian distribution, the corresponding $R(D)$ function is given by $\frac{1}{2}\log_2\frac{1}{D}$. From this figure, several observations can be made:

1. For the same image fidelity, exploiting 2-D correlation in the image leads to lower $R(D)$ (more compression) than if the image were assumed to be correlated only along one direction. Specifically, for the assumptions on which the $R(D)$ calculations were based, a doubling of compression ratio is achieved.

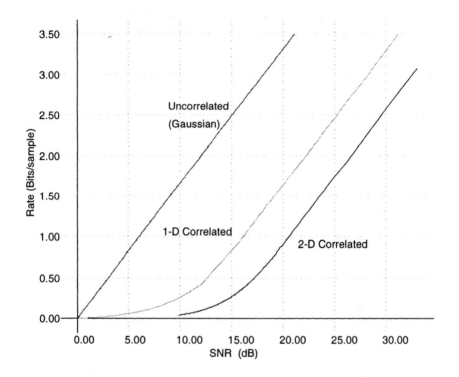

Figure 3.1 Rate-distortion performance for various image models.

2. If the image is uncorrelated or if the correlation is weak, then there is a significant degradation in compression, as seen by comparing the $R(D)$ plots for the Gaussian source model and the 2-D correlation source model. A quadrupling of the compression ratio is obtained in the latter case for the same image fidelity.

3. In regions of low image fidelity (low SNR), exploiting spatial redundancy in both dimensions of the image does not offer any significant reductions in the compression ratio. In these regions, the coding gain provided by the better correlation model is offset by the distortion introduced in the lossy compression scheme.

The $R(D)$ function has several practical uses:

- For sources where the correlation coefficients ρ_1 and ρ_2 can be estimated, the equations described above can be used to determine whether the exploitation of spatial redundancies would be beneficial. In a practical coder, a compression method that exploits spatial redundancy has a higher complexity than a coder that simply assumes that the source is uncorrelated. Thus, the $R(D)$ function can provide some insight into the R versus D versus complexity tradeoffs discussed in Chapter 1.

- For practical coders, one can measure D and the corresponding rate R and compare them against the theoretical limit $R(D)$. This comparison will determine the efficacy of the practical coder. In general, it is difficult to compare one coder against another, given the different parameters used by each coder. However, one can always compare the performance of each coder against the theoretical limit $R(D)$. These comparisons can then be used to compare the two coders. Note that, in Figure 3.1, the SNR is computed for a unit variance source. When comparing the results of a practical coder against this plot, the SNR axis has to be scaled appropriately to account for the estimated variance of the source being compressed. Instead of SNR, peak SNR (PSNR) may also be used, provided that the horizontal axis in Figure 3.1 is scaled accordingly.

- In a practical setting, the $R(D)$ function can be used to optimize the design of functional blocks within the encoder and decoder.

For example, assume we want to compress an 8-bit grayscale image with a maximum distortion of 20 dB. From Figure 3.1, using the 2-D correlated model, the lowest bit rate is close to 0.9 bits per sample, which corresponds to a compression ratio of 8.88 to 1. Similarly, if we want a compression ratio of 20 to 1 (or 0.4 bits per sample), the best SNR we can achieve is close to 16 dB.

In the calculations for $R(D)$, we have assumed a stochastic image model. For images, using a perceptual model in addition to the stochastic model is more appropriate. This can be done quite simply by modifying the power spectral density $S_X(f_x, f_y)$ of (3.4) as $W(f_x, f_y)S_X(f_x, f_y)$, where $W(f_x, f_y)$ is an estimate of the frequency response of the human visual system.

The reader must exercise caution when comparing the $R(D)$ plots in Figure 3.1 with the measured data from a specific coder, since

- in deriving the closed-form expressions for D and R, some assumptions were made regarding the bandwidth of $S_X(f_x, f_y)$,

- the $R(D)$ function was computed for specific values for the correlation coefficients ρ_1 and ρ_2, and these values may change from image to image, and

- the distortion metric used in the specific coder may not be the mean-squared error (MSE) metric which we have used here.

Later in this chapter we consider two of the lossy compression schemes that are the basis for the image compression standards, and we compare their performance against the theoretical rate-distortion bound we have described here. In another chapter, we will revisit the $R(D)$ formulation and extend it to video coding.

3.3   BASIC CODING SCHEMES FOR LOSSY COMPRESSION

There are two classes of lossy compression schemes for images: sample-based coding and block-based coding.

3.3.1   Sample-Based Coding

In sample-based coding, the image samples are compressed on a sample-by-sample basis. The samples can be either in the spatial domain or in the frequency domain. A simple sample-based compression method is a scalar predictive coder such as the differential pulse code modulation (DPCM) method shown in Figure 3.2. This scheme is very similar to the one used in lossless JPEG (described in Chapter 2), except that in lossless JPEG there is no quantization. For each input sample x_{ij}, a residual signal e_{ij} is formed using the prediction signal P_{ij}. P_{ij} is typically a weighted sum of three to four previously decoded pixels near $\hat{x}_{ij}$, where

$$\hat{x}_{ij} = P_{ij} + q_{ij} \tag{3.13}$$

denotes the decoder's estimate for x_{ij}.

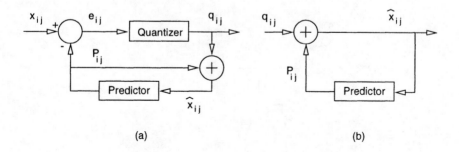

(a)   (b)

Figure 3.2 DPCM compression method. (a) Encoder; (b) decoder.

For example, for the prediction kernel shown in Figure 3.3,

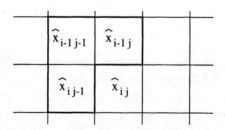

Figure 3.3 Example of a prediction kernel for DPCM.

$$P_{ij} = w_1\hat{x}_{ij-1} + w_2\hat{x}_{i-1j-1} + w_3\hat{x}_{i-1j}, \tag{3.14}$$

where w_1, w_2, and w_3 are weights that sum to one. Both the encoder and the decoder use the same predictor.

If the image is highly correlated, P_{ij} will track x_{ij}, and e_{ij} will consequently be quite small. From Figure 3.2a, the residual signal e_{ij} is quantized. A typical uniform quantization function is shown in Figure 3.4. The quantizer maps several of its inputs into a single output. This process is irreversible and is the main cause of information loss. For a uniform quantizer, the quantization

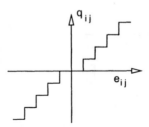

Figure 3.4   Input-output characteristics of a typical quantizer.

process can be expressed as

$$q_{ij} = round\left(\frac{e_{ij}}{\Delta}\right), \qquad (3.15)$$

where Δ is the quantization step. For 8-bit image data, Δ is usually an integer between 1 and 255.

Since the variance of e_{ij} is lower than the variance of x_{ij}, quantizing e_{ij} will not introduce significant distortion. Furthermore, the lower variance corresponds to lower entropy and thus to higher compression. As mentioned earlier, for typical images, the covariance function decays rapidly at distances longer than eight pixels. This implies that there is no benefit to using more than an eight-pixel neighborhood when forming P_{ij}.

If x_{ij} represents 8-bit pixels, then e_{ij} will be in the range [-255, 255]. The quantizer remaps e_{ij} into the quantized data q_{ij}, which may occupy the same dynamic range as e_{ij} but require fewer bits in their representation. For example, if there are only 16 quantization levels, then the quantized data can be represented by only four bits.

DPCM coders do not yield performance close to the $R(D)$ bound that we described in the previous section. Let P_{ij} be given by (3.14). If we assume an isotropic covariance model with correlation coefficient ρ and if the effects of quantization noise are ignored, then the rate-distortion function for DPCM can be expressed as

$$R(D)_{DPCM} = R(D)_{uncorrelated} - \frac{1}{2}log_2 G_p, \qquad (3.16)$$

where G_p is the gain due to prediction and is given by

$$G_p = 1 - \rho^2 \left(\frac{2}{1 + \rho^{\sqrt{2}}} \right). \tag{3.17}$$

For $\rho = 0.95$, DPCM coding yields a bit rate that is lower by 1.97 bits per sample than the bit rate of the uncorrelated source. To visualize this plot, in Figure 3.1 simply shift down the R versus D curve labeled *Uncorrelated* by 1.97. In practice, the effects of quantization noise cannot be ignored, and the true DPCM performance is nearly 6 to 10 dB worse than the $R(D)$ bound shown in Figure 3.1 for the 2-D correlated image model.

The computational complexity of DPCM is quite reasonable. For example, if P_{ij} is computed using three prior samples, then computing P_{ij} requires three multiplications and two additions, computing e_{ij} requires one subtraction, and forming q_{ij} requires one multiplication, for an overall cost of four multiplications and three additions per pixel. Since most of the multiplications involve small constants, lookup-table-based multiplication or multiplication using shift and add operations is sufficient.

3.3.2 Block-Based Coding

From rate-distortion theory, the source-coding theorem shows that, as the size of the coding block increases, we can find a coding scheme of rate $R > R(D)$ whose distortion is arbitrarily close to D. In other words, if a source is coded as an infinitely large block, then it is possible to find a block-coding scheme with rate $R(D)$ that can achieve distortion D. This implies that a sample-based compression method, such as DPCM, cannot attain the $R(D)$ bound and that block-based coding schemes yield better compression ratios for the same level of distortion. Many compression techniques have been developed to operate in a blockwise manner. These techniques fall into one of two classes: spatial domain block coding and transform-domain block coding.

In spatial-domain block coding, the pixels are grouped into blocks, and the blocks are then compressed in the spatial domain. Vector-quantization-based methods fall into this category. Vector-quantization (VQ) methods require far more processing in the encoder than in the decoder and are described in more detail in Chapter 17. From a rate-distortion viewpoint, for the same rate, such coders yield images with at least 3 dB better SNR than DPCM methods.

In transform-domain block coding, the pixels are grouped into blocks, and the blocks are then transformed to another domain, such as the frequency domain. Let us define two $N \times N$ transformation matrices

$$T_c = \{t_c(u, i)\}, u, i = 0, 1, .., N - 1, \qquad (3.18)$$
$$T_r = \{t_r(v, j)\}, v, j = 0, 1, .., N - 1. \qquad (3.19)$$

If the 2-D image in the spatial domain is X, then the $N \times N$ linear transformation process can be expressed as $Y = T_c X T_r^t$, where T_r^t denotes the matrix transpose of T_r. The T_c and T_r matrices are also referred to as transformation kernels or basis functions. For symmetric kernels, $T_c = T_r = T$ and $Y = TXT^t$.

The motivation for transforming X into Y is to obtain in Y a more compact representation of the data in X. Lossy transform-based compression methods achieve compression by first performing the transformation from X to Y and then by discarding the less important information in Y.

In any practical system, the compression process is followed at some point by decompression. Thus, the transformation process has to be reversible so that X can be reconstructed from Y. It follows that, for a specific T, there should be a U such that $X = UYU^t$.

Some of the most commonly used basis functions for T include the basis functions of the discrete Fourier transform (DFT), the discrete cosine transform (DCT), the discrete sine transform (DST), the discrete Hadamard transform (DHT), and the Karhunen-Loeve transform (KLT). In Figure 3.5, we show the output of the transform process for a typical image. Dark regions indicate less energy.

From this figure, it is seen that the KLT basis is the most efficient in terms of compaction efficiency, since all the energy is compacted into the top left corner. The DHT basis has poor compaction efficiency; its basis functions are rectangular compared to the smoother functions for the DCT or DST basis. The KLT is considered to be the optimum transform in the sense that

1. it packs the most energy in the least number of elements in Y,

2. it minimizes the total entropy of the sequence, and

3. it completely decorrelates the elements in X.

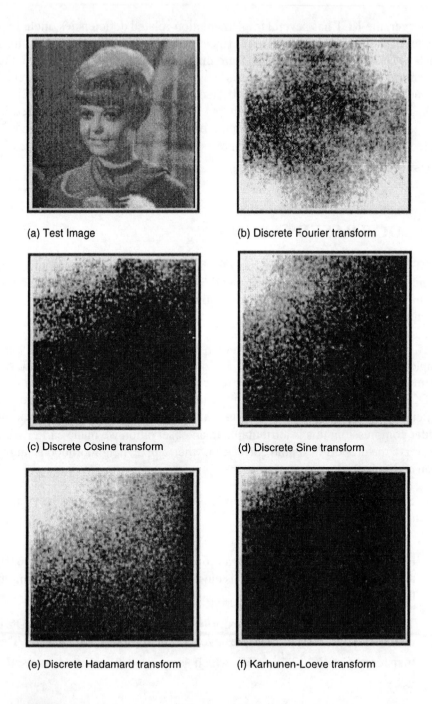

(a) Test Image

(b) Discrete Fourier transform

(c) Discrete Cosine transform

(d) Discrete Sine transform

(e) Discrete Hadamard transform

(f) Karhunen-Loeve transform

Figure 3.5 Compaction efficiency for various image transforms.

However, the KLT has several implementation-related deficiencies, including the fact that the basis functions are image dependent. The other basis functions such as DFT, DCT, DST, and DHT are image independent. Among them, the transform output for the DCT basis is seen to be close to the output produced by the KLT; that is, its compaction efficiency is close to that of the KLT. Since it has a performance very close to the KLT, the DCT basis is widely used in image and video compression and is the basis of choice for all the image and video compression standards. In the next section, we will discuss the basics of image coding using the DCT basis.

3.4   DCT-BASED CODING

DCT-based image coding is the basis for all the image and video compression standards. The basic computation in a DCT-based system is the transformation of an $N \times N$ image block from the spatial domain to the DCT domain. For the image compression standards, $N = 8$.

An 8×8 blocksize is chosen for several reasons.[1] From a hardware or software implementation viewpoint, an 8×8 blocksize does not impose significant memory requirements; furthermore, the computational complexity of an 8×8 DCT is manageable on most computing platforms.[2] From a compaction efficiency viewpoint, a blocksize larger than 8×8 does not offer significantly better compression; this is attributable to an observation we made earlier with regard to the dropoff in spatial correlation when a pixel neighborhood is larger than eight pixels.

The choice of the DCT in the standards is motivated by the many benefits it offers:

- [3] For highly correlated image data (correlation coefficients $\rho_1, \rho_2 > 0.7$), the DCT compaction efficiency is close to that obtained with the optimum transform, namely, the KLT.

- [4] The DCT is an orthogonal transform. Thus, if in matrix form the DCT output is $Y = TXT^t$, then the inverse transform is $X = T^tYT$. The transformation $X \overset{8 \times 8 DCT}{\longrightarrow} Y$, which is commonly referred to as the forward DCT or simply the DCT, is expressed as

$$y_{kl} = \frac{c(k)c(l)}{4} \sum_{i=0}^{7} \sum_{j=0}^{7} x_{ij} \cos\left(\frac{(2i+1)k\pi}{16}\right) \cos\left(\frac{(2j+1)l\pi}{16}\right),$$

(3.20)

where $k, l = 0, 1, ..., 7$ and

$$c(k) = \begin{cases} \frac{1}{\sqrt{2}} & \text{if } k = 0 \\ 1 & \text{otherwise} \end{cases}.$$

(3.21)

The DCT transformation can also be expressed in vector-matrix form as

$$y = Tx,$$

(3.22)

where $x = \{x_{00}, x_{01}, \cdots, x_{07}, x_{10}, x_{11}, \cdots, x_{17}, \cdots, x_{70}, x_{71}, \cdots, x_{77}\}$, T is a 64×64 matrix whose elements are the product of the cosine functions defined in (3.20), and $y = \{y_{00}, y_{01}, \cdots, y_{07}, y_{10}, y_{11}, \cdots, y_{17}, \cdots, y_{70}, y_{71}, \cdots, y_{77}\}$.

The DCT transformation decomposes each input block into a series of waveforms, each with a particular spatial frequency. The 64 waveforms composing the DCT basis functions are depicted in Figure 3.6. The DCT transformation can be viewed as the process of finding for each waveform shown in Figure 3.6 the corresponding weight y_{kl} so that the sum of the 64 waveforms scaled by the corresponding weights y_{kl} yields the reconstructed version of the original 8×8 matrix X.

From (3.20) and the orthogonality property of the DCT, the 8×8 inverse DCT transform (commonly referred to as IDCT) can be derived as

$$x_{ij} = \sum_{k=0}^{7} \sum_{l=0}^{7} y_{kl} \frac{c(k)c(l)}{4} \cos\left(\frac{(2i+1)k\pi}{16}\right) \cos\left(\frac{(2j+1)l\pi}{16}\right), \quad (3.23)$$

where $i, j = 0, 1, ..., 7$. Note that, the computations for the forward and inverse DCT are nearly the same. Thus, from a hardware implementation viewpoint, the same computation unit can be used for both the forward and the inverse DCT.

■ An important property of the 2-D DCT and IDCT transforms is separability. The 1-D DCT is computed as

$$z_k = \frac{c(k)}{2} \sum_{i=0}^{7} x_i \cos\left(\frac{(2i+1)k\pi}{16}\right), \quad k = 0, 1, \cdots, 7.$$

(3.24)

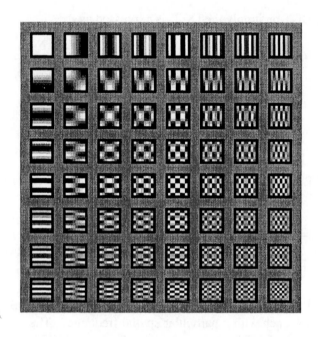

Figure 3.6 The 64 8 × 8 DCT basis functions.

This equation can also be expressed in vector-matrix form as $z = Tx^t$, where T is an 8 × 8 matrix whose elements are the cosine function values defined in (3.24), $x = [x_0, x_1, \cdots, x_7]$ is a row vector, and z is a column vector. From (3.20), the output of the 2-D DCT can be expressed as

$$y_{kl} = \frac{c(k)}{2} \sum_{i=0}^{7} \left[\frac{c(l)}{2} \sum_{j=0}^{7} x_{ij} \cos\left(\frac{(2j+1)l\pi}{16}\right) \right] \cos\left(\frac{(2i+1)k\pi}{16}\right).$$

(3.25)

Let

$$z_{il} = \frac{c(l)}{2} \sum_{j=0}^{7} x_{ij} \cos\left(\frac{(2j+1)l\pi}{16}\right), \quad i = 0, 1, \cdots, 7 \qquad (3.26)$$

denote the output of the 1-D DCTs of the *rows* of x_{ij}. The above equations imply that the 2-D DCT can be obtained by first performing 1-D DCTs of the rows of x_{ij} followed by 1-D DCTs of the *columns* of z_{il}. In matrix notation, $Y = TXT^t$, and this can also be expressed as

$$Z = TX^t, \tag{3.27}$$
$$Y = TZ^t = TXT^t. \tag{3.28}$$

From an implementation viewpoint, this row-column approach may also simplify the hardware requirements at the expense of a slight increase in the overall operations count. This issue is covered in more detail in a later section of this chapter.

- The DCT basis is image independent. This is an important issue in compression, since an image-dependent basis implies that additional computations need to be performed to determine the basis. Image independence, however, will result in some loss of performance.

- As observed earlier, the spatial-domain block is decomposed by the DCT in terms of the 64 waveforms shown in Figure 3.6. Since each y_{kl} represents the contribution of the corresponding kl-th waveform, the characteristics of the human visual system could be easily incorporated by suitably modifying y_{kl}. Compare this with a spatial domain coding technique such as DPCM. In DPCM, it is not intuitively obvious how the prediction residual can be modified to account for the characteristics of the human visual system.

- The DCT computations as expressed in (3.20) or (3.24) can be performed with fast algorithms that require fewer operations than the computations performed directly from these equations. Fast algorithms are desirable from both a hardware and a software implementation viewpoint. These fast algorithms also tend to be parallelizable and thus can be efficiently implemented on parallel architectures. We will describe some fast DCT algorithms later in this chapter.

3.4.1   A Generic DCT-based Image Coding System

Figure 3.7 shows the key functional blocks in a generic DCT-based image coding system. This diagram represents the core computation pipeline employed in all the lossy image and video compression standards discussed in this text. In the encoder, the DCT process transforms each 8×8 block X into a set of DCT coefficients Y. As noted earlier, each of these coefficients

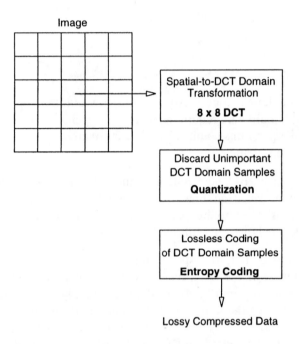

Figure 3.7   Generic DCT-based coding system.

is the weight associated with the corresponding DCT basis waveform. In the lossy compression mode, some of the weights will be deleted and thus the corresponding waveforms will not be used during decompression. The process of deleting some of the weights is referred to as the quantization process in Figure 3.7. The quantization process is an irreversible process and is the only source of loss in a DCT coding scheme. Strictly speaking, even with no quantization, there may be additional losses related to the implementation accuracy of the DCT and IDCT. Furthermore, in a transmission application, there may be additional losses due to noise in the transmission link.

After quantization, the nonzero quantized y_{kl} values are then compressed in a lossless manner using an entropy coder. In most applications, the entropy coder combines a run-length coder with a Huffman coder. Entropy coding specifics are discussed in detail in a later chapter on the JPEG still-image compression standard.

We provide two examples that illustrate the various steps performed during compression and decompression in a DCT-based coding system.

Example 1: The DCT coding for a typical image is shown in Figure 3.8. Here,

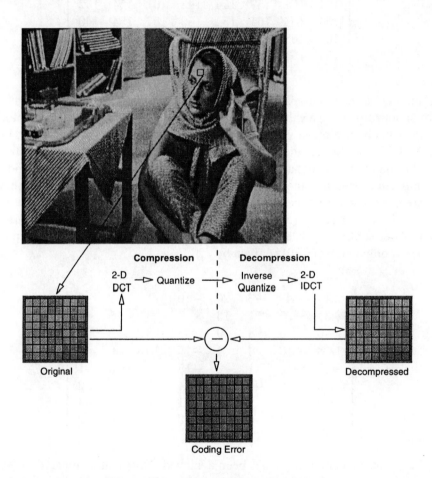

Figure 3.8 DCT coding - an example.

the input 8×8 block (labeled *original*) is taken from a low activity region; that is, there are very small differences among pixel values in that area. The pixel values for this block are given by

$$X = \begin{bmatrix} 168 & 161 & 161 & 150 & 154 & 168 & 164 & 154 \\ 171 & 154 & 161 & 150 & 157 & 171 & 150 & 164 \\ 171 & 168 & 147 & 164 & 164 & 161 & 143 & 154 \\ 164 & 171 & 154 & 161 & 157 & 157 & 147 & 132 \\ 161 & 161 & 157 & 154 & 143 & 161 & 154 & 132 \\ 164 & 161 & 161 & 154 & 150 & 157 & 154 & 140 \\ 161 & 168 & 157 & 154 & 161 & 140 & 140 & 132 \\ 154 & 161 & 157 & 150 & 140 & 132 & 136 & 128 \end{bmatrix}.$$

Depending on the color space, image pixels of a color component may have zero or nonzero average values. For example, in the RGB color space, all color components have a mean value of 128 (assuming 8-bit pixels). However, in the YCbCr color space, the Y component has an average value of 128, but the chroma components have an average value of zero. In order to provide for uniform processing, most standard DCT coders require that image pixels are preprocessed so that their expected mean value is zero. The subtracted (or added) bias is then added (or subtracted) back by the decoder after the inverse DCT. After subtracting 128 from each element of X, the 8×8 DCT output block as computed by (3.20) is given by

$$Y = \begin{bmatrix} 214 & 49 & -3 & 20 & -10 & -1 & 1 & -6 \\ 34 & -25 & 11 & 13 & 5 & -3 & 15 & -6 \\ -6 & -4 & 8 & -9 & 3 & -3 & 5 & 10 \\ 8 & -10 & 4 & 4 & -15 & 10 & 6 & 6 \\ -12 & 5 & -1 & -2 & -15 & 9 & -5 & -1 \\ 5 & 9 & -8 & 3 & 4 & -7 & -14 & 2 \\ 2 & -2 & 3 & -1 & 1 & 3 & -3 & -4 \\ -1 & 1 & 0 & 2 & 3 & -2 & -4 & -2 \end{bmatrix}.$$

At this point, no compression has been achieved. Note that, compared to X, the DCT-transformed datum Y has large amplitudes clustered close to y_{00}, commonly referred to as the DC coefficient. In general, for a low-activity block, most of the high-amplitude data will be in the low-order coefficients, as is the case in this example.

It is the process of quantization which leads to compression in DCT domain coding. The process of quantization of y_{kl} is expressed as

$$z_{kl} = round\left(\frac{y_{kl}}{q_{kl}}\right) = \left\lfloor \frac{y_{kl} \pm \lfloor \frac{q_{kl}}{2} \rfloor}{q_{kl}} \right\rfloor, k, l = 0, 1, ..., 7, \qquad (3.29)$$

where q_{kl} denotes the kl-th element of an 8×8 quantization matrix Q. ($\lfloor x \rfloor$ denotes the largest integer smaller or equal to x.) In order to ensure that the same type of clipping is performed for either positive or negative valued y_{kl}, in (3.29), if $y_{kl} \geq 0$, then the two terms in the nominator are added; otherwise they are subtracted. In this example, if the 8×8 quantization matrix is given by

$$Q = \begin{bmatrix} 16 & 11 & 10 & 16 & 24 & 40 & 51 & 61 \\ 12 & 12 & 14 & 19 & 26 & 58 & 60 & 55 \\ 14 & 13 & 16 & 24 & 40 & 57 & 69 & 56 \\ 14 & 17 & 22 & 29 & 51 & 87 & 80 & 62 \\ 18 & 22 & 37 & 56 & 68 & 109 & 103 & 77 \\ 24 & 35 & 55 & 64 & 81 & 104 & 113 & 92 \\ 49 & 64 & 78 & 87 & 103 & 121 & 120 & 101 \\ 72 & 92 & 95 & 98 & 112 & 100 & 103 & 99 \end{bmatrix}, \qquad (3.30)$$

then the quantized DCT output is given by

$$Z = \begin{bmatrix} 13 & 4 & 0 & 1 & 0 & 0 & 0 & 0 \\ 3 & -2 & 1 & 1 & 0 & 0 & 0 & 0 \\ 0 & 0 & 1 & 0 & 0 & 0 & 0 & 0 \\ 1 & -1 & 0 & 0 & 0 & 0 & 0 & 0 \\ -1 & 0 & 0 & 0 & 0 & 0 & 0 & 0 \\ 0 & 0 & 0 & 0 & 0 & 0 & 0 & 0 \\ 0 & 0 & 0 & 0 & 0 & 0 & 0 & 0 \\ 0 & 0 & 0 & 0 & 0 & 0 & 0 & 0 \end{bmatrix}.$$

The process of quantization has resulted in the zeroing out of many of the DCT coefficients y_{kl}. The specific design of Q depends on psychovisual characteristics and compression-ratio considerations. All compression standards provide default values for Q. In a later chapter, we describe some of the techniques used to develop quantization tables.

At this point, the quantized DCT domain representation, Z, has resulted in significant savings, since only 11 values are needed to represent Z, compared to the 64 values needed to represent X; this represents a compression ratio of 5.8. The matrix Z can be efficiently represented using a combination of a run-length coding scheme and a Huffman coding scheme; we will describe the specifics of such an entropy coder in a later chapter.

Decompression begins with the entropy decoding of the coded bitstream. Since entropy coding is a lossless compression scheme, the decoder should be able to reconstruct an exact version of Z. Inverse quantization on Z is simply performed as

$$\hat{z}_{kl} = z_{kl} q_{kl} . \tag{3.31}$$

In this example, the inverse quantizer output matrix $\hat{Z}$ is

$$\hat{Z} = \begin{bmatrix}
208 & 44 & 0 & 16 & 0 & 0 & 0 & 0 \\
36 & -24 & 14 & 19 & 0 & 0 & 0 & 0 \\
0 & 0 & 16 & 0 & 0 & 0 & 0 & 0 \\
14 & -17 & 0 & 0 & 0 & 0 & 0 & 0 \\
-18 & 0 & 0 & 0 & 0 & 0 & 0 & 0 \\
0 & 0 & 0 & 0 & 0 & 0 & 0 & 0 \\
0 & 0 & 0 & 0 & 0 & 0 & 0 & 0 \\
0 & 0 & 0 & 0 & 0 & 0 & 0 & 0
\end{bmatrix}.$$

After the inverse quantization, the decoder computes an 8×8 IDCT of $\hat{Z}$ using (3.23). The 8×8 IDCT output is given by

$$\tilde{X} = \begin{bmatrix}
171 & 160 & 149 & 149 & 158 & 166 & 166 & 162 \\
174 & 164 & 155 & 154 & 160 & 164 & 161 & 156 \\
171 & 164 & 157 & 156 & 158 & 158 & 151 & 145 \\
161 & 157 & 154 & 154 & 155 & 151 & 144 & 137 \\
156 & 155 & 155 & 156 & 156 & 152 & 145 & 140 \\
159 & 160 & 160 & 160 & 157 & 153 & 148 & 145 \\
161 & 161 & 160 & 156 & 150 & 144 & 141 & 139 \\
159 & 158 & 155 & 148 & 139 & 132 & 129 & 128
\end{bmatrix}.$$

Observe that $\tilde{X} \neq X$ and that the only cause of this coding error is the quantization of the DCT coefficients. However, in this example, as evidenced in Figure 3.8, the coding error is not perceptually significant.

Example 2: To better illustrate the properties of the DCT, we include another example as depicted in Figure 3.9. Here, we have chosen an 8 × 8 input block

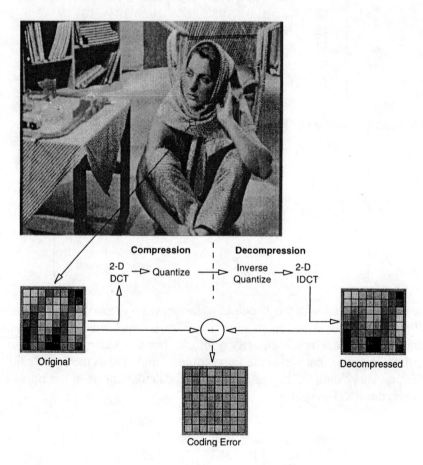

Figure 3.9 DCT coding example for a block in a high-activity region.

from a high-activity region. Zooming into this block, from Figure 3.9, we note that the block has two edge regions. The pixel values for this block are given by

$$X = \begin{bmatrix} 197 & 184 & 144 & 103 & 130 & 133 & 70 & 51 \\ 200 & 158 & 111 & 141 & 179 & 151 & 70 & 73 \\ 172 & 110 & 111 & 179 & 192 & 135 & 95 & 144 \\ 118 & 77 & 139 & 193 & 156 & 102 & 128 & 193 \\ 73 & 75 & 151 & 163 & 110 & 84 & 154 & 197 \\ 54 & 84 & 142 & 122 & 73 & 90 & 160 & 162 \\ 50 & 95 & 130 & 71 & 52 & 101 & 146 & 117 \\ 68 & 115 & 106 & 55 & 63 & 116 & 118 & 72 \end{bmatrix},$$

and the corresponding DCT output is

$$Y = \begin{bmatrix} -60 & -5 & -14 & -38 & 17 & 15 & 15 & 7 \\ 127 & 139 & -40 & 103 & 102 & -41 & 12 & -13 \\ -76 & 123 & 22 & 110 & -105 & -46 & 1 & -8 \\ -20 & -5 & 29 & -53 & -54 & 18 & -1 & 11 \\ -4 & 4 & 5 & -25 & -6 & 4 & 2 & 5 \\ -3 & -10 & 9 & -19 & -5 & 8 & 7 & 6 \\ 3 & 0 & 1 & 1 & 4 & 0 & -1 & -2 \\ -1 & -4 & 5 & -4 & -2 & -2 & 1 & 4 \end{bmatrix}.$$

Note that, compared with the Y output of the previous example, the dominant DCT values are not clustered close to y_{00}. This is usually the case when the spatial-domain block is a high-activity block. For this example, we chose a quantization matrix that yields the same compression ratio as the one obtained in the previous example. Repeating the same calculations as in the previous example, the IDCT output is

$$\tilde{X} = \begin{bmatrix} 198 & 182 & 153 & 136 & 145 & 145 & 95 & 32 \\ 182 & 159 & 146 & 153 & 152 & 129 & 98 & 81 \\ 153 & 124 & 135 & 174 & 159 & 105 & 104 & 150 \\ 120 & 95 & 125 & 180 & 153 & 86 & 112 & 203 \\ 88 & 84 & 120 & 159 & 130 & 81 & 121 & 211 \\ 62 & 93 & 120 & 114 & 92 & 92 & 131 & 173 \\ 45 & 112 & 123 & 64 & 52 & 110 & 139 & 114 \\ 37 & 127 & 126 & 31 & 27 & 123 & 143 & 72 \end{bmatrix}.$$

A comparison of the coding error images in Figure 3.8 and Figure 3.9 indicates that, for the same compression ratio, we have incurred more error for the high-activity block. However, from a perceptual viewpoint, the viewer may find the decompressed image block in Figure 3.9 to be the same as the original image block; this is because the eye is less sensitive to quantization noise on edges.

Another way to visualize the DCT coding process for our two examples is to look at which of the DCT kernels of Figure 3.6 are retained after the quantization process. Figure 3.10a shows the main DCT kernels for the low-activity block, and Figure 3.10b shows the main DCT kernels for the high-activity block. Note that, for the high-activity block of Example 2, some

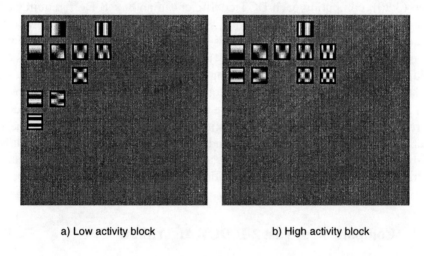

a) Low activity block   b) High activity block

Figure 3.10 DCT kernels retained for reconstruction of 8 × 8 block.

of the high-frequency information-bearing kernels need to be retained for good image fidelity.

Unlike DPCM, in DCT-based coding, higher lossy compression results in blurred decompressed images. This is because an increase in compression ratio implies using larger values for the quantization coefficients q_{kl}. Since most of the larger values are often concentrated in the high frequencies, these frequencies are zeroed out and cause loss of detail in the reconstructed image.

In these examples, we have illustrated the basic steps in a DCT-based coding scheme. From a computational point of view, quantization and inverse quantization require 64 multiplications each. On the other hand, brute-force evaluation of an 8 × 8 DCT or IDCT requires close to 4,096 multiply-accumulate operations. In the next section, we describe several algorithms for the fast computation of the 8 × 8 DCT or IDCT.

3.5   FAST ALGORITHMS FOR THE DCT

From (3.20), computing each DCT coefficient in an 8 × 8 DCT requires 64 multiplications and 64 additions. Thus, 4,096 multiply-accumulate operations are needed for each 8 × 8 block. This number can easily be reduced if we use a row-column approach. From (3.24), an eight-point 1-D DCT requires 64 multiplications and 64 additions. Using the row-column decomposition, we need to compute 16 1-D DCTs (eight for the rows and eight for the columns) for a total of 1,024 multiply-accumulate operations. The separability property has reduced the computational complexity by a factor of four; however, these numbers are still quite prohibitive and need to be reduced if we require real-time completion of the DCT process for a whole image. In recent years, several fast algorithms for the computation of the DCT have been developed.

3.5.1   Comparison of Fast 2-D DCT Algorithms

Most of the fast algorithms for the DCT exploit the properties of the transformation matrix T. Essentially, it has been shown that the matrix T of either (3.22) or (3.27) can be factorized so that $T = T_1 T_2 ... T_k$, where each of the matrices $T_1, T_2, ..., T_k$ is sparse. Sparseness implies that most of the elements of the matrix are zero. Thus, the transform TX can be expressed as $T_1 T_2 ... T_k X$. If the calculations are performed in a sequential manner; that is, if we compute $s_k = T_k X$ first and then use the result to perform the next matrix product, $s_{k-1} = T_{k-1} s_k$, then it can be shown that there is an overall reduction in the number of operations needed to perform the 2-D DCT or IDCT. We avoid the details of this factorization; instead, we refer the reader to various references cited in the last section of this chapter. As an example, Figure 3.11 shows the flowgraph for the fast eight-point Chen DCT, where each stage shows the operations required for the transformation of the input data with

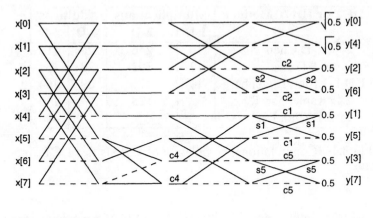

Flowgraph notations

Figure 3.11   Flowgraph for the eight-point 1-D Chen DCT.

the corresponding T_i matrix. Note that, in this flowgraph, $cj = cos(\frac{\pi j}{16})$ and $sj = sin(\frac{\pi j}{16})$.

The computational complexity of some of the most commonly used fast DCT algorithms is shown in Table 3.1. The first three methods provide a fast algorithm for an eight-point 1-D DCT from which the 2-D DCT can be computed using the separability property of the DCT. The last two methods were developed exclusively for a 2-D DCT. All these methods offer a significant reduction in complexity compared with the naive matrix multiplication approach of the previous section. At this stage, the reader might conclude that a true 2-D method, such as the Cho approach, yields lower complexity than the separable extensions of the 1-D DCT. However, from an implementation viewpoint, many of the true 2-D DCT methods have several disadvantages:

DCT or IDCT Method	Multiplications		Additions	
	1-D	**2-D**	**1-D**	**2-D**
1-D Chen	16	256	26	416
1-D Lee	12	192	29	464
1-D Loeffler, Ligtenberg	11	176	29	464
2-D Kamangar, Rao		128		430
2-D Cho, Lee		96		466

Table 3.1 Computational complexity of various fast DCT algorithms.

1. In a software implementation, storage for up to 128 elements is needed. In a computer system that is register limited such storage is not always feasible.

2. Data addressing is highly irregular. In a software or hardware implementation, this irregularity leads to additional overhead for address calculations, which are not included in the original number of multiplications and additions.

Most practical software and hardware implementations of DCT-based coders use the row-column extension of the 1-D DCT. In some DCT-based coding schemes, a further reduction in complexity for the 1-D DCT approach can be obtained, as we discuss in the next section.

3.5.2   The Fast Scaled DCT

From Figure 3.7, we observe that the output of the DCT computation unit is always processed by the quantizer. Excluding clipping effects, the quantization process of (3.29) can be expressed as $z_{kl} = \frac{y_{kl}}{q_{kl}}$. Let us assume that, in the matrix factorization of $T = T_1 T_2 \cdots T_k$, the last (in processing order) matrix T_1 can be expressed as $T_1 = H\hat{T}_1$, where H is a diagonal matrix. Then it can been shown that using $\hat{T}_1$ instead of T_1 will yield a scaled version of y_{kl}. By suitably scaling q_{kl}, the entries of the quantization matrix, by the entries of the diagonal matrix H, one can reproduce the true y_{kl} during the quantization process. Through this restricted factorization, the resulting $\hat{T}_1$ is more sparse than T_1 and thus reduces the overall operations count when calculating the scaled version of y_{kl}.

As an example, consider the following DCT butterfly operations at the last stage of a DCT algorithm:

$$y_i = t_j b + t_i a , \qquad (3.32)$$

$$y_j = t_i b - t_j a, \qquad (3.33)$$

where a and b are constants. The above operations require four multiplications and two additions. If q_i and q_j denote the quantization coefficients for y_i and y_j, then after quantization $z_i = \frac{y_i}{q_i}$ and $z_j = \frac{y_j}{q_j}$. Let $c = \frac{b}{a}$. Then the above operations can be rewritten as

$$y_i = a(t_j c + t_i), \qquad (3.34)$$

$$y_j = a(t_i c - t_j). \qquad (3.35)$$

At first glance, there is no reduction in the number of operations; however, since y_i and y_j will be quantized, the two multiplications by a can be absorbed by the quantizer. Let

$$\hat{y}_i = t_j c + t_i, \qquad (3.36)$$

$$\hat{y}_j = t_i c - t_j, \qquad (3.37)$$

be the output of the scaled DCT butterfly. The above operations require only two multiplications and two additions. If $\hat{q}_i = \frac{q_i}{a}$ and $\hat{q}_j = \frac{q_i}{a}$ are the scaled quantization coefficients, then $z_i = \frac{\hat{y}_i}{\hat{q}_i} = \frac{y_i}{q_i}$.

If the above procedure is applied to all the butterfly operations in the last stage of a DCT algorithm, then the output vector y of a 1-D DCT can be expressed as

$$y = H\hat{y}, \qquad (3.38)$$

where $\hat{y}$ is the output of the scaled 1-D DCT and H is a diagonal matrix.

For example, Figure 3.12 shows the flowgraph for the scaled eight-point Chen DCT. From the discussion above, in this flowgraph $dj = \frac{sj}{cj}$, where $cj = cos(\frac{\pi j}{16})$ and $sj = sin(\frac{\pi j}{16})$. In this example,

$$H = diag(\sqrt{0.5}, \frac{c1}{2}, \frac{c2}{2}, \frac{c5}{2}, \sqrt{0.5}, \frac{c1}{2}, \frac{c2}{2}, \frac{c5}{2}). \qquad (3.39)$$

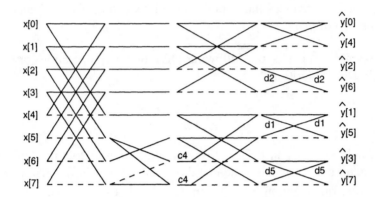

Figure 3.12   Flowgraph for the scaled eight-point Chen DCT.

From Figure 3.12, the scaled Chen DCT requires eight multiplications, whereas, from the flowgraph of Figure 3.11, the non-scaled Chen DCT requires 16 nontrivial multiplications. Similar scaling techniques can be applied to other DCT algorithms as well.

Let a scaled 2-D DCT be computed using row-column decomposition and 1-D scaled DCTs. If $\hat{Y}$ denotes the output of the scaled 2-D DCT, then the output of the 2-D DCT is given by

$$Y = H\hat{Y}H^t. \tag{3.40}$$

As before, the pre- and post-multiplications by H in (3.40) can be absorbed by the quantizer. From (3.40), using matrix algebra and taking into account that H is diagonal,

$$Y_{ij} = h_i h_j \hat{Y}_{ij}, \tag{3.41}$$

where h_i is the i-th element of H. From (3.41), the scaled quantization coefficients are given by

$$\hat{q}_{ij} = \frac{q_{ij}}{h_i h_j}. \tag{3.42}$$

If we assume that the elements of the quantization matrix do not change during the coding process, then the scaling of the quantization matrix has a one-time cost of 64 multiplications, and thus its contribution to the overall operations

DCT or IDCT Method	Multiplications		Additions	
	1-D	**2-D**	**1-D**	**2-D**
1-D Chen	8	128	26	416
1-D Winograd	5	80	29	464
1-D Lee	11	176	29	464
2-D Kamangar, Rao		92		430
2-D Feig, Winograd		54		462

Table 3.2 Computational complexity of scaled DCT algorithms.

count is minimal. In Table 3.2, we show the computation complexity of the scaled DCT approach.

The scaled DCT method of Feig and Winograd is the most efficient scheme. The main disadvantage of the scaled DCT method is that it is efficient only when the quantization matrix is fixed, which is the case for the still-image compression standard; however, in the video compression standards, the quantization matrix is typically varying at a rate of nearly 600 times per second.

3.5.3   DCT on a Multiply-Accumulate-based Architecture

In our discussion of fast DCT methods, we have represented the computational complexity in terms of the number of multiplications and additions. If in the underlying computation unit the basic operation is a single-cycle multiply-accumulate operation, expressed as $a = bc + d$, then the computational complexity of a fast DCT can be further reduced. We illustrate the basic ideas using the DCT computational flowgraph of Figure 3.11. If we examine this flowgraph, we note that several calculations can be recast in the multiply-accumulate form. These are shown in Figure 3.13.

For example, in Figure 3.13a, a multiplication by a constant ($c4$) followed by two additions can be recast as two multiply-accumulate operations, for a savings of one operation. In a similar manner, if $t2 = \frac{c2}{s2}$, then a calculation requiring four multiplications and two additions, as shown in Figure 3.13b, can be remapped into only two scaled multiply-accumulate operations. Since the

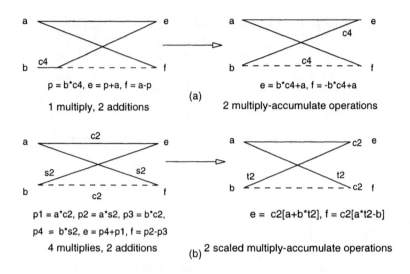

Figure 3.13 Flowgraph remapping for multiply-accumulate operations.

postmultiply by $c2$ is done in the last stage of the DCT calculations, the notion of the scaled DCT, which was discussed earlier, can be used to eliminate the postmultiply by $c2$. We can simply scale the appropriate quantization matrix entry q_{kl} by $c2$ to account for this postmultiply. These flowgraph modifications lead to a row-column approach for a 2-D 8×8 Chen-based DCT that requires only 416 multiply-accumulate operations. A comparison with the other transforms shown in Table 3.1 and Table 3.2 indicates that the multiply-accumulate approach is the best scheme.

A similar exercise can be conducted for all the DCT methods that we have listed in Tables 3.1 and 3.2, and in all cases the multiply-accumulate approach yields the smallest number of operations. For the IDCT, the operations in the flowgraph of Figure 3.11 are simply reversed; that is, operations are performed from right to left. In this case, the scaling operations are combined with the operations in the inverse quantizer.

This approach lends itself to very efficient DCT implementations on conventional digital signal processors, since such architectures include a special multiply-accumulate unit. Additional implementations for the DCT are described in a later chapter.

3.5.4 Multiplication-Free DCTs

A careful examination of the flowgraphs associated with the fast algorithms for the DCT or IDCT indicates that the multiplications are of the type $y = cx$, where c is a constant and x is image dependent. The c are fractional-valued constants, and, in an integer implementation of the DCT, these constants have to be appropriately scaled. Recently, several approximations to the DCT basis functions have been developed. The objective of these approximations is to use very small integer-valued constants and to replace multiplications by simple shift and add operations. These are the so-called multiplication-free DCT and IDCT implementations. Multiplication-free implementation is desirable for low-cost processors that do not have a dedicated multiplier.

We give a brief overview regarding the construction of a set of basis functions that approximate the 8×8 IDCT and can be implemented without any multiplications. A 2-D IDCT can be implemented using the approximate 1-D IDCT transformation matrix using the usual row-column approach. The 1-D DCT computation is of the form

$$y_k = \frac{c(k)}{2} \sum_{i=0}^{7} x_i \cos\left(\frac{(2i+1)k\pi}{16}\right).$$ (3.43)

We refer to $\cos\left(\frac{(2i+1)k\pi}{16}\right)$ as the true DCT basis functions. For $i, k = 0, 1, ...,$ 7, the corresponding 8×8 matrix, denoted as T_C, has values

$$\begin{bmatrix}
1.000 & 1.000 & 1.000 & 1.000 & 1.000 & 1.000 & 1.000 & 1.000 \\
0.981 & 0.831 & 0.556 & 0.195 & -0.195 & -0.556 & -0.831 & -0.981 \\
0.924 & 0.383 & -0.383 & -0.924 & -0.924 & -0.383 & 0.383 & 0.924 \\
0.831 & -0.195 & -0.981 & -0.556 & 0.556 & 0.981 & 0.195 & -0.831 \\
0.707 & -0.707 & -0.707 & 0.707 & 0.707 & -0.707 & -0.707 & 0.707 \\
0.556 & -0.981 & 0.195 & 0.831 & -0.831 & -0.195 & 0.981 & -0.556 \\
0.383 & -0.924 & 0.924 & -0.383 & -0.383 & 0.924 & -0.924 & 0.383 \\
0.195 & -0.556 & 0.831 & -0.981 & 0.981 & -0.831 & 0.556 & -0.195
\end{bmatrix}.$$

Our objective is to express this matrix as a product of a diagonal matrix and a matrix with integer weights, which we denote as $\tilde{T}_C$. As observed in the section on scaled DCT, the diagonal matrix can be incorporated into the quantization matrix. Thus, the DCT calculations can now be performed using the integer-valued matrix $\tilde{T}_C$.

For the design of $\tilde{T}_C$, we impose the following constraints:

1. The row elements of $\tilde{T}_C$ must have sign alternations identical to that of T_C.

2. If we examine T_C, we find that within a scale factor all elements of the first row and fifth row are unity. We will preserve this feature in $\tilde{T}_C$. Furthermore, by a suitable choice of a diagonal matrix, T_C has only seven unique values. We will preserve this feature in $\tilde{T}_C$.

3. If we examine values in, say, the second row of T_C, we see an ordered data arrangement. $\tilde{T}_C$ will retain a similar ordering.

4. DCT is an orthogonal transform. This feature must be preserved in $\tilde{T}_C$ as well.

Based on these constraints, $\tilde{T}_C$ can be expressed in terms of seven constants $a, b, c, d, e, f,$ and g as

$$
\tilde{T}_C = \begin{bmatrix}
g & g & g & g & g & g & g & g \\
a & b & c & d & -d & -c & -b & -a \\
e & f & -f & -e & -e & -f & f & e \\
b & -d & -a & -c & c & a & d & -b \\
g & -g & -g & g & g & -g & -g & g \\
c & -a & d & b & -b & -d & a & -c \\
f & -e & e & -f & -f & e & -e & f \\
d & -c & b & -a & a & -b & c & -d
\end{bmatrix}.
$$

Without any loss of generality, we require $a \geq b \geq c \geq d$. Furthermore, we set $g = 1$ so as to mimic the first row of T_C. To preserve orthogonality, we require $ab = ac + bd + cd$. The constraints above can be used in a computer search procedure to find small integers $a, b, ..., f$; however, this will yield many feasible sets. In order to find a good choice among these candidate sets, we will choose the set or sets that have a high decorrelation efficiency. The decorrelation efficiency is calculated as follows:

- Let $Y = \tilde{T}_C X$ be the DCT of the 1-D vector X.

- We assume that the 1-D source X is correlated with correlation coefficient ρ and that the corresponding covariance function is the 8×8 matrix $Cov_x = \rho^{|i-j|}$.

- We compute the covariance matrix for Y as $Cov_y = \tilde{T}_C Cov_x (\tilde{T}_C)^t$.

- We define the decorrelation efficiency as $\dfrac{\sum diagonal\ entries\ of\ COV_y}{\sum all\ entries\ of\ COV_y}$.

We seek the values $a, b, ..., f$ for which the resulting decorrelation efficiency approaches one. One set of values is $a = 10$, $b = 9$, $c = 6$, $d = 2$, $e = 3$, and $f = 1$. For this set, $\tilde{T}_C$ can be factorized into a product of sparse integer-valued matrices, and the corresponding 8×8 DCT or IDCT would require 576 additions and 192 arithmetic shift operations. According to published results, a hardware implementation of the above scheme requires far less hardware than conventional implementations. The primary disadvantage of the above scheme is that the coding error is increased, since $\tilde{T}_C$ is an approximation to the true DCT basis.

3.6 RATE-DISTORTION PERFORMANCE OF THE DCT

In the previous sections, we have described the usage of DCT in an image compression system. Several computationally efficient approaches for a low-complexity implementation were also discussed. In an earlier section in this chapter, we also provided the rate-distortion function for a typical, highly correlated image. Does a DCT-based coding scheme approach the rate-distortion function of the source?

If we neglect the quantization process in a DCT-based coding system, for the same SNR, the resulting bit rate is 2.37 bits per sample lower than the corresponding rate for the uncorrelated source shown in Figure 3.1. In low SNR regions or for a source with correlation coefficient less than 0.75, the rate improvement will be around 1 bit per sample. From a rate-distortion viewpoint, at very low distortions, a DPCM system is 0.4 bits per sample worse than the DCT system.

3.7 TO PROBE FURTHER

In this chapter, we introduced the concept of rate-distortion. In [22], the rate-distortion function has been developed for uncorrelated sources and for sources with correlation along one dimension. The rate-distortion function for images based on a separable covariance function is described in [173]. The isotropic model is considered to be a better choice for images, and the development of the rate distortion function for the isotropic model is given in [184]. Much of the work reported in this chapter on the rate-distortion function for 2-D images is based on the developments reported in [249], [76], and [35].

Over the years, many lossy compression methods have been developed for still-images. Here, we restricted our discussion to DPCM, a sample-based coding scheme, and to a DCT-based block-coding scheme. Motivation for using block-based coding can be found in the source-coding theorem, which suggests that block-based coding can approach the rate-distortion bound; a recent development of this theorem is in [128]. All the image and video compression standards discussed in this book employ both DPCM and DCT coding methods. A detailed discussion of the DPCM coding scheme can be found in [121].

The basis of many block-coding schemes is the transformation of images from a spatial domain to the frequency domain. Many transformation methods exist and are extensively discussed in [54]. The DCT has received close attention among the transformation methods, due to its use in all the current and emerging standards in image and video compression. A detailed discussion on the DCT can be obtained from [216]. Among the fast DCT algorithms we referred to in this chapter, Chen's DCT algorithm is described in [216], Lee's DCT can be found in [140], Winograd's DCT can be found in [11], the 2-D Kamangar and Rao DCT is described in [125], and the 2-D Cho DCT is described in [43]. The development of fast DCT methods using the concept of a scaled DCT is provided in [56]. A split-radix fast DCT is reported in [234]. In case one needs to compute only the low-frequency components of a DCT, a pruning algorithm for the fast DCT is described in [233].

The theory of DCT implementations on a multiply-accumulate architecture was developed in [156]. DCT coding systems might be used in the future in low power devices and in systems that use fairly simple processors. Multiplication-free DCT implementations would be quite useful in such cases, and one such development can be found in [6]. The development of fast DCT methods for software as well as hardware implementations is currently being researched extensively. In our discussion on the rate-distortion function for DCT coding, we ignored quantization effects. A detailed analysis that takes into account the quantization effects as well as the transmission link imperfections can be found in [174]. A more detailed discussion on the hardware implementation of the DCT is presented in Chapter 10.

4

FUNDAMENTALS OF LOSSY
VIDEO COMPRESSION

4.1 INTRODUCTION

In the previous chapter, we introduced some of the basic concepts in the theory of lossy compression of still-images. These concepts can be readily applied to the compression of image sequences (such as video) as well, by simply treating each image in the sequence as a still-image. This approach is inherently simple; however, it does not provide significant compression. For instance, consider an uncompressed image sequence at a data rate of 166 Mbits per second. Using a DCT-based coding system on an image-by-image basis, for good image fidelity, one can achieve close to 12:1 data compression. Thus, the compressed bit rate is close to 14 Mbits per second, which is too high for most practical uses. For instance, the bandwidth for digital TV broadcasting is close to 4 to 6 Mbits per second. In this case, we need a compression ratio around 41:1. In another example, the typical output rate of a CD-ROM drive is close to 1.5 Mbits per second. In this case, we need a compression ratio around 110:1. Hence, in practical applications with image sequences, we require at least four times better compression ratios than those we can achieve with coders for still-images.

Quadrupling the compression ratio of a still-image coder is nontrivial. In this chapter, we introduce some of the basic concepts for the compression of image sequences. These are essentially extensions of the DCT-based coding techniques employed in still-image coding and form the basis of the video coding standards that will be described in the ensuing chapters.

4.2 VIDEO CODING BASICS

4.2.1 Image Sequence Model

In the still-image case, we observed that image data tend to have a high degree of spatial redundancy. Consider now the problem of capturing the movements of a 3-D object through time (Figure 4.1). In the first image, we capture a spatial projection of the object, say, in region A. Since this projection is comprised of pixels from the object, we expect correlation within the image. If the object is moving, it will yield a spatial projection in the next image as well, say in region X. Thus, we would expect a high degree of temporal redundancy between neighbor images as well; that is, there is a strong correlation between pixels in region A in one image and pixels in region X in the next image. The

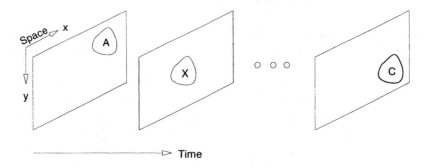

Figure 4.1   Temporal correlation in an image sequence.

goal of video compression algorithms is to exploit both the spatial and the temporal redundancy within an image sequence for optimum compression.

In Chapter 3 we described how the 2-D DCT can be used to exploit the spatial redundancy within an image. In order to exploit both the spatial and the temporal redundancy, one might suggest using a 3-D DCT instead of the 2-D DCT. This approach has been shown to be quite effective from a compression viewpoint; however, the excessive complexity of a 3-D DCT renders this approach impractical. Instead, most video coders use a two-stage process to achieve good compression. The first stage uses a method that exploits the temporal redundancy between frames. The output of this stage is followed by a coding method that exploits spatial redundancy within the frame. The basic two-stage process is illustrated in Figure 4.2.

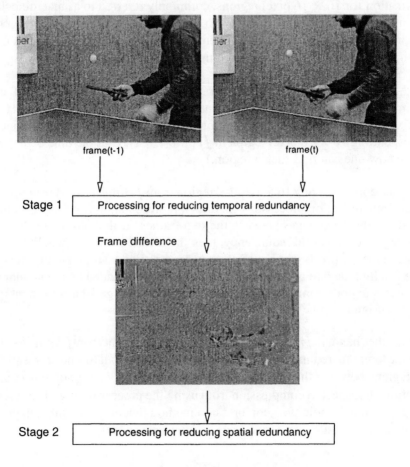

Figure 4.2   Two-stage video coding process.

4.2.2   Reducing Temporal Redundancy

Intuitively, one might expect that the ideal processor for reducing temporal redundancy is one that tracks every pixel from frame to frame. This is computationally intensive, and such methods do not provide reliable tracking due to the presence of noise in the frames. Instead of tracking individual pixels from frame to frame, video coding standards only allow tracking information for 16×16 pixel regions, commonly referred to as macroblocks. The macroblock dimension of 16×16 is chosen because it provides a good compromise between providing efficient temporal redundancy reduction and requiring moderate computational requirements.

Let the two contiguous frames in Figure 4.2 be denoted as $frame(t-1)$ and $frame(t)$. In the first stage, we segment $frame(t)$ into nonoverlapping 16×16 pixel regions (macroblocks), and for each 16×16 block we determine a corresponding 16×16 pixel region in $frame(t-1)$. (For the time being, we ignore how one can find such a region.)

Using the corresponding 16×16 pixel region from $frame(t-1)$, the temporal redundancy reduction processor generates a representation for $frame(t)$ that contains only the changes between the two frames. If the two frames have a high degree of temporal redundancy, then the difference frame would have a large number of pixels that have values near zero. For example, in Figure 4.2, there is a high degree of temporal redundancy, as evidenced by the similarity of features in both frames. Thus, the output image of stage 1 has lower energy than the original frame and is more amenable to compression.

On the other hand, if $frame(t)$ were completely different than $frame(t-1)$, then the temporal redundancy reduction processor may fail to find corresponding regions between the two frames. In this case, one would not expect any benefit with respect to compression from using the process in stage 1. In video compression terminology, a compression method that employs only temporal redundancy reduction is referred to as an *interframe* coder.

4.2.3   Reducing Spatial Redundancy

The output of stage 1 is a difference frame in which the pixels are spatially correlated. Thus, one can use a processor that can exploit this spatial redundancy and thereby yield a compressed representation for $frame(t)$. This is the process performed in stage 2 of Figure 4.2. Video coding standards use a DCT coding method for reducing spatial redundancy. This DCT coding technique is essentially the same as the one described in the previous chapter.

In video coding terminology, a compression method that employs only spatial redundancy reduction is referred to as an *intraframe* coder. The combination of interframe coding and intraframe coding, as depicted in Figure 4.2, is referred to as a hybrid (intraframe/interframe) coding method.

4.2.4 Motion Compensation

The process of computing changes among frames by establishing correspondence between frames is referred to as temporal prediction with *motion compensation*. We define motion compensation as the process of compensating for the displacement of moving objects from one frame to another. In practice, motion compensation is preceded by *motion estimation*, the process of finding corresponding pixels among frames. If the temporal redundancy reduction processor employs motion compensation, then we can express its output as

$$e(x, y, t) = I(x, y, t) - I(x - u, y - v, t - 1), \qquad (4.1)$$

where $I(x, y, t)$ are pixel values at spatial location (x, y) in $frame(t)$ and $I(x - u, y - v, t - 1)$ are *corresponding* pixel values at spatial location $(x - u, y - v)$ in $frame(t - 1)$. The output of the motion estimator, the coordinates (u, v), defines the relative motion of a block from one frame to another and is referred to as the motion vector for block at (x, y). Later we describe various schemes for computing (u, v). $I(x - u, y - v, t - 1)$ is referred to as the motion-compensated prediction of $I(x, y, t)$, and $e(x, y, t)$ is the prediction residual for $I(x, y, t)$.

Note that, the notion of temporal prediction and the formation of the difference signal $e(x, y, t)$ is very similar to the differential coding scheme (DPCM) described in the previous chapter. The primary difference with respect to the previous discussion on DPCM is that here we form the temporal prediction using temporally adjacent samples, which are determined through the process of motion estimation. Thus, the hybrid coding scheme for video can be viewed as a DPCM method followed by a DCT coding method. A generic block diagram of such a hybrid coding scheme is shown in Figure 4.3. Notice that motion compensation is performed in both the encoder and the decoder, but motion estimation is needed only in the encoder. A practical coder will also include buffer memory and additional preprocessing and postprocessing functional blocks, which are not shown in this figure. Detailed schematics of video coders and decoders will be provided in subsequent chapters.

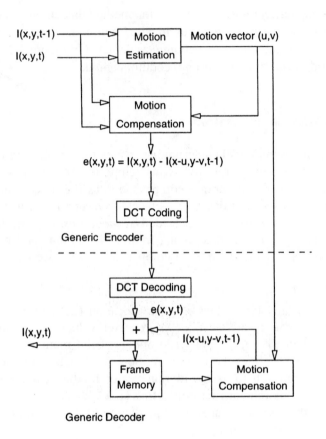

Figure 4.3 Generic hybrid video coder and decoder.

Intuitively, one would expect that interframe coding combined with intraframe coding should yield better compression than either interframe or intraframe coding alone. In the next section, we attempt to confirm this intuitive expectation by applying results from rate-distortion theory to different video coding models. This analysis will provide us in a relative sense the efficacy of temporal prediction versus direct DCT coding of frames. Even though the analysis is not very rigorous, readers that are not interested in the mathematical models of video coding may comfortably skip the next section.

4.3 RATE-DISTORTION FUNCTIONS IN VIDEO CODING

As in the lossy coding of still-images, in a practical lossy compression scheme for video, there is a tradeoff between the distortion D and the rate R at the output of the encoder. The D versus R relationship can be measured for specific implementations of the compressor and decompressor. Under various assumptions for the video signal, it is possible to derive an information theoretic bound for the D versus R relationship, which we refer to as the rate-distortion function $R(D)$ for the video source. $R(D)$ is independent of the particular compression method and depends only on the underlying stochastic model for the source. The practical significance of $R(D)$ is that, for a given distortion D, it provides a bound on the best compression rate that can be achieved, regardless of the compression method. Thus, $R(D)$ can be used as a measure of the effectiveness of a particular video coding methodology.

Our motivation for determining $R(D)$ is to enable us to provide answers to these questions:

- How does intraframe coding compare to interframe coding?

- How do hybrid coding schemes compare to intraframe or interframe coding alone? Does the added complexity of motion compensation warrant its use?

In particular, we compute the rate-distortion function for three video coding models, namely, (1) frame differencing with no motion compensation, (2) motion compensation with no intraframe coding, and (3) motion compensation followed by intraframe coding.

4.3.1 Rate-Distortion Function for Frame Differencing

For the purposes of computing $R(D)$, we simplify the generic encoder-decoder tandem of Figure 4.3 as shown in the model of Figure 4.4. The motion-estimation and motion-compensation blocks are replaced by a single filter function $h(x, y)$ and the coding noise $n(x, y, t)$, induced by the DCT coding of $e(x, y, t)$, is modeled as an additive and uncorrelated white noise

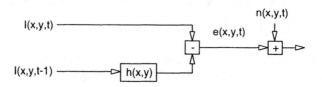

Figure 4.4   Simplified model for interframe coding.

with constant spectral density D. We further assume that the blocks $I(x, y, t)$ and $I(x, y, t - 1)$ are temporally correlated, and we denote by ρ the temporal correlation coefficient. This is a simplified model in that a single correlation coefficient is assumed for the entire frame.

We will also assume that the impulse response of $h(x, y)$ is a single impulse at $x = u$ and $y = v$; that is, we assume that each block $I(x, y, t)$ corresponds to a shifted (by (u, v)) block in the prior frame. Note that, in this case (u, v) is not a true motion vector but an arbitrary estimate that remains the same for all blocks in the frame. We refer to this case as *frame differencing without motion compensation*. Note that there is no added complexity in evaluating $R(D)$ for nonzero values of u and v, even though in practice by frame differencing we often imply that $u = 0$ and $v = 0$.

For this impulse response, the corresponding frequency response is $H(f_x, f_y)$ $= e^{j2\pi(f_x u + f_y v)}$. From (3.6), $R(D)$ is computed as the integral of the power spectral density. From the model in Figure 4.4 and the frequency response of $h(x, y)$ we can express the power spectral density of the signal $e(x, y, t)$ + $n(x, y, t)$ as

$$S_e(f_x, f_y) = 2(1 - \rho \cos(2\pi(f_x u + f_y v)))S_I(f_x, f_y) + D . \qquad (4.2)$$

For $S_I(f_x, f_y)$, the power spectral density of a frame, we use the isotropic covariance model established in (3.1) of the previous chapter. As before, the power spectral density is bandlimited such that $\sqrt{f_x^2 + f_y^2} \leq \frac{1}{\sqrt{\pi}}$.

The computation of $R(D)$ for this model follows along the same lines as the computations outlined in (3.7) to (3.10) of the last chapter. $D(\theta)$ and $R(\theta)$ can be expressed in terms of $S_e(f_x, f_y)$ by replacing $S_X(f_x, f_y)$ with $S_e(f_x, f_y)$. Let $R(D)_{diff}$ denote the rate-distortion function for the case of frame differencing without motion compensation. The integral expression for this function can be computed in closed form as

$$R(D)_{diff} = \frac{1}{2} log_2 \left(\frac{\sigma_e^2}{D} + 1 \right), \tag{4.3}$$

$$\sigma_e^2 = 2\sigma_I^2 (1 - \rho e^{-2\pi f_0 \sqrt{u^2+v^2}}), \tag{4.4}$$

where σ_I^2 is the variance of $I(x, y, t)$. From the previous chapter, the rate-distortion function for an uncorrelated source with Gaussian distribution was given by

$$R(D)_{uncorrelated} = \frac{1}{2} log_2 \frac{\sigma_I^2}{D}. \tag{4.5}$$

Thus, for a video compression scheme based on frame differencing to be effective, we require $R(D)_{diff} < R(D)_{uncorrelated}$. Using (4.3) and (4.5), this inequality implies $\sigma_e^2 + D < \sigma_I^2$. If we ignore the coding noise term D and use (4.4), the variance inequality reduces to

$$2\sigma_I^2 (1 - \rho e^{-2\pi f_0 \sqrt{u^2+v^2}}) < \sigma_I^2. \tag{4.6}$$

With some simple algebra, this inequality is satisfied if $\sqrt{u^2 + v^2} < \frac{1}{2\pi f_0} ln\, 2\rho$. Let $f_0 = \frac{0.05}{\pi \sqrt{\pi}}$ and $\rho = 0.95$, then the inequality is achieved if $\sqrt{u^2 + v^2} < 11.38$. The implication of this result is that temporal prediction with frame differencing is effective only if two neighbor frames are displaced by at most ± 2 pixels. In a typical video sequence it is not uncommon to find displacements that are ± 7 to ± 15 pixels; thus, frame differencing is not effective in such cases.

In Figure 4.5, we plot the rate-distortion functions of (4.3) for several settings of $K = \sqrt{u^2 + v^2}$. For comparison we also plot the rate-distortion functions for an uncorrelated source (4.5) and a 2-D correlated source (as defined in the previous chapter). Observe that for frame displacements greater than eight pixels, frame differencing offers no compression improvement compared with a simplistic compression method that treats the source as a memoryless source. Furthermore, even for a displacement that is at most one pixel (that is, $K = 1$), there is only a 25 to 30 percent improvement in the compression ratio compared with the memoryless case. In all cases, the 2-D correlated model (that is, intraframe coding alone on a frame-by-frame basis) yields better performance than frame differencing alone. Thus, one can conclude that temporal prediction based on simple frame differencing is not a viable compression method except for regions in a scene that have very little motion.

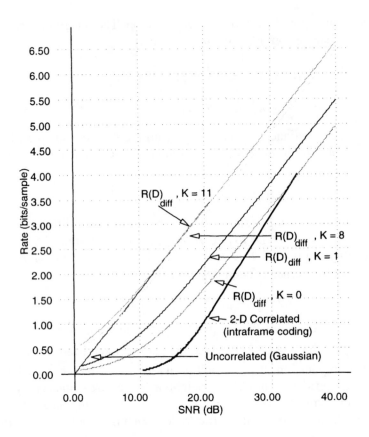

Figure 4.5   Rate-distortion functions for frame differencing with no motion compensation, a Gaussian source, and a 2-D correlated source (intraframe coding).

In the next section, we develop the rate-distortion functions for video coders that track motion of regions across frames and employ a motion-compensated prediction to exploit the temporal redundancy.

4.3.2   Rate-Distortion Function for Motion-Compensated Video

Let us assume that regions in block $I(x, y, t)$ have corresponding regions in block $I(x, y, t - 1)$ at displacement (u, v); that is, (u, v) is the true motion vector associated with the pixel location (x, y). Motion estimation attempts to estimate (u, v), and this estimation process will not be error free. The reason for the estimation error is that in practice $I(x, y, t)$ and $I(x, y, t-1)$ are available only on integer-sampled positions for (x, y) even though the change in position of an object from $frame(t-1)$ to $frame(t)$ might be less than one pixel distance. Thus, due to the sampling grid limitations for the frame, the motion vector accuracy may be limited to ± 1 pixel. Methods are known for reducing the estimation error, but it has been shown by Girod that for typical video imagery and a frame resolution corresponding to broadcast TV, even if the motion-estimation error can be reduced below ± 0.1 pixels, there is a negligible decrease in the variance of the prediction error, that is, the variance of the signal $e(x, y, t)$. Since reduction in prediction-error variance is a good measure of the compressibility of $e(x, y, t)$, Girod observed that for broadcast TV it is sufficient to attain a motion-estimation error of ± 0.25 pixels.

Procedures for motion estimation will be described in a later section of this chapter. For now, let us assume a motion-estimation procedure that yields a motion vector (u, v) with estimation errors denoted by d_u and d_v. The temporal prediction signal can be expressed as $e(x, y, t) = I(x, y, t) - I(x - u - d_u, y - v - d_v, t - 1)$. Note that, if (u, v) is the true motion vector, that is, $d_u = d_v = 0$, then $e(x, y, t)$ will be zero valued. Referring to Figure 4.4, the estimation error can be easily incorporated into this model if we view $h(x, y)$ as a combination of two filters $h_1(x, y)$ and $h_2(x, y)$, where $h_1(x, y)$ has the same impulse response as the $h(x, y)$ in the frame differencing case and the frequency response of $h_2(x, y)$ can be written as $H_2(f_x, f_y) = e^{j2\pi(f_x d_u + f_y d_v)}$. Note that, the estimation error vector (d_u, d_v) is random. To compute $R(D)$, we need to determine the power spectral density of $e(x, y, t)$. This requires a simple modification of (4.2), and the power spectral density is given by

$$S_e(f_x, f_y) = 2(1 - E[\rho \cos(2\pi(f_x d_u + f_y d_v))])S_I(f_x, f_y) + D, \quad (4.7)$$

where $S_I(f_x, f_y)$ is bandlimited as noted earlier. The expectation operator is needed due to the randomness of (d_u, d_v). Assume an isotropic displacement error probability density function for (d_u, d_v) as

$$p(d_u, d_v) = \frac{2}{\pi \sigma_d^2} e^{\left(-\frac{2\sqrt{d_u^2 + d_v^2}}{\sigma_d}\right)}, \quad (4.8)$$

where σ_d^2 is the variance of the motion-estimation error. The prediction error variance for $e(x, y, t)$ is computed as the integral of $S_e(f_x, f_y)$ and is given by

$$\sigma_e^2 = 2\sigma_I^2 \left(1 - \frac{\rho}{(1 + \pi \sigma_d f_0)^2} \right), \tag{4.9}$$

where f_0 is the same as before. From this equation, we claim that temporal prediction with motion compensation is effective if $\sigma_e^2 < \sigma_I^2$. Applying this condition to (4.9) leads to the following requirement:

$$\sigma_d < \frac{\sqrt{2\rho} - 1}{\pi f_0} . \tag{4.10}$$

Thus, for accurate motion-compensation-based temporal prediction, the variance of the motion-estimation error needs to satisfy the above inequality. This equation suggests that effective motion compensation requires a high degree of motion-estimation accuracy relative to spatial variations (f_0) and the temporal correlation coefficient ρ. For typical values of f_0 and ρ, as defined before, effective motion compensation requires that $\sigma_d < 4.26$. This suggests that motion estimators with inaccuracies in excess of ± 4 pixels will render motion compensation ineffective.

If the motion-compensated prediction signal $e(x, y, t)$ is coded as if it were an uncorrelated source (that is, with no additional intraframe coding), then its rate-distortion function is the same as in (4.5) but with σ_I^2 replaced by σ_e^2:

$$R = \frac{1}{2} log_2 \frac{\sigma_e^2}{D}. \tag{4.11}$$

In Figure 4.6, we plot the rate-distortion function of (4.11) for various motion-estimation error settings of σ_d. For comparison, we also show the rate-distortion function for a Gaussian uncorrelated source. Observe from this figure that even for a high motion-estimation error of ± 1 pixel, we get a 40 percent improvement in the compression ratio when motion compensation is employed. We note that reducing the motion-estimation error below $\pm \frac{1}{4}$ pixel does not provide noticeable improvement in the compression ratio. In the video coding standards for broadcast TV, the motion-estimation error is at most $\pm \frac{1}{2}$ pixel. Thus, one can conclude that motion compensation followed by a simple coding scheme that considers $e(x, y, t)$ as being uncorrelated can yield a lower rate compared to the case where no motion compensation is used.

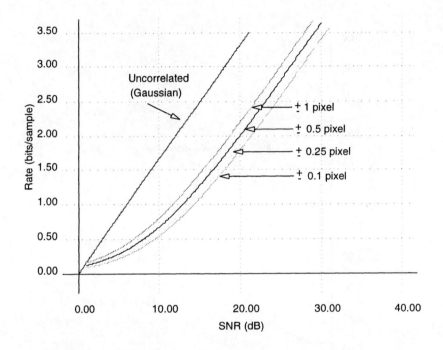

Figure 4.6   Rate-distortion performance for motion-compensated prediction with no intraframe coding.

Does this interframe-only coding method outperform an intraframe coding method? In Figure 4.7, we plot the rate-distortion function for the motion-compensated prediction case (interframe-only coding) and a motion-estimation error setting of $\pm\frac{1}{2}$ pixel, against the rate-distortion function for 2-D intraframe coding that was developed in the previous chapter. From this figure, rate-distortion theory suggests that a video coding method that uses motion-compensated prediction and does not exploit the spatial redundancies in $e(x,y,t)$ will not outperform an intraframe-only coding technique.

In practice, $e(x,y,t)$ is spatially correlated, and hence treating it as such can lead to a lower rate. For instance, in the video coding standards, the signal $e(x,y,t)$ is further processed by a DCT-based coder. We now define the rate-distortion function for the case of spatially correlated $e(x,y,t)$. This corresponds to the performance achieved by an ideal compression scheme that

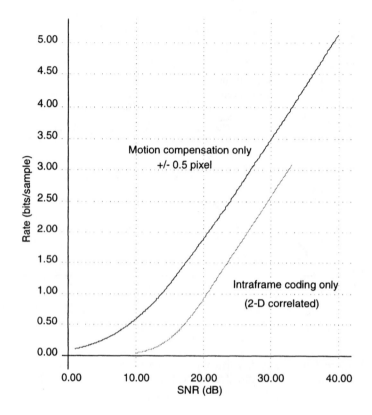

Figure 4.7 Rate-distortion performance for motion-compensated prediction versus intraframe coding.

uses motion compensated prediction followed by additional processing of the prediction residual $e(x, y, t)$.

4.3.3 Rate-Distortion Function for Motion-Compensated Video Followed by Intraframe Coding

We write (4.7) as

$$S_e(f_x, f_y) = 2(1 - \rho\phi(f_x, f_y))S_I(f_x, f_y) + D \ , \qquad (4.12)$$

where the characteristic function $\phi(f_x, f_y)$ is derived from (4.8) as

$$\phi(f_x, f_y) = \frac{1}{((\pi\sigma_d f)^2 + 1)^{\frac{3}{2}}}, \qquad (4.13)$$

and $f = \sqrt{f_x^2 + f_y^2}$. We denote $\phi(f_x, f_y)$ as $\phi(f)$. Using the basic expression for rate, we can express $R(D)$ in integral form as

$$R = \pi \int_0^{\frac{1}{\sqrt{\pi}}} f \, log_2\left(\frac{\sigma^2(f)}{D} + 1\right) df, \qquad (4.14)$$

$$\sigma^2(f) = 2(1 - \rho\phi(f))S_0(f), \qquad (4.15)$$

where $S_0(f)$ is given by (3.4). Note that, if the motion-estimation method yields no estimation errors (that is, $\sigma_d = 0$), we modify (4.15) to $\sigma^2(f) = 2(1 - \rho)S_0(f)$ and compute $R(D)$ as per the integral in (4.14). Computing a closed-form expression for R in (4.14) is quite cumbersome; instead, we resort to a numerical integration procedure to provide the rate R for a specific choice of D.

In Figure 4.8, we plot this rate-distortion function for various settings of the motion-estimation error σ_d. This rate-distortion function yields the rate that is achievable with a hybrid interframe/intraframe coding method. For comparison, we plot the rate-distortion function for the interframe-only case as per (4.11); for this plot, we have set the motion-estimation error to $\pm\frac{1}{4}$ pixel. We note from this figure that exploiting spatial redundancy in the motion-compensated prediction residual $e(x, y, t)$ can result in nearly 40 percent improvement in the compression ratio for a moderate SNR of 20 dB.

The key question to ask at this point is whether this hybrid interframe and intraframe coding method can outperform an intraframe-only coding method. To address this question, in Figure 4.9, we depict a series of rate-distortion plots. Here, we plot the rate-distortion function for the hybrid coding method by computing (4.14); this plot is for a motion-estimation error of $\frac{1}{2}$ pixel. For comparison, we also show the rate-distortion function for the interframe-only case as developed earlier in this section; this plot is for a motion-estimation error of $\frac{1}{2}$ pixel. We also show the rate-distortion function for the 2-D intraframe-only coding case as developed in the previous chapter. From this figure, we observe that exploiting spatial redundancy and temporal redundancy lowers the rate by a factor of two compared with the interframe-only case. In

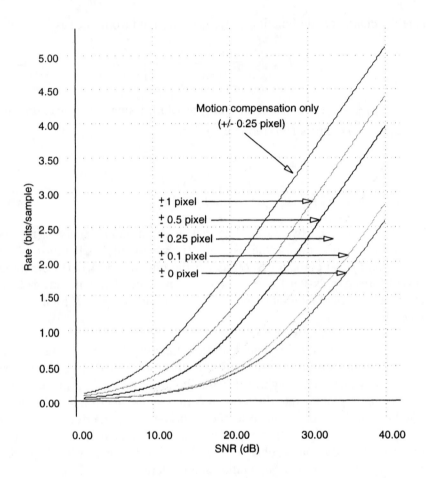

Figure 4.8 Rate-distortion functions for hybrid video coding.

the literature the factor-of-two improvement has often been cited, but these observations have been based largely on experiments. It is informative to note that rate-distortion theory is able to predict this gain. Furthermore, at moderate SNR (say, 25 dB), there is nearly a 30 percent improvement in the compression ratio for the interframe/intraframe approach compared with the intraframe-only coding method.

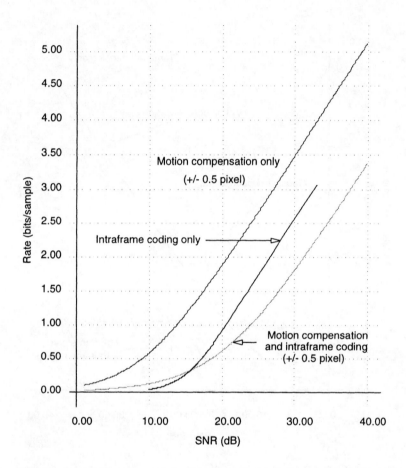

Figure 4.9 Rate-distortion functions for hybrid, interframe, and intraframe video coders.

4.3.4 Rate-Distortion Functions for Video-A Summary

We summarize all the rate-distortion functions developed in this chapter in Figure 4.10. This plot indicates the relative performance improvements that can be obtained through various approaches to video coding, ranging from the simple case of treating the video source as a memoryless source, to schemes that exploit spatial correlation and finally to schemes that employ both spatial and temporal correlation.

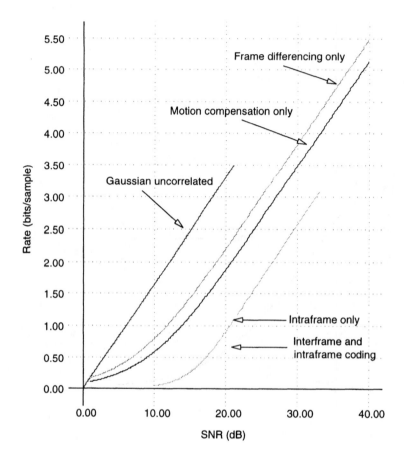

Figure 4.10   Rate-distortion functions for various video coding schemes.

Caution must be exercised in using the data in these plots, since many assumptions were used when developing the rate-distortion functions. An astute reader might observe that at moderate SNR (say, 25 dB), we only obtain a 4- to 5-fold reduction in the rate compared with the memoryless coding scheme, while in practice, hybrid coding schemes such as those used in the video coding standards report much higher rate-reduction factors. Rate-distortion theory as we have outlined it here does not take into account various additional techniques used in the video coding standards, including spatial subsampling of frames, skipping of regions in frames, entropy coding, and switching between intra- and interframe coding.

In an information-theoretic sense, these plots provide the motivation for why hybrid coding methods have been used for video. In the following sections, we provide details on motion compensation and then describe the hybrid coding scheme that is the basis for the video coding standards.

4.4 MOTION-COMPENSATED PREDICTION

As noted in the previous section, a hybrid (interframe/intraframe) coding method is quite effective for video compression. A rudimentary video coding scheme based on this method was shown in Figure 4.3. One of the most compute-intensive operations in interframe coding is the motion-estimation process. In this section we will describe in more detail the motion-estimation process and associated algorithms.

Figure 4.11 illustrates the motion-estimation problem as it is posed in the video coding standards. Given a reference picture and an $N \times M$ macroblock in a current picture, the objective of motion estimation is to determine the $N \times M$ block in the reference picture that better matches (according to a given criterion) the characteristics of the macroblock in the current picture. As current picture, we define an image or frame at time t. As reference picture, we define an image or frame either at past time $t - n$, for forward motion estimation, or at future time $t + k$, for backward motion estimation. In the more general case of motion estimation, the geometry of the matching block at the reference picture need not be the same as the geometry of the block in the current picture, since objects in the real world undergo scale changes as well as rotation and warping. However, in the video coding standards, only the translatory motion model is assumed for objects in the scene, and thus a rectangular geometry is sufficient.

The location of the macroblock regions is given usually by the (x, y) coordinates of their left-top corner. Ideally, we would like to search the whole reference picture for the best match; however, this is impractical. Instead, we restrict the search to a $[-p, p]$ search region around the original location of our macroblock in the current picture. (Many implementations restrict

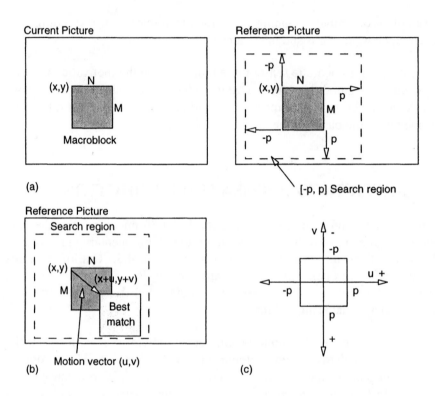

Figure 4.11 Motion-estimation process.

the search range to $[-p, p - 1]$. Both definitions are equally common.) Let $(x + u, y + v)$ be the location of the best matching block in the reference picture (Figure 4.11b). In motion-estimation terminology, the vector from (x, y) to $(x + u, y + v)$ is referred to as the motion vector associated with the macroblock at location (x, y). Often, the motion vector is expressed in relative coordinates; that is, we assume that (x, y) is at location (0,0), and thus the motion vector is simply expressed as (u, v).

Note that, our assumption for a common displacement (u, v) for all pixels in the macroblock implies that we are essentially imposing a local smoothness constraint on the motion vector field. The local smoothness constraint is only satisfied for small macroblock sizes. The choice of the dimensions of the macroblock is the result of tradeoffs among three conflicting requirements. Specifically,

1. small values for N and M (from four to eight) are preferable, since the smoothness constraint would be easily met at this resolution;

2. small values for N and M reduce the reliability of the motion vector (u, v), since few pixels participate in the matching process; and

3. fast algorithms for finding motion vectors are more efficient for larger values of N and M.

In the video coding standards, $N = M = 16$.

The coordinate system associated with the motion vector is shown in Figure 4.11c. For the search region shown in Figure 4.11a, $-p \le u \le p$ and $-p \le v \le p$. For broadcast TV, good performance is obtained at $p = 15$ for head-and-shoulders-type video scenes, and at $p = 63$ for sporting events (high motion).

4.4.1   The Matching Criterion

Let the pixels of the macroblock in the current frame be denoted as $C(x+k, y+l)$ and the pixels in the reference picture be denoted as $R(x + i + k, y + j + l)$. We define a cost function

$$MAE(i,j) = \frac{1}{MN} \sum_{k=0}^{M-1} \sum_{l=0}^{N-1} |C(x+k, y+l) - R(x+i+k, y+j+l)|, \quad (4.16)$$

where i and j are defined in $-p \le i \le p$ and $-p \le j \le p$. This is referred to as the mean absolute error (MAE) or mean absolute difference (MAD) criterion. We define as the best matching block, the block $R(x + i, y + j)$ for which $MAE(i, j)$ is minimized. Thus, the coordinates (i, j) for which MAE is minimized define also the motion vector. In (4.16), we use the absolute value function due to to its computational simplicity. Squaring the difference (mean squared error criterion) or explicitly computing the correlation between blocks are also valid choices, but within a typical coding system, the cost function of (4.16) performs just as well. In video coding terminology, since the match is being performed between rectangular regions, this is referred to as a block matching criterion, and search techniques to find the (u, v) that yield the smallest MAE are referred to as block matching algorithms (BMA). In the next section, we describe several algorithms for block matching motion estimation.

4.5 ALGORITHMS FOR MOTION ESTIMATION

4.5.1 Full-Search Method

Given (4.16), the simplest method to find the motion vector for each macroblock is to compute $MAE(i, j)$ at each location in the search space. This is referred to as the full-search algorithm. We estimate the computational complexity of the full-search (FS) algorithm as follows.

For each motion vector there are $(2p + 1)^2$ search locations. At each search location (i, j) we compare $N \times M$ pixels. Each pixel comparison requires three operations, namely, a subtraction, an absolute-value calculation, and one addition. We ignore the cost of accessing the pixels $C(x + k, y + l)$ and $R(x + i + k, y + j + l)$. Thus, the total complexity per macroblock is $(2p+1)^2 \times MN \times 3$ operations. For a picture resolution of $I \times J$ and a picture rate of F pictures per second, the overall complexity is $\frac{IJF}{MN}(2p+1)^2 \times MN \times 3$ operations per second. Let $N = M = 16$. For typical values for broadcast TV ($I = 720$, $J = 480$, and $F = 30$), motion estimation based on the full-search algorithm requires 29.89 GOPS (Giga operations per second) for $p = 15$ and 6.99 GOPS for $p = 7$.

Full-search is computationally expensive but guarantees finding the minimum MAE value. Due to the high computational complexity of full-search, alternative search methods are desirable. In the next section, we describe several heuristic search strategies for motion estimation. These schemes achieve suboptimum performance at significantly reduced complexity compared to the full-search method. By suboptimum, we imply that the heuristic search strategies do not guarantee that we will find the minimum MAE value. The complexity is significantly reduced either by decreasing the number of search locations or by computing (4.16) using fewer than $M \times N$ pixel differences per search location. Combinations of these two schemes are also possible.

4.5.2   Two-Dimensional Logarithmic Search

Two-dimensional logarithmic search is very similar to binary search. In the first step, the $[-p, p]$ search rectangle is divided into two areas: one inside a $[\frac{-p}{2}, \frac{p}{2}]$ (at integer pixel resolution) rectangle and one outside it. Furthermore, instead of searching the whole $[\frac{-p}{2}, \frac{p}{2}]$ area, we only compute the MAE function for nine locations: at $(0,0)$ and at the eight major points in the perimeter of the $[\frac{-p}{2}, \frac{p}{2}]$ area. That is, if the distance between these points is d_1, we compute the minimum MAE from the MAE computed at $(0,0)$, $(0, d_1)$, $(0, -d_1)$, $(-d_1, 0)$, $(d_1, 0)$, (d_1, d_1), $(d_1, -d_1)$, $(-d_1, d_1)$, and $(-d_1, -d_1)$. The distance d_1 is given by

$$d_1 = 2^{k-1}, \tag{4.17}$$

where $k = \lceil log_2 p \rceil$. For example, for $p = 7$, $k = 3$ and $d_1 = 4$ pixels. Using the best match location as the starting point, we then look for the best match in the eight perimeter points at distance $d_2 = \frac{d_1}{2}$. We continue this process until the k-th search, where the eight perimeter search locations are spaced by one point. After these eight locations have been examined, we determine the location that yields the smallest MAE.

Overall, logarithmic search examines $8k + 1$ search locations and computes the MAE at each search location using (4.16). For pictures at resolution $I \times J$ at F pictures/s, the computational complexity of logarithmic search is $IJF \times (8k + 1) \times 3$. For $I = 720$, $J = 480$, $F = 30$, and $p = 15$ ($k = 4$), logarithmic search requires around one GOP. The complexity of logarithmic search is only 3.3 percent of the complexity of full-search.

In Figure 4.12, we illustrate the logarithmic search procedure in its more popular form, referred to as the three-step search (TSS), where $k = 3$ and $p = 7$. In videoconferencing applications, $p = 7$ is found to be sufficient for good performance. Search locations corresponding to each of the steps in the three-step search procedure are labeled as 1, 2, and 3.

- In the first step, starting from $(0,0)$, we compute the MAE for the the nine search locations labeled 1. The spacing between these search locations is $d_1 = 4$. Assume that MAE is a minimum for the search location $(-4,0)$.

- In the second step, using $(-4,0)$ as the center, we search among the eight search locations around it labeled 2; the spacing between locations is now

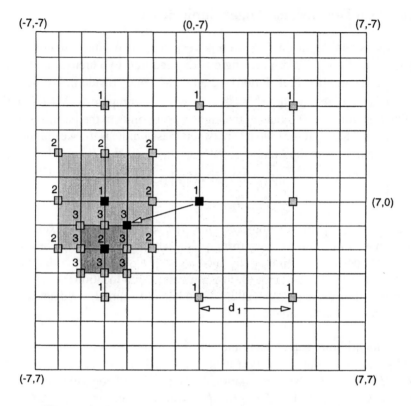

Figure 4.12 Example of a three-step logarithmic search.

$d_2 = \frac{d_1}{2} = 2$ pixels. Let us assume that MAE is a minimum at search location (-4,2).

■ In the third step, from (-4,2) as center, we search among the eight search locations labeled 3, and as in the previous step, the spacing between search locations is halved to $d_3 = 1$. Assume that the MAE is a minimum at search location (-3,1). The search process is terminated at this point, since no further subdivision of the search space is possible, and the output motion vector is defined by the coordinates (-3,1).

The computation complexity associated with these 25 search locations is 777.6 MOPS (Million operations per second).

4.5.3   Parallel Hierarchical One-Dimensional Search (PHODS)

Unlike the logarithmic search, the search in this search strategy is done independently along the two dimensions. The search algorithm is as follows:

1. For a $[-p, p]$ search region as shown in Figure 4.11, let $S = 2^{\lfloor \log_2 p \rfloor}$ and set the origin of the search space at search location (0, 0). Denote the origin as (di, dj).

2. In parallel, compute the

   ■ i-axis local minimum: Among the three locations $(di - S, dj)$, (di, dj), $(di + S, dj)$, find the location that yields the smallest MAE. Set di to the i coordinate of this location.

   ■ j-axis local minimum: Among the three locations $(di, dj - S)$, (di, dj), $(di, dj + S)$, find the location that yields the smallest MAE. Set dj to the j coordinate of this location. Set $S = \frac{S}{2}$.

Repeat step 2, until $S = 0$. The final (di, dj) is the motion vector that yields the best match for the macroblock in the current picture.

This procedure is illustrated in Figure 4.13 for the case of $p = 7$. The reader may want to compare this search process against the search process in Figure 4.12.

First, $S = 4$.

1. In the first step, for the i-axis minimum, we compute MAE for the three search locations labeled $x1$. Assume that the minimum is obtained at $i = 0$. In parallel, we compute the j-axis minimum using search locations labeled $y1$. Assume that the MAE minimum is obtained at $y = 0$. Thus the new origin is again (0,0).

2. The spacing is reduced to 2. The i-axis minimum is obtained from the search locations centered at the origin obtained in the previous step, and these search locations are labeled $x2$. Assume that the MAE minimum is attained at $i = -2$. For the j-axis minimum, as in the i-axis case, we compute MAE for the j-axis search locations labeled $y2$. Assume the MAE minimum is attained at $j = 2$. Thus the new search origin is (-2,2).

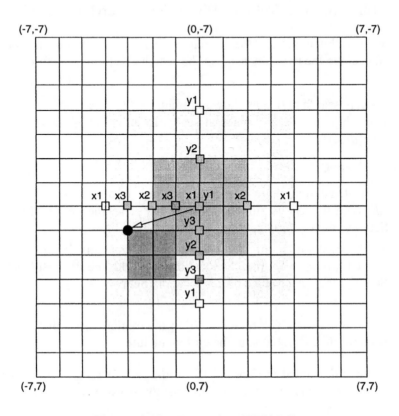

Figure 4.13   Example of PHODS strategy.

3. For this step, $S = 1$, and the minimum MAE along the i-axis is determined from MAE computations at search locations labeled $x3$. We can determine in parallel the minimum MAE along the j-axis by computing the MAE at locations labeled $y3$. Assume that the MAE minimum along the i- and j-axis is attained at (-3,1). Since $S = 1$ at the start of this step, no further reduction in spacing is possible, and this terminates the search algorithm. We declare (-3,1) as the location yielding the smallest MAE and therefore the motion vector for the macroblock in the current picture.

For the case of $p = 7$, we need to examine 13 search locations, which for frames at 720×480 resolution and 30 frames/s corresponds to 404.35 MOPS.

PHODS has two distinct advantages over TSS: (1) the MAE calculations are parallelizable, and (2) it has a regular data flow, since the search locations are always along the i-axis and the j-axis.

Both the logarithmic and the PHODS methods belong to the class of fast algorithms that reduce motion-estimation complexity by reducing the number of search locations that are used in determining the minimum MAE. For p = 7, compared to a full-search method, the complexity is reduced from 6.99 GOPS to 404.35 MOPS. Fast algorithms that work in the reduced search space assume that $MAE(i, j)$ of (4.16) increases monotonically as the search area moves away from the best matched location. Such algorithms perform as well as the full-search method if this assumption holds; however, in practice the assumption often fails, since not all the search locations are visited, and the search for a global minimum may get trapped into a local minimum. This is illustrated in Figure 4.14, which shows hypothetical MAE values for different search points.

In Figure 4.14a, there is a single global minimum. Thus, the result of the fast algorithm coincides with the best solution obtainable with full-search. In Figure 4.14b, there is a global minimum and a local minimum. The fast algorithm gets trapped in the local minimum and declares X as the location of the lowest MAE, even though the full-search method would have found the true minimum at location Y. Thus, fast algorithms such as those described here may yield a larger motion-compensated prediction error $e(x, y, t)$ than a full-search scheme.

4.5.4 Pixel Subsampling and Pixel Projections for MAE Calculations

An alternative approach to reduce the motion-estimation complexity is to reduce the number of pixels that are used in computing block differences in (4.16). The motivation for using a reduced number of pixels is this: since block matching implies that all pixels within the macroblock have the same motion, an estimate of the motion can also be obtained with fewer pixels. Pixel decimation must be done in a careful manner so as not to reduce the motion estimation accuracy. A strategy for pixel decimation in the context of motion estimation is shown in Figure 4.15.

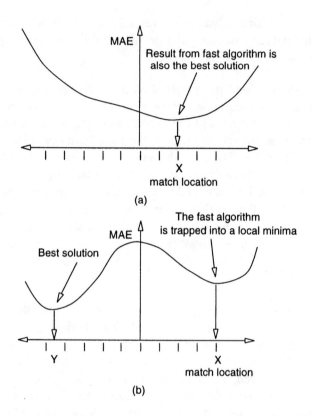

Figure 4.14 MAE versus search location. (a) MAE function is monotonic; (b) search result is trapped into a local minimum.

This decimation process is explained as follows. Assume we are using a full-search method. For each MAE value, we use only one-fourth of the pixels. Pixels labeled 1 are used for computing $MAE(i, j)$ at (0,0) and all search locations $(2i, 2j)$. Pixels labeled 2 are used at all search locations $(2i, 2j + 1)$. Pixels labeled 3 are used at search locations $(2i + 1, 2j + 1)$, and pixels labeled 4 are used at all search locations $(2i + 1, 2j)$. By alternating the pixel patterns and associating a different pattern for each neighboring search location, all the pixels within the block are covered within four search locations, and hence, all pixels in the picture are used in motion estimation. Furthermore, this alternation minimizes the possibility of not considering one-pixel-wide

1	2	1	2	1	2	1	2
3	4	3	4	3	4	3	4
1	2	1	2	1	2	1	2
3	4	3	4	3	4	3	4
1	2	1	2	1	2	1	2
3	4	3	4	3	4	3	4
1	2	1	2	1	2	1	2
3	4	3	4	3	4	3	4

Figure 4.15 Pixel decimation for block matching in an 8 × 8 block.

horizontal, vertical and diagonal lines. With this scheme, the computational complexity of a full-search method is reduced by 25 percent.

Yet another approach to reduce the pixels used in the MAE calculations is to form a projection of the macroblock along several directions and use the samples along the projections instead of the pixels themselves. This is depicted in Figure 4.16. Each element of the column projection vector is the sum of pixels along the corresponding column of the 8 × 8 block. The row projection vector is formed in a similar manner. For a 16 × 16 macroblock, the operations are identical. In the MAE calculations in (4.16), we replace the sum of absolute values of $N \times M$ pixel differences with the sum of absolute values of the differences of the corresponding row and column projections. This reduces the complexity by approximately a factor of $\frac{MN}{M+N}$. For improved motion-estimation accuracy, we can also form projections along various diagonals. In practice, for videoconferencing applications that require a very high compression ratio, two projections have been found to provide accuracy only slightly inferior to that of the full-search method. On the other hand, this approach requires only 12.5 percent of the computations required in the full-search method. For additional computation savings, the projection method can be combined with any of the fast-search strategies discussed in this chapter.

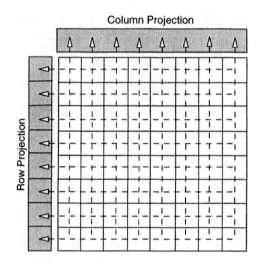

Figure 4.16 Row and column projection of pixels in an 8 × 8 block.

The main disadvantage in using fewer pixels than 16 × 16 is that the motion vector estimate may be unreliable due to noise in the current and reference pictures. A summation over 16 × 16 pixel differences has the side benefit of masking this noise by noise averaging. This effect is reduced when we use fewer pixels in the MAE calculations.

4.5.5   Hierarchical Motion Estimation

Several fast motion-estimation schemes use a combination of search strategies that use both fewer search locations and fewer pixels in computing the MAE. A scheme that combines these two features is widely referred to as the hierarchical search method. There are many variations of this method, but one approach is depicted in Figure 4.17.

The motion vector search proceeds as follows:

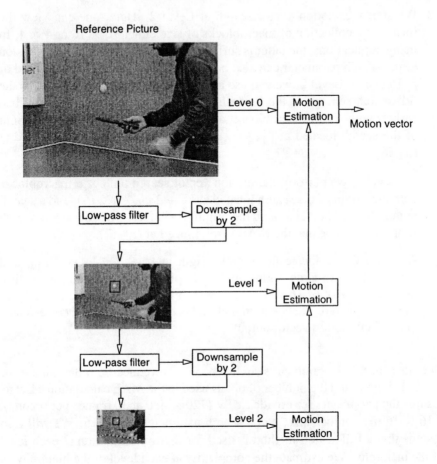

Figure 4.17   A generic hierarchical motion vector search strategy.

1. We form several low-resolution versions of the current picture and the reference picture. In Figure 4.17, we show two low-resolution versions at Level 1 and Level 2 for the reference picture. A similar set of operations is performed to obtain low-resolution versions for the current picture. Let us assume that the macroblock in the current picture is located at (x, y). In Level 1 and Level 2, the corresponding macroblock is at $\left(\frac{x}{2}, \frac{y}{2}\right)$ and $\left(\frac{x}{4}, \frac{y}{4}\right)$, respectively. Let us also assume that the size of the macroblock at Level 0 is 16×16 and that the motion vector has a dynamic range of $\pm p$ pixels.

2. We start the motion vector search at Level 2. Here, one can view the image as a collection of macroblocks of size either 16×16 or 4×4. In many applications, the latter is sufficient. Furthermore, the scaled version of the search parameter can also be used; that is, we can use $\frac{p}{4}$ instead of p. Hence, at Level 2, we can use any motion vector strategy, such as the full-search scheme or the logarithmic search scheme that we described earlier, with 4×4 macroblocks, a search parameter of $\frac{p}{4}$, and the origin of the search located at $(\frac{x}{4}, \frac{y}{4})$. Assume that the MAE is minimized at (u_2, v_2).

3. At Level 1, we perform the motion vector search on 8×8 macroblocks with the origin of the search located at $(\frac{x}{2} + 2\ u_2, \frac{y}{2} + 2\ v_2)$, and a search region of [-1, 1] pixel around the origin. As before, we can use any search strategy. Assume that the MAE is minimized at (u_1, v_1).

4. At Level 0, we locate the search origin at $(x + 2\ u_1, y + 2\ v_1)$, and we search in a search region of [-1, 1] pixel around the origin, with a macroblock size of 16×16. Any search method, including full-search, can be employed. The location that yields the smallest MAE corresponds to the final motion vector output.

Note that, in this hierarchical search scheme, we reduce both the number of search locations and the number of pixels used in the MAE calculations. Let us assume the parameters for broadcast TV (720×480 at 30 frames per second), a 16×16 macroblock size, and a search parameter of $p = 15$. We will also assume that a full-search method is used for motion estimation at each level of the hierarchy. We estimate the complexity at each level of the hierarchy as follows:

1. **Level 2**

   - Picture size = 180×120, and macroblock size = 4×4.
   - Number of macroblocks = $\frac{Picture\ size}{Macroblock\ size}$ = 1,350. At 30 frames/s, we have 40,500 macroblocks.
   - Search parameter = $\lceil \frac{p}{4} \rceil$ = 4.
   - Number of search locations = $(2 \times 4 + 1)^2 = 81$.
   - Number of operations per search location = macroblock size $\times$ 3 = 48.

Complexity for Level 2 = 40,500 $\times$ 81 $\times$ 48 = 157.46 MOPS.

2. **Level 1**

 - Picture size = 360 $\times$ 240, and macroblock size = 8 $\times$ 8.
 - Number of macroblocks = 1,350. At 30 frames/s, we have 40,500 macroblocks.
 - Search parameter = 1.
 - Number of search locations = 9.
 - Number of operations per search location = macroblock size $\times$ 3 = 192.

Complexity for Level 1 = 40,500 $\times$ 9 $\times$ 192 = 69.98 MOPS.

3. **Level 0**

 - Picture size = 720 $\times$ 480, and macroblock size = 16 $\times$ 16.
 - Number of macroblocks = 1,350. At 30 frames/s, we have 40,500 macroblocks.
 - Search parameter = 1.
 - Number of search locations = 9.
 - Number of operations per search location = macroblock size $\times$ 3 = 768.

Complexity for Level 0 = 40,500 $\times$ 72 $\times$ 48 = 279.9 MOPS.

The total complexity is the sum of complexities in Levels 0 through 2 and is 507.38 MOPS. This is a significant reduction over the 29.89 GOPS needed for a full-search method.

From a complexity viewpoint, the hierarchical search method is very efficient; however, such a method requires increased storage due to the need to keep pictures at different resolutions. Furthermore, this scheme may yield inaccurate motion vectors for regions containing small objects. Since the search method starts at the lowest resolution of the hierarchy, regions containing small objects may be completely eliminated and thus fail to be tracked. On the other hand, the creation of low-resolution pictures does provide some immunity to noise.

4.5.6 Performance Comparison of Motion-Estimation Methods

For the MAE criterion, Table 4.1 summarizes the computational requirements of the four major motion-estimation techniques we described in this chapter. In all cases, we assume 16×16 macroblocks and a search range of $[p, -p]$ pixels.

Search Method	Operations per Macroblock	Operations for pictures 720 x 480 at 30 frames/s	
		$p = 15$	$p = 7$
Full-search	$(2p + 1)^2 NM3$	29.89 GOPS	6.99 GOPS
Logarithmic	$(8\lceil log_2 p \rceil + 1)NM3$	1.02 GOPS	777.60 MOPS
PHODS	$(4\lceil log_2 p \rceil + 1)NM3$	528.76 MOPS	404.35 MOPS
Hierarchical	$\left[(2\lceil \frac{p}{4} \rceil + 1)^2 + 180\right] \frac{NM}{16}3$	507.38 MOPS	398.52 MOPS

Table 4.1 Computational complexity and MOPS requirements for various motion-estimation algorithms using the MAE criterion and a [-p, p] search range.

In Figure 4.18, we show the performance of several motion-estimation methods. In all cases, $p = 15$, and we use 16×16 macroblocks.

Figures 4.18c to 4.18f, show the motion-compensated prediction error. If motion estimation were perfect, then we would expect an error picture that would be completely black. Figure 4.18c shows the prediction error for the case of a zero-valued motion vector (simple frame differencing). The full-search method performs the best in that the prediction error is close to zero in most regions of the picture. The logarithmic search scheme is inferior to the full-search and the hierarchical methods, probably because the search strategy gets trapped into local minima. Due to lowpass filtering, the probability of getting trapped into local minima is reduced in the hierarchical scheme, and Figure 4.18f indicates that the performance is nearly as good as using full-search.

All our discussions have considered macroblock sizes of 16×16. In practice, the performance among the search methods is similar for block sizes of 16×8 or 8×8. At smaller block sizes, such as 8×4 and 4×4, the hierarchical scheme performs much worse than the logarithmic scheme. This is probably

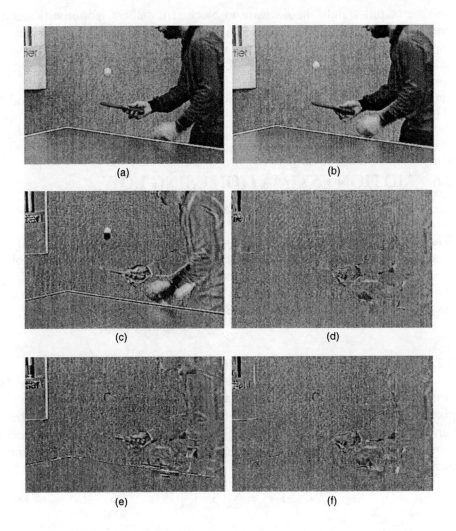

Figure 4.18 Performance comparison of motion-estimation methods. (a) Reference picture; (b) current picture; (c) simple frame differencing; (d) full-search; (e) logarithmic search; (f) three-level hierarchical search.

due to an insufficient number of pixels within the small blocks to obtain accurate estimates of MAE. For the fast-search methods, visually, the decrease in block size manifests as block flickering in the reconstructed sequence. In the case of full-search, the decrease in block size reduces motion artifacts, since we assign motion vectors to smaller regions of the picture. This also improves the smoothness of the motion vector field and thus makes the vectors amenable to compression.

The video coding standards do not specify which motion-estimation method to use. Any of the methods discussed here is applicable.

4.6 MOTION ESTIMATION USING LOW COMPLEXITY MATCHING CRITERIA

Whatever the search strategy, evaluating the MAE criterion is computationally expensive, with overall complexity being proportional to the bit resolution of the pixel values. Therefore, the complexity of motion estimation can also be reduced if one operates on pixel data of lower resolution, or if one applies a simpler matching criterion than MAE. Such approaches are examined next.

4.6.1 The Pixel Difference Classification (PDC)

In this method, each pixel is classified as either a *matching* or a *mismatching* pixel. Following our previous notation, let the pixels of the macroblock in the current frame be denoted as $C(x + k, y + l)$ and the pixels in the reference picture be denoted as $R(x + i + k, y + j + l)$. The pixel difference classification metric is defined as

$$PDC(i, j) = \sum_k \sum_l T_{i,j}(k, l), \qquad (4.18)$$

where, given a threshold t

$$T_{i,j}(k, l) = \begin{cases} 1 & \text{if } |C(x + k, y + l) - R(x + i + k, y + j + l)| \leq t \\ 0 & \text{otherwise} \end{cases}$$

$$(4.19)$$

In effect, PDC counts the number of *matching* pixels between two blocks and the motion vector is defined by the (i, j) values for which $PDC(i, j)$ is maximum.

Assuming that $t = 2^{t_1}$, the binary representation of PDC is given by

$$BPDC(i,j) = \sum_k \sum_l \text{ and}\{\textbf{xnor}(C_{t_1}(x+k,y+l), R_{t_1}(x+i+k,y+j+l))\},$$
(4.20)

where $C_{t_1}(x+k,y+l)$ and $R_{t_1}(x+i+k,y+j+l)$ are the $8-t_1$ most significant bits of $C(x+k,y+l)$ and $R(x+i+k,y+j+l)$, respectively. From (4.20), to evaluate $T_{i,j}(k,l)$ we only need to compare the $8-t_1$ most significant bits of the two pixels. If all these bits are exactly the same, then the absolute difference of the full-resolution pixels will be less than t, and we can count them as matching pixels.

4.6.2 The Binary Level Matching Criterion (BPROP)

The BPDC criterion simply counts the number of matching pixels, thus all matching pixels have equal weight. Alternatively, one can assign more weight to the more significant bits. This is the motivation behind the binary level matching criterion (BPROP) defined by

$$BPROP(i,j) = \sum_k \sum_l \textbf{xor}(C_{t_1}(x+k,y+l), R_{t_1}(x+i+k,y+j+l)).$$
(4.21)

Under this criterion, the motion vector is defined by the values of (i,j) for which $BPROP(i,j)$ is minimum. Experimental results show that $t = 2^{t_1} = 16$ yields the best output signal to noise ratio. In terms of output PSNR, BPROP is only about 0.5 to 1 dB worse than MAE.

4.6.3 The Difference Pixel Count Criterion (DPC)

The difference pixel count criterion (DPC) can be considered a variation of the BPDC metric. Let $C_q(x+k,y+l)$ and $R_q(x+i+k,y+j+l)$ be quantized values of $C(x+k,y+l)$ and $R(x+i+k,y+j+l)$, respectively; then

$$DPC(i,j) = \sum_k \sum_l \text{ and}\{\textbf{xnor}(C_q(x+k,y+l), R_q(x+i+k,y+j+l))\}.$$
(4.22)

In other words, DPC counts the number of *matching quantized* pixels. Since quantized pixels have lower resolution than the original pixels, evaluation of the DPC criterion requires less hardware than the evaluation of the MAE criterion. Experimental results show that quantizing the original 8-bit pixels to 2-bit pixels yields the best tradeoff between performance and computational cost. Given input pixels $p(k, l)$, let

$$p_q(k, l) = \begin{cases} 01 & \text{if } p(k, l) \geq t \\ 00 & \text{if } t > p(k, l) \geq 0 \\ 11 & \text{if } 0 > p(k, l) \geq -t \\ 10 & \text{if } p(k, l) < -t \end{cases}, \tag{4.23}$$

denote the output of a four-level quantizer. For each macroblock, the value of t can be derived using

$$t = \frac{3}{2MN} \sum_{k=0}^{M-1} \sum_{l=0}^{N-1} |p(k, l) - m|, \tag{4.24}$$

where m denotes the mean of the block.

4.6.4   The Bit-plane Matching Criterion (BPM)

In this method, both the current and the reference frames are transformed first into frames of binary-valued pixels. For two binary blocks, the MAE cost function of (4.16) reduces to

$$BPM(i, j) = \frac{1}{MN} \sum_k \sum_l \mathbf{xor}(\hat{C}(x + k, y + l), \hat{R}(x + i + k, y + j + l)), \tag{4.25}$$

where $\hat{C}(x + k, y + l)$ and $\hat{R}(x + i + k, y + j + l)$ denote pixel values after the transformation to one-bit frames. The motion vector is defined by the (i, j) values for which $BPM(i, j)$ is minimum.

One way to compute the one-bit data is the following. Let F denote a frame (current or reference), and let G denote the filtered version of F obtained by applying a convolution kernel K to F. The pixels of the binary frame $\hat{F}$ are given by

$$\hat{F}(i, j) = \begin{cases} 1 & \text{if } F(i, j) \geq G(i, j) \\ 0 & \text{otherwise} \end{cases}. \tag{4.26}$$

For example, a 17×17 convolution kernel K is given by

$$K(i, j) = \begin{cases} \frac{1}{25} & \text{if } i, j \in [1, 4, 8, 12, 16] \\ 0 & \text{otherwise} \end{cases}. \tag{4.27}$$

The motivation behind this approach rests on the observation that the edges in an image are key to accurate motion estimation. A simple way to extract the edges is to carry out a high-pass thresholding, that is, compare the frame pixel by pixel to a high-pass filtered version of the frame, and threshold the pixels to 0 or 1, depending on the outcome of the comparison. Unfortunately, this would also cause the thresholded frame to track the high-frequency noise in the original frame. To overcome this, the convolution kernel defined above performs band-pass filtering, so that the thresholded frame represents the mid-frequency content of the original frame. One could also quantize macroblocks based on their mean value. That is, set $\hat{F}(i,j) = 1$ if $F(i,j)$ is larger than or equal to the mean of the block and set $\hat{F}(i,j) = 0$ otherwise. This scheme corresponds to using low-pass thresholding.

From an implementation point of view, the number of operations required for evaluating the BPM criterion can be computed as follows.

- **Thresholding:** Using the K kernel defined above, the transformation of each pixel to binary data requires 24 additions, one shift (for scaling), and one comparison, for a total of 26 operations per pixel. For frames of size 720×480 pixels, at 30 frames/s, thresholding requires 269.56 MOPS.

- **Pixel Matching:** Assuming 16×16 macroblocks, from (4.25), each search position requires 256 exclusive-or (xor) operations and 255 additions. Assuming a 32-bit architecture, since we operate on binary pixels, we can perform 32 bit-wise xor operations per xor instruction, requiring a total of eight such instructions. Assuming a native instruction that counts the number of ones in a register, we need eight such instructions plus another seven add instructions to compute the total sum of ones, for a total of 23 operations per search position.

  If there is no native instruction that counts the number of ones in a register, then this operation can be replaced by table lookups and additions. Assuming a table with 256 entries, where each entry specifies the number of ones for an 8-bit input, then we can compute the total number of ones in a 32-bit register using four table lookups and three additions. In that case, the total will be $8 + 8 \times (4+3) + 7 = 71$ operations per search position. For comparison, the MAE criterion requires $3 \times 256 = 768$ operations per search position.

Table 4.2 summarizes the computational requirements for the MAE and the BPM metrics for full-search and logarithmic search. We assume frame sizes 720 × 480 pixels at 30 frames per second, 16 × 16 macroblocks, and a [-15, 15] search range. BPM-32 refers to a system with a native instruction to compute the number of ones in a 32-bit register. The operation count for the BPM criterion also includes the overhead for the binary thresholding of the frames. From Table 4.2, for the logarithmic search, there is little difference between BPM and BPM-32, since most of the time (269.56 MOPS) is spent for the thresholding operation.

Search Method	MAE	BPM	BPM-32
Full-search	29.89 GOPS	3.03 GOPS	1.16 GOPS
Logarithmic	1.02 GOPS	364.45 MOPS	300.30 MOPS

Table 4.2 Computation requirements for full and logarithmic searches using the MAE and the BPM criteria, on frames with 720 × 480 pixels, at 30 frames/s, and a [-15, 15] search range.

In a fixed-rate coder, quantization noise can mask the prediction error resulting from a sub-optimum search strategy. This is illustrated in Figure 4.19, which shows the output quality of an H.263 decoder, measured in terms of the PSNR, versus bit rate, for various search strategies used in macroblock motion estimation. The results were obtained using the Telenor TMN software coder (version 1.5) on the "Foreman" test sequence, with a fixed quantizer, fixed frame skipping (one), in arithmetic coding mode, with no advanced prediction modes, and a motion vector search range of 15 pixels (full-search in all cases). From Figure 4.19, as expected, the MAE criterion at full pixel resolution performs the best. However, at bit rates above 50 kbits/s the difference in PSNR between MAE and the other criteria is less than 0.3 dB. Among the low-complexity criteria, DPC performs the best and BPDC (with $t_1 = 4$) performs the worst. However, in terms of PSNR, there is no much of a difference among all four low-complexity matching criteria.

All of the above matching criteria can be used with any of the search schemes described in the previous section. In addition, they can also be used in combination with the MAE criterion. For example, one can use a low-

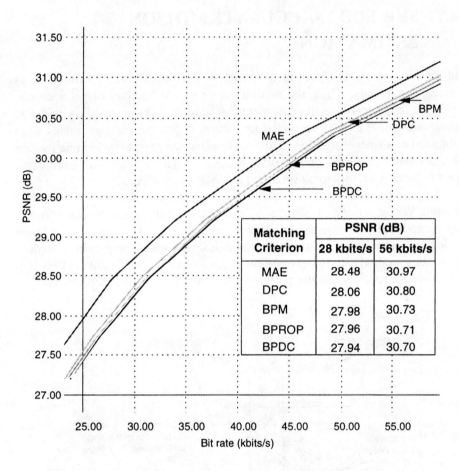

Figure 4.19   PSNR versus bit rate for an H.263 coder and various full-search strategies. Test sequence: Foreman, QCIF, 12 frames/s, fixed quantizer.

complexity matching criterion to find a first set of candidate motion vectors and then refine this search using the full-resolution data and the MAE criterion.

4.7 SUB-PIXEL-ACCURATE MOTION ESTIMATION

In all our discussions, we have restricted the motion vector estimation to integer pixel grids. Thus, the motion vector would be pixel or pel-accurate. The true frame-to-frame displacements are unrelated to the sampling grid, and thus, one would expect improved prediction if displacement estimates were obtained at a finer resolution. This implies that we need to determine motion vectors with fractional or sub-pixel accuracy. The video coding standards permit motion vectors to be specified to a half-pixel accuracy.

Motion vector estimation with, say, half-pixel accuracy can be easily found by interpolating the current and reference pictures by a factor of two and then using any of the motion-estimation methods described in the previous section. Since this method uses excessive storage requirements, the following two-step alternate approach is usually prefered. We illustrate the two-step approach in Figure 4.20.

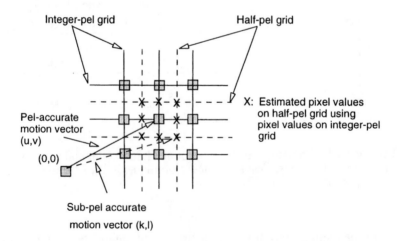

Figure 4.20   Half-pel accurate motion vector estimation.

1. In the first step, we find a motion vector with integer-pel accuracy using any of the motion-estimation methods described in the previous section.

The resulting integer-pel accurate motion vector (u, v) that yields the smallest MAE is as shown in Figure 4.20. Let us denote the corresponding MAE as $MAE(u, v)$.

2. In the second step, we refine the results of the first step to obtain the motion vector with the desired half-pixel accuracy. To refine the (u, v) motion vector to half-pixel accuracy, we do the following:

   - We form eight new search blocks of 16×16 pixels, with the top-left corner of each block being at the locations marked X in Figure 4.20. These locations are denoted as $(u + \frac{m}{2}, v + \frac{n}{2})$, where $m, n \in [-1, 0, 1]$, but m and n cannot both be zero at the same time. For each of these eight new search blocks, we know only those pixel values with coordinates matching our original integer-pel grid. The remaining values on the half-pel grid of Figure 4.20 can be estimated using simple bilinear interpolation techniques.

   - We compute new MAE values by comparing each of these eight search blocks with the original macroblock. Denote by $MAE(m, n)$ the MAE value associated with the search block whose top-left corner is located at $(u + \frac{m}{2}, v + \frac{n}{2})$.

   - Denote by (k, l) the position of the block for which $MAE(m, n)$ is minimum. Note that, $MAE(u, v)$ is also included in finding the minimum. Thus, (k, l) is the half-pel refinement to the integer-pel accurate motion vector (u, v), and the final half-pixel accurate motion vector is given by $(u + \frac{k}{2}, v + \frac{l}{2})$. Note that, the k, l values do not have to be nonzero values. Depending on the nature of the video sequence, there will be regions in the picture where $MAE(u, v)$ is minimum among the nine MAE values. In such cases, the integer-pel accurate motion vector is the right choice.

A similar procedure can be adopted if motion vectors are needed at, say, quarter-pel accuracy.

The pixel interpolation step is compute intensive. In recent years, fast methods that do not perform the interpolation step to determine sub-pixel accurate motion vectors have been developed. We briefly describe one such fast method. The various steps in the fast method are depicted in Figure 4.21.

- Let us assume that we found the integer-pel accurate motion vector using any of the methods described earlier. The location corresponding to this

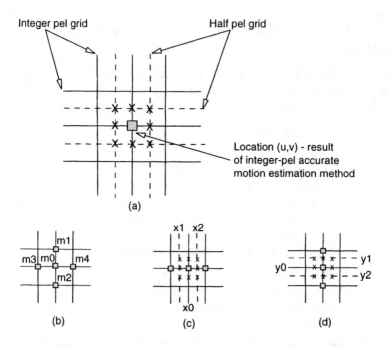

Figure 4.21 Fast method for half-pel accurate motion vector estimation. (a) Integer-pel accurate motion vector estimate; (b) computation of MAE at neighboring integer-pel displacements; (c) integer-pel locations used in computing the MAE minimum along the horizontal direction; (d) integer-pel locations used in computing the MAE minimum along the vertical direction.

motion vector is shown in Figure 4.21a. Let us denote the corresponding MAE value as $m0$. Since pixel values are known on the integer-pel grid, we can compute MAE at the neighboring four locations on the integer-pel grid as shown in Figure 4.21b. These MAE values are denoted as $m1, m2, m3$, and $m4$.

- In (4.16), around the location corresponding to $m0$, we will model MAE as a separable function.

- Along the horizontal (i) direction, this function is of the form $p(i) = a|i - b| + c$. Here, a, b, and c are constants and can be explicitly derived using $p(0) = m0, p(1) = m4$, and $p(-1) = m3$. Using simple algebra, it can be shown that for the motion vector to yield the smallest MAE on the sub-pixel grid in the i direction, one of the following conditions must be met:

1. If $2(m3 - m0) < (m4 - m0)$, the i coordinate of the motion vector is one of the three values on the $x1$ line shown in Figure 4.21c.

2. If $(m3 - m0) > 2(m4 - m0)$, the i coordinate of the motion vector is one of the three values on the $x2$ line shown in Figure 4.21c.

If neither of these conditions is met, the i coordinate of the motion vector is on the $x0$ line shown in Figure 4.21c. Note that, the $x1$ line corresponds to a displacement of $(u - 0.5)$, the $x2$ line corresponds to a displacement of $(u + 0.5)$, and the $x0$ line corresponds to a displacement of u.

■ In a similar manner, we model the MAE along the vertical (j) direction as a function of the form $p(j) = a|j - b| + c$. Here, $a, b,$ and c are constants and can be explicitly derived using $p(0) = m0, p(1) = m2,$ and $p(-1) = m1$. Using simple algebra, it can be shown that for the motion vector to yield the smallest MAE on the sub-pixel grid in the j direction, one of the following conditions must be met:

1. If $2(m1 - m0) < (m2 - m0)$, the j coordinate of the motion vector is at one of the three locations on the $y1$ line in Figure 4.21d.

2. If $(m1 - m0) > 2(m2 - m0)$, the j coordinate of the motion vector is at one of the three locations on the $y2$ line in Figure 4.21d.

If neither of these conditions is met, the j coordinate of the motion vector is at one of the three locations on the $y0$ line in Figure 4.21d.

For example, if after these tests we find that the i coordinate of the motion vector is on the $x2$ line and that the j coordinate is on the $y1$ line, then the half-pel accurate motion vector is the vector $(u + \frac{1}{2}, v - \frac{1}{2})$.

Thus, one can determine the half-pel accurate motion vector without the explicit computation of pixel values on the half-pel grid. Note that such computations are not completely avoidable. If the motion estimator yields a motion vector with coordinates at the half-pel grid, then we still need to compute pixel values at the half-pel grid in order to compute the motion-compensated prediction error.

4.8   MULTIPICTURE MOTION ESTIMATION

In our discussions on motion estimation and motion-compensated prediction, we have restricted attention to the case where the motion estimation is performed using only one reference picture. In instants of smooth motion, there is a great deal of correlation between the current picture at time instant t and pictures prior to and after time instant t. Thus, it is beneficial to compute a prediction of a macroblock in the current picture using motion-compensated regions from the pictures prior to and after time instant t. This is called bidirectional prediction. The motion-compensated prediction for the macroblock X is simply the average of the motion-compensated prediction between X and A and between X and C. This form of prediction is also referred to as interpolative prediction.

In Figure 4.22, we show the current picture and two pictures denoted as reference pictures. The motion vector estimated using the reference picture at

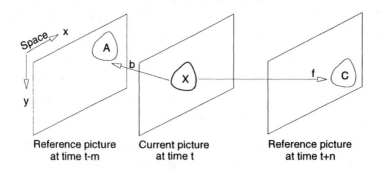

Reference picture   Current picture   Reference picture
at time t-m   at time t   at time t+n

Figure 4.22   Multiframe motion-compensated prediction.

time $t - m$ is shown as the vector b, and the motion vector estimated using the reference picture at time $t + n$ is shown as the vector f. Motion vectors b and f can be found using the methods for motion estimation described in the previous sections. Since bidirectional prediction doubles the complexity of motion vector estimation, several fast methods have been devised; for instance, one might use b as a hint to search for f.

In the video coding standards, for certain types of pictures, two motion vectors can be used for a macroblock X. The specific details on when one uses one motion vector instead of two will be described in Chapter 6.

4.9 TO PROBE FURTHER

In this chapter we introduced the basic concepts associated with the compression and decompression of image sequences (video). The main differences between the spatial-domain coding methods discussed in the previous chapter and the video coding techniques discussed in this chapter are the temporal extensions, which comprise two functions: motion estimation and motion compensation. From an information-theoretic viewpoint, the benefits of temporal extension were determined through the development of rate-distortion functions for video coding. Our approach to computing the rate-distortion functions parallels the developments in [196], [249], and [76]; we have omitted many of the details, and we direct the reader to these references for further details. The modeling of the motion-compensated prediction error, which is useful in the calculation of the rate-distortion functions, is also discussed in [230] and [77].

In this chapter, we have focused on motion-estimation methods that find the best match between a 16×16 macroblock in the current picture and a 16×16 pixel region in the reference picture. In recent years, it has been observed that the translatory-motion model assumption of such motion-estimation methods may not be appropriate for many video sequences. A generalized block-matching method that maps a 16×16 macroblock to a deformed (quadrilateral-shaped) region in the reference picture is described in [227]. This scheme outperforms the square-region-based block matching discussed in this chapter. In most motion-estimation methods, for the best match, the mean absolute difference (MAE) cost function is used. An alternative criterion based on classifying each pixel is described in [75]. This method is claimed to yield good performance for large search windows.

Over the years, several heuristic motion-vector search schemes have been developed; descriptions of some of these schemes can be found in [216], [74], [145], [264], and [211]. A description of the 2-D logarithmic search discussed in this chapter can be found in [118]. Its more popular variation, namely, the three-step search scheme, is described in [126]. The PHODS method of motion vector search discussed in this chapter was first presented in [37], and comparisons with the three-step search method are provided in this paper. Motion vector search methods that employ pixel decimation are developed in [157], and the approaches adopted here yield performance very close to the full-search method. There have been several motion vector search schemes that employ a hierarchical decomposition of the current and the reference pictures, and one of the implementations is discussed in [26]. A hierarchical block-matching method with specific emphasis on hardware implementation is described in [206]. Additional hardware implementations for motion estimation are also described in a later chapter.

The PDC matching criterion was presented first by Gharavi and Mills [75]. Its binary representation (BPDC) and comparisons with the BPROC criterion are described in [268]. In [148], Lee et al. present the DPC criterion and an architecture for motion estimation that combines the DPC with the MAE criterion. The discussion on the BPM criterion is based on work by Natarajan et al. [183]. In [58], Feng et al. describe an adaptive block-matching estimation scheme that combines the MAE and BPM criteria.

The video coding standards allow motion vectors to be represented with sub-pixel (half-pixel) accuracy. In this chapter, we provided a fast method for determining such motion vectors using a model for the MAE function. A discussion of the performance using this model versus a model that employs a squaring function is provided in [123]. There it is shown that the squaring function, though mathematically tractable, is not necessarily a good fit to the MAE.

5

THE JPEG STANDARD

5.1 INTRODUCTION

Until recently, the Group 3 and Group 4 standards for facsimile transmission were the only international standard methods for the compression of images. However, these standards deal only with bilevel images and do not address the problem of compressing continuous-tone color or grayscale images.

Since the mid-1980s, members from both the International Telecommunication Union (ITU) and the International Organization for Standardization (ISO) have been working together to establish a joint international standard for the compression of multilevel *still* images. This effort has been known as JPEG, the Joint Photographic Experts Group. Officially, JPEG corresponds to the ISO/IEC international standard 10918-1, *Digital compression and coding of continuous-tone still images* or to the ITU-T Recommendation T. 81. The text in both these ISO and ITU-T documents is identical.

In recent years, there have been many new developments in the field of image compression. These include new compression schemes based on transform coding, vector quantization, subband filtering, wavelets, and fractals. The goal of JPEG has been to develop a general method for image compression that meets a number of diverse requirements, including the following:

- Be as close as possible to the state of the art in image compression.
- Allow applications (or a user) to tradeoff easily between desired compression and image quality.

- Work independently of the image type. That is, the method should not be restricted by the type of image source, image content, color spaces, dimensions, pixel resolution, etc.

- Have modest computational complexity that would allow software-only implementations even on low-end computers. Low-complexity hardware implementations should also be feasible.

- Allow both sequential (single scan) and progressive coding (multiple scans).

- Offer the option for hierarchical encoding, in which a low-resolution version of the image can be accessed without a need to decompress the image at full resolution.

After evaluating a number of coding schemes, the JPEG members selected a DCT-based method in 1988. From 1988 to 1990, the JPEG group continued its work by simulating, testing, and documenting the algorithm. JPEG became a draft international standard in 1991 and an international standard in 1992. The JPEG group has not yet completed its mission. It continues to work on future enhancements. ISO 10918-3, which specifies recent extensions to JPEG, has already been approved as an international standard (ITU-T Recommendation T.84), and the JPEG group has already started soliciting proposals for JPEG-2000, the next generation of compression algorithms.

JPEG includes two basic compression methods: a DCT-based lossy compression method and a predictive method for lossless compression. We have already examined the lossless compression method in Chapter 2. In this chapter, we provide an overview of the lossy JPEG method.

5.2   DCT-BASED CODING

The JPEG standard specifies four modes of operation: sequential DCT-based, progressive DCT-based, lossless, and hierarchical. Under the lossless mode, a predictive coder followed by either a Huffman or an arithmetic coder is used instead of a DCT-based scheme. The details of operation under the lossless mode were discussed in Chapter 2. To better understand the other modes of operation, we need first to review the DCT-based coder.

5.2.1 JPEG Encoding

Figure 5.1 shows a block diagram of the DCT-based JPEG encoder for an image with a single color component (grayscale). For color images, the process is repeated for each of the color components.

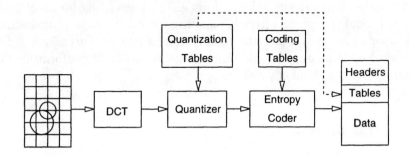

Figure 5.1 Block diagram of a JPEG encoder.

From Figure 5.1, the image is first divided into nonoverlapping blocks. Each block has 8 × 8 pixels. If any of the dimensions of the image is not a multiple of eight, then the pixels of the last row or the last column in the image are duplicated appropriately.

Each block is transformed into the frequency domain by a 2-D DCT. The standard does not specify a unique DCT algorithm. Consequently, users may choose the algorithm that is best suited for their applications. The DCT output coefficients are quantized and entropy coded.

The entropy coder consists of two stages. The first stage is either a predictive coder for the DC (or [0,0]) coefficients or a run-length coder for the AC coefficients. The second stage is either a Huffman coder or an arithmetic coder. Arithmetic coding provides better compression than Huffman coding; however, there are very few JPEG implementations that support arithmetic coding. There are three main reasons for this. First, the improvement in compression (2 percent to 10 percent) does not justify the additional complexity (especially for hardware implementations). Second, many of the algorithms on arithmetic coding are covered by patents in the United States and Japan. Therefore, most implementors are reluctant to pay license fees for minimal gains in performance. Third, the *baseline implementation*, that is, the implementation with the minimum set of requirements for a JPEG compliant decoder, uses only Huffman coding.

In order to facilitate the acceptance of JPEG as an international standard and because of the various options available during JPEG encoding, the JPEG committee also defined an interchange format. This interchange format embeds image and coding parameters (type of compression, coding tables, quantization tables, image size, etc.) within the compressed bit stream. This allows JPEG compressed bit streams to be interchanged among different platforms and to be decompressed without any ambiguity.

5.2.2   JPEG Decoding

Figure 5.2 shows a block diagram of a JPEG decoder. After extracting the coding and the quantization tables from the compressed bit stream, the compressed data passes through an entropy decoder. The DCT coefficients are

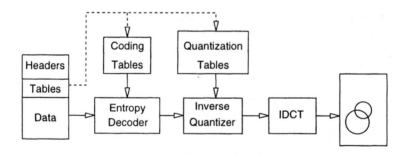

Figure 5.2   Block diagram of a JPEG decoder.

first dequantized and then translated to the spatial domain via a 2-D inverse DCT. After a block-to-raster translation, the image is fully decoded.

5.3 PROCESSING OF COLOR IMAGES

Thus far, we have considered JPEG compression of images with a single component. In practice, images may be represented by multiple color components, each at a different resolution. For example, most color scanners generate images with red, green, and blue color components (RGB).

JPEG sets no restrictions on the type of the input color space. Instead, it views each image as a collection of image components. The maximum number of color components in JPEG is 255. Each component consists of a rectangular array of samples (pixels). Each sample may be represented by P-bits of precision. In JPEG, P can be either 8 or 12 for DCT-based coders and from 2 to 16 for lossless coders. Images with other pixel resolutions can still be coded using JPEG; however, pixel values have to be shifted to be within the resolutions supported by JPEG.

It is not necessary that all color components have the same dimensions. Let x_i and y_i denote the horizontal and vertical dimensions for the i-th component. If $\frac{x_i}{x_j} = \frac{H_i}{H_j}$ and $\frac{y_i}{y_j} = \frac{V_i}{V_j}$, then the relative horizontal and sampling factors for the components (that is, H_i and V_i) are allowed only the integer values 1 to 4. Under the JPEG interchange format, a color image is specified by the maximum horizontal and vertical dimensions ($X = \max(x_i)$ and $Y = \max(y_i)$), the relative sampling factors for each component, and the maximum relative sampling factors (H_{max} and V_{max}). The maximum dimension supported by JPEG is $2^{16} = 65,536$ samples. Given the above information, the decoder extracts the dimensions of each component using the following equations:

$$x_i = \lceil X \times \frac{H_i}{H_{max}} \rceil, \tag{5.1}$$

$$y_i = \lceil Y \times \frac{V_i}{V_{max}} \rceil. \tag{5.2}$$

Example. We want to compress a color 512×512 RGB image. Since JPEG requires each color component to be handled separately, one approach would be to compress the RGB image by applying the JPEG compression method on the R, G, and B components separately. In typical RGB images, there is significant correlation between the color components; thus, from a compression viewpoint, the RGB space is not an efficient representation for the JPEG compression method. The ideal approach would be to transform the RGB image into another component representation, where each of the components is decorrelated with respect to each other, and then apply the JPEG method. In practice, this is done by first transforming the RGB representation into a luminance and chrominance (or color difference) representation, such as YUV, YCbCr, or CIELAB. Assume that the image is translated from the RGB color space into the YCbCr color space. (The reader may want to refer to Appendix A for the RGB to YCbCr color transformation matrix). There is very little correlation among the components in this representation. Furthermore, since most of the spatial information is in the luminance (Y) component, we lose little information if we subsample the Cb and Cr components by a factor of two in both dimensions. Thus, even before applying JPEG compression, we have reduced the image size in half by this simple color-space transformation.

After color-space transformation and subsampling, the image color components have the following dimensions: for the luminance component, $x_0 = y_0 = 512$; for Cb and Cr, $x_1 = x_2 = y_1 = y_2 = 256$. Thus, $X = Y = \max(256, 512) = 512$. The relative sampling factors are $H_0 = V_0 = 2$ and $H_1 = H_2 = V_1 = V_2 = 1$. Thus, $H_{max} = V_{max} = 2$.

5.3.1   Data Interleaving

Images with multiple components can be stored and processed either with or without data interleaving. When there is no data interleaving, each component is stored and processed separately. For example, a 512×512 noninterleaved RGB image would be stored as three separate 512×512 8-bit images: one with the red pixels, one with the green pixels, and one with the blue pixels.

For more efficient storage and processing, the color components can be interleaved. A data unit is defined as the smallest logical unit of source data that can be processed by JPEG. For lossless JPEG, a data unit corresponds to a single sample. For lossy JPEG, a data unit is a single 8×8 block of data. Under color interleaving, the i-th color component is partitioned into small rectangular blocks of dimension $H_i \times V_i$ data units. If we select one $H_i \times V_i$ data block from each of the color components, then we form what is referred to by JPEG as the minimum coded unit (MCU), that is, the smallest group of interleaved data. Figure 5.3 shows an example of interleaved data ordering for a YCbCr image. In this example, the Y component is divided into blocks of 2×2 data units, and the Cb and Cr components are divided into blocks of

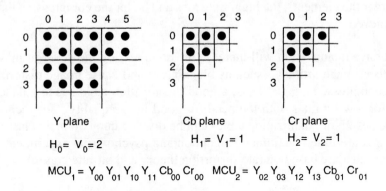

Figure 5.3 Example of data interleaving.

single data units. An MCU is created by taking the four data units from Y, one data unit from Cb, and one data unit from Cr. Within one data unit, pixels are ordered from left to right, top to bottom.

In general, the maximum number of interleaved components is four, and the maximum number of data units within an MCU is ten. The last restriction is expressed as

$$\sum_{i=1}^{N_s} H_i V_i \leq 10, \tag{5.3}$$

where N_s is the total number of color components. If the above equation is not satisfied, the data can still be coded, but in a noninterleaved format.

5.4   DESIGN OF QUANTIZATION TABLES

The principle source of compression in DCT-based coders is the quantizer. JPEG requires an 8×8 quantization matrix for each image component to be compressed; however, multiple components can share the same quantization matrix. Each element of the quantization matrix can be any integer value between 1 and 255 and defines the quantization step for the corresponding DCT coefficients. A quantization matrix with all ones will result in nearly lossless compression (all the loss will be due to round-off errors). In general, the larger the elements, the bigger the loss and better the compression that can be achieved.

Aggressive quantization will increase the compression; however, it may also introduce image artifacts (such as blockiness) and cause image distortion in the decompressed image. Thus, a good quantization table has to balance the need for low bit rates with the need for good image quality. The techniques for the design of quantization tables can be divided into two main classes: (1) those that are based on human perception and psychovisual experiments, and (2) those that are based on rate-distortion theory and bit-rate control.

5.4.1 Techniques Based on Psychovisual Experiments

Techniques based on human perception define the elements of the quantization matrix from visibility thresholds for the DCT basis functions so that any loss due to quantization is perceptually irrelevant. These visibility thresholds are determined from a series of psychovisual experiments that can be described analytically as

$$u_r = u_o + \gamma u_t. \tag{5.4}$$

Given a viewing image u_r, a uniform background u_o, and a stimulant function u_t, the test person has to determine the value of the variable γ that yields a just-noticeable difference between u_o and u_t. In practice, a DCT basis function is superimposed on a uniform background, and the intensity (γ) of the basis function is increased until it becomes just noticeable.

Lohscheller was among the first to determine visibility thresholds for the 2-D DCT basis functions. His work led to the definition of the two quantization tables listed in Annex K of the standard and shown here in Figure 5.4. One table can be used for either grayscale images or for luminance components, and the other can be used for chrominance components. Use of these tables is optional; however, experience has shown that they are quite robust and applicable to a wide range of images and applications.

As mentioned before, quantization tables based on psychovisual experiments will yield decompressed images of very good quality; however, image quality or overall compression may not be satisfactory for certain applications. In those cases, one can tradeoff between image quality and data compression by uniformly scaling the original quantization table by a *quality factor*.

16	11	10	16	24	40	51	61
12	12	14	19	26	58	60	55
14	13	16	24	40	57	69	56
14	17	22	29	51	87	80	62
18	22	37	56	68	109	103	77
24	35	55	64	81	104	113	92
49	64	78	87	103	121	120	101
72	92	95	98	112	100	103	99

(a)

17	18	24	47	99	99	99	99
18	21	26	66	99	99	99	99
24	26	56	99	99	99	99	99
47	66	99	99	99	99	99	99
99	99	99	99	99	99	99	99
99	99	99	99	99	99	99	99
99	99	99	99	99	99	99	99
99	99	99	99	99	99	99	99

(b)

Figure 5.4 JPEG (Annex K) quantization tables. (a) Luminance quantization table; (b) chrominance quantization table.

For the two quantization tables listed in the standard, a quality factor of one-half will yield decompressed images indistinguishable from the originals. For additional compression, quality factors from one to five are typical. Figure 5.5 shows the effects on image quality and compression ratio for quality factors (q) of two and four. At $q = 4$, the compression ratio doubles; however, there is significantly more blockiness in the background of the image.

5.4.2 Techniques Based on Bit-Rate Control

In contrast to the design techniques that are based on general psychovisual experiments, techniques based on bit-rate control generate quantization tables based on the statistical properties of an image. Results from the theory of human perception may also influence the final design; however, the emphasis here is on bit-rate control. A simple design technique is described next.

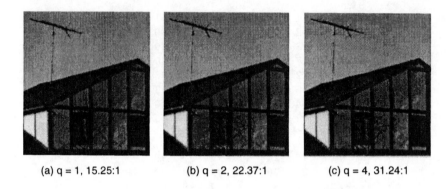

(a) q = 1, 15.25:1   (b) q = 2, 22.37:1   (c) q = 4, 31.24:1

Figure 5.5   Quality of reconstructed images for different quality factors.

Let r denote the desired average compressed bit rate. (Note that, this is the bit rate before the Huffman coder.) If we allocate b_{ij} bits per quantized DCT coefficient, then

$$\frac{1}{64}\sum_{i=1}^{8}\sum_{j=1}^{8} b_{ij} = r. \tag{5.5}$$

Given an image, let B be the number of 8×8 blocks in the image and let $y_k[i, j]$ denote the (i, j) output DCT coefficient for the k-th block. Let

$$\mu_{i,j} = mean(y[i,j]) = \frac{1}{B}\sum_{k=1}^{B} y_k[i,j], \tag{5.6}$$

and let

$$\sigma_{i,j}^2 = Var(y[i,j]) = \frac{1}{B}\sum_{k=1}^{B} [y_k[i,j] - \mu_{i,j}]^2, \tag{5.7}$$

denote the mean and the variance for each of the output DCT coefficients. The key idea of the rate-control method is to allocate more bits to the coefficients with larger variances. From rate-distortion theory, the optimum bit allocation is given by

$$b_{ij} = r + \frac{1}{2}log_2 \frac{\sigma_{i,j}^2}{\left[\prod_{i,j}\sigma_{i,j}^2\right]^{1/64}}. \tag{5.8}$$

For images with eight bits per pixel, the maximum range of the AC coefficients is from -1023 to 1023. Given that this range needs to be divided into $2^{b_{ij}}$ equal quantization levels, the elements of the quantization matrix are given by

$$Q[i,j] = \frac{2046}{2^{b_{ij}}}. \qquad (5.9)$$

For the same perceptual quality, the above procedure generates quantization tables that yield higher compression ratios than the quantization tables listed in Annex K of the standard. Recent proposals try to combine the two design techniques; however, the design of quantization tables is still an open problem. In a way, the design of quantization tables is related to the very difficult problem of developing a mathematical model for the human visual system.

5.5 ENTROPY CODING

The last processing block in JPEG is the entropy coder. This block improves overall performance by performing lossless coding on the quantized DCT coefficients. The entropy coder employed in the JPEG standard is not a straightforward implementation of the Huffman or arithmetic coding methods described in Chapter 2; instead, the quantized data are preprocessed by a run-length coder whose operation will be described later in this section. If the entropy coder employs Huffman coding, then one or more sets of Huffman tables (the maximum number of Huffman tables is eight) need to be specified by the application. There are no default tables, but most applications use the Huffman tables listed in the standard. JPEG imposes only two restrictions on the Huffman tables: (1) no codeword may exceed 16 bits, and (2) no codeword may be the all-ones sequence (that is, FF_{16}).

The arithmetic coding option in JPEG requires no external table specifications, since it is able to adapt to the image characteristics. However, for improved performance, optional statistical tables can be used.

The baseline JPEG implementation uses Huffman coding only. It also restricts the number of Huffman tables to four: two for the AC components and two for the DC components. This is not a major constraint, since most applications operate on luminance-chrominance data and use one set of Huffman tables for the luminance component and one set of tables for the chrominance components. This restriction can also be bypassed if one uses noninterleaved data. In that case, a new set of tables may be loaded before the decompression of each color component. Details for the baseline Huffman coder are presented next.

5.5.1   Huffman Coding of the DC Coefficients

Figure 5.6 shows a block diagram of the Huffman coder in baseline JPEG. Let

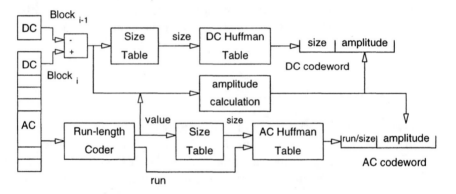

Figure 5.6   Huffman coding in baseline JPEG.

DC_i and DC_{i-1} denote the DC coefficients of blocks i and $i-1$. Due to the high correlation of DC values among adjacent blocks, JPEG uses differential coding for the DC coefficients. For 8-bit-per-pixel data, DC differentials $(DC_i - DC_{i-1})$ can take values in the range [-2,047, 2,047]. This range is divided into 12 size categories, where the i-th category includes all differentials that can be represented by i bits. The entries for these categories are the same as the first 12 categories in Table 2.9. Thus, after a table lookup, each DC differential can be described by the pair (size, amplitude), where *size* defines the number of bits required to represent the amplitude, and *amplitude* is simply the amplitude of the differential. Given a DC residual value, its amplitude is computed as follows: if the residual is positive, then the amplitude is simply its binary representation with *size* bits of precision; and if the residual is negative, then we take the one's complement of its absolute value. From this pair of values, only the first (the size) is Huffman coded.

For example, if the DC differential has an amplitude of 195, then, from Table 2.9, size = 8. Thus, 195 is described by the pair (8, 11000011). If the Huffman codeword for size = 8 is 111110, then 195 is coded as 11111011000011. Similarly, -195 would be coded as 11111000111100. Huffman decoding is quite simple. From the input bit stream, we first decode the size = 8 information. Then, the next eight bits in the input bit stream directly give the amplitude of the DC differential, which we decode according to the value of its most significant bit.

5.5.2   Huffman Coding of the AC Coefficients

For 8-bit pixels, AC coefficients may take any value in the range [-1023, 1023]. As before, this range is divided into 10 size categories, and each AC coefficient can be described by the pair (size, amplitude). After quantization, most of the AC coefficients will be zero; thus, only the nonzero AC coefficients need to be coded.

AC coefficients are processed in zig-zag order. Figure 5.7 shows the conventional and the zig-zag ordering of elements in an 8×8 matrix. Zig-zag ordering allows for a more efficient operation of the run-length coder.

A run-length coder yields the *value* of the next nonzero AC coefficient and a *run*, that is, the number of zero AC coefficients preceding this one. Hence, each nonzero AC coefficient can be described by the pair (run/size, amplitude). The value of run/size is Huffman coded, and the value of the amplitude (computed as in the case of the DC differentials) is appended to that code.

For example, assume an AC coefficient is preceded by six zeros and has a value of -18. From Table 2.9, -18 falls into category 5. The one's complement of -18 is 01101. Hence, this coefficient is represented by (6/5, 01101). The pair (6/5) is Huffman coded, and the 5-bit value of -18 is appended to that code. If the Huffman codeword for (6/5) is 1101, then the codeword for -18 is 110101101.

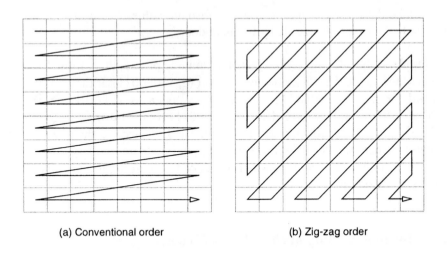

(a) Conventional order   (b) Zig-zag order

Figure 5.7   Conventional and zig-zag ordering in an 8 × 8 matrix.

There are two special cases in the coding of AC coefficients, as follows: (1) The run-length value may be larger than 15. In that case, JPEG uses the symbol (15/0) to denote a run-length of 15 zeros followed by a zero. Such symbols can be cascaded as needed; however, the codeword for the last AC coefficient must have a nonzero amplitude. (2) If after a nonzero AC value all the remaining coefficients are zero, then the special symbol (0/0) denotes an end of block (EOB).

A Coding Example. Assume that the values of a quantized DCT matrix are given (in zig-zag order) by

$$
\begin{array}{rrrrrrrr}
42 & 16 & -21 & 10 & -15 & 0 & 0 & 0 \\
3 & -2 & 0 & 2 & -3 & 0 & 0 & 0 \\
0 & 0 & 2 & -1 & 0 & 0 & 0 & 0 \\
0 & 0 & 0 & 0 & 0 & 0 & 0 & 0 \\
0 & 0 & 0 & 0 & 0 & 0 & 0 & 0 \\
0 & 0 & 0 & 0 & 0 & 0 & 0 & 0 \\
0 & 0 & 0 & 0 & 0 & 0 & 0 & 0 \\
0 & 0 & 0 & 0 & 0 & 0 & 0 & 0
\end{array}
\tag{5.10}
$$

If the DC value of the previous block is 40, then $DC_i - DC_{i-1} = 2$. This can be expressed as the (size, amplitude) pair (2, 2). If the Huffman codeword for size = 2 is 011, then the codeword for the DC value is 01110.

Table 5.1 shows the codewords for the AC values. For Huffman codewords, we use the tables given in Annex K of the JPEG standard document for luminance AC coefficients (Table K.5). For this example, we need 82 bits to

Value	Run/Size	Huffman Code	Amplitude	Total Bits
16	0/5	11010	10000	10
-21	0/5	11010	01010	10
10	0/4	1011	1010	8
-15	0/4	1011	0000	8
3	3/2	111110111	11	11
-2	0/2	01	01	4
2	1/2	11011	10	7
-3	0/2	01	00	4
2	5/2	11111110111	10	13
-1	0/1	00	0	3
EOB	0/0	1010		4

Table 5.1 Example for the Huffman coding of AC coefficients.

encode the AC coefficients and five bits to encode the DC coefficient, for a total of 87 bits or an average bit rate of $\frac{87}{64} = 1.36$ bits per pixel. If the input resolution was eight bits per pixel, then the compression ratio is $\frac{8}{1.36} = 5.88$.

5.5.3 Compression Efficiency of Entropy Coding in JPEG

From the discussion on entropy coding of the DC and AC coefficients, we note that the entropy coder employed in the JPEG standard is not a straightforward implementation of the Huffman or arithmetic coding methods described in Chapter 2. In JPEG, Huffman or arithmetic coding is preceded by a run-length coder. Furthermore, entropy coding in the JPEG standard includes these features:

1. The DC and AC coefficients are treated separately. This is motivated by the fact that the statistics for the DC and AC coefficients are quite dissimilar; hence, better coding efficiencies can be obtained using different Huffman tables.

2. For typical values of the quality factor q, many of the AC coefficients within an 8×8 block will be zero-valued. Zig-zag scanning of the AC coefficients leads to an efficient (run-length, value) representation for the nonzero AC coefficients.

3. Values for the DC differentials range between -2047 and 2047, and for the AC coefficients range between -1023 and 1023. Direct Huffman coding of these values would require code tables with 4,095 and 2,047 entries, respectively. By Huffman coding only the *size* or the *(run/size)* information, the size of these tables is reduced to 12 and 162 entries, respectively.

To illustrate the benefits of run-length coding, Figure 5.8 shows for a typical grayscale image the output bit rate with and without a run-length coder. The top plot shows the output bit rate when an ideal Huffman or arithmetic coder is applied directly to the output of the DCT quantizer in Figure 5.1. The bottom plot shows the output bit rate when the ideal Huffman or arithmetic coder is preceded by a run-length coder. Bit rates are measured for various settings of the quality factor used to scale the quantization table. For a quality factor of one, the bit rate with a run-length coder is nearly four bits per pixel lower than the bit rate of an entropy coder alone. This is largely attributable to the efficient run-length representation of the zig-zag ordered AC coefficients. As the quality factor increases, more of the quantized AC values will be zero, and as expected, the benefits from a run-length coder are even higher.

5.6   JPEG MODES OF OPERATION

As mentioned before, in addition to the lossless mode of operation, JPEG defines the following other modes: sequential, progressive, and hierarchical.

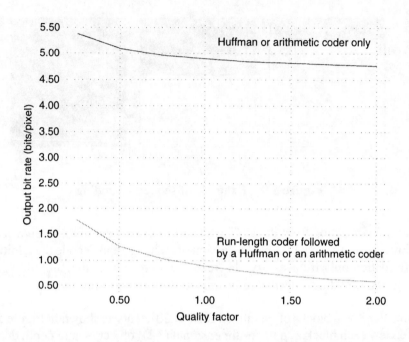

Figure 5.8   Effects of run-length coding on data compression.

5.6.1   Sequential Coding

Sequential coding is the most common mode of operation. Image blocks are coded in a scan-like sequence, from left to right and from top to bottom. Transformed and encoded blocks can be transmitted before the end of the image. Similarly, the decoder may begin sequential decoding before it receives the complete compressed image. Figure 5.9 shows an example of sequential coding.

5.6.2   Progressive Coding

In progressive mode, image blocks are also processed sequentially, but the coding is completed in multiple *scans*. The first scan yields the full image but without all the details, which are provided in successive scans. This mode requires that the output of the DCT be buffered so that during each scan only

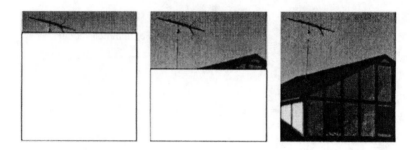

Figure 5.9   Example of sequential coding.

partial information from the DCT coefficients is encoded. Progressive coding allows a user to preview a rough version of an image and decode the additional information only if necessary. There are two procedures that are allowed for progressive coding: *spectral selection* and *successive approximation*.

Consider 8×8 blocks of quantized DCT coefficients as shown in Figure 5.10. We view each block as a three-dimensional (3-D) object, where depth denotes

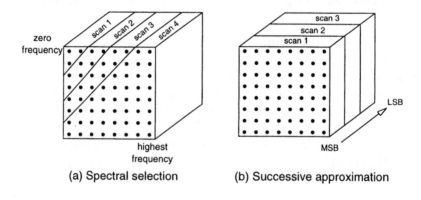

Figure 5.10   Description of progressive coding in JPEG.

the arithmetic precision of the quantized coefficients. Under spectral selection, each block is divided into frequency bands, and each band is transmitted during a different scan. For example, in Figure 5.10a, the DCT output is divided into four scans. Scan 1 includes the DC coefficient and the first two AC coefficients (counted in zig-zag order); Scan 2 includes the next seven AC coefficients; Scan 3 includes another 11 AC coefficients; and Scan 4 includes the remaining AC coefficients. For most images, most of the information is contained in the DC and the first few AC coefficients. Thus, encoding and transmission of the first scan of coefficients will provide adequate information for a rough preview of the image. Encoding and transmission of the remaining scans just adds progressively additional detail.

Figure 5.11 shows an example of progressive coding based on spectral selection. Figure 5.11a shows the output image after decoding only the DC

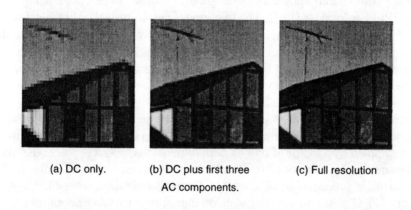

(a) DC only.   (b) DC plus first three   (c) Full resolution
                      AC components.

Figure 5.11 Example of JPEG progressive coding using spectral selection. (a) Decoding of the DC coefficient only; (b) decoding of the DC and the first three (in zig-zag order) AC coefficients; (c) decoded image at full spectral resolution.

coefficients. The image is rather blocky, but we can still get a rough preview of the image. Figure 5.11b shows the output image after decoding the DC and the first three AC coefficients. The diagonal edges of the house are still blocky. Figure 5.11c shows the decoded image at full spectral resolution.

Under successive approximation, given a frequency band, the DCT coefficients are divided by a power of two before encoding. This scheme allows the encoder to transmit the most significant bits first, for a rough preview, and the least significant bits later, for decoding at full resolution. For example, in Figure 5.10b, the DCT output is encoded using three successive approximation scans.

In the decoder, the coefficients are scaled back by the same power of two before computing the IDCT. The two progressive schemes may be combined or used separately.

5.6.3   Hierarchical Coding

In hierarchical mode, each image component is encoded as a sequence of *frames*. The first frame is usually a low-resolution version of the original image (possibly downsampled). Subsequent frames are differential frames between source components (possibly downsampled) and reference reconstructed components (possibly upsampled). Frames can be coded using either lossy JPEG or lossless JPEG. The two modes can be mixed only when lossless JPEG is used for the last stage of DCT-based hierarchical process. Hierarchical coding is useful when there are multiresolution requirements. For example, an application may support both high-resolution displays on workstations and low resolution displays on personal computers.

Figure 5.12 shows an example of a three-level hierarchical coder. From the original image X, we generate two subsampled versions: X_2, where the image is subsampled by a factor of two on both dimensions; and X_4, where the image is subsampled by a factor of four on both dimensions. Note that, subsampling may also be preceded by a low-pass filtering operation to reduce aliasing effects. JPEG poses no requirements on these preprocessing operations.

The encoded image consists of three frames: $S1, S2$, and $S3$. Frame $S1$ is simply the X_4 image compressed. Using only $S1$, the decoder can extract a low-resolution estimate of the original image (X_l'). $S2$ (uncompressed) is the difference image between X_2 and an estimate of X_2 (X_2') after upsampling X_4 by a factor of two. Using $S1$ and $S2$, the decoder can extract a medium resolution estimate of the input X (X_m'). Similarly, $S3$ (uncompressed) is the difference image between X and an estimate of X based on X_2 and X_4. The reader can verify that, under lossless compression, $X' = X$.

Figure 5.13 shows an example of hierarchical coding. The first image (a) is the original image shown at full resolution (200 dpi). Images (b) and (c) are subsampled versions of the original (subsampled by factors of two and four, respectively).

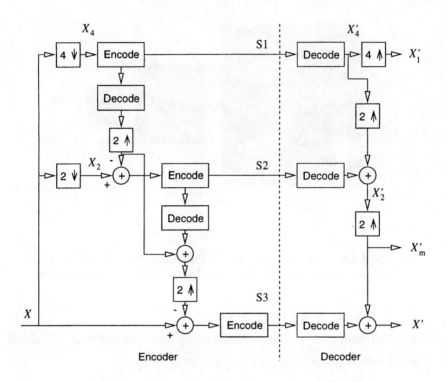

Figure 5.12   Three-level hierarchical coder.

To summarize, the essential characteristics of the main JPEG coding processes are given as follows.

1. Baseline Process

- Coding: DCT-based, sequential, one to four color components.
- Resolution: eight bits per pixel.
- Huffman coding: two AC and two DC tables.
- Interleaved and noninterleaved scans.

2. Extended DCT-based Process

- Coding: DCT-based, sequential or progressive.

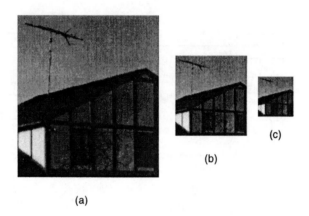

(b)

(c)

(a)

Figure 5.13 Example of hierarchical coding. (a) Original; (b) subsampled by a factor of two; (c) subsampled by a factor of four.

- Resolution: 8 or 12 bits per pixel.
- Huffman coding or arithmetic coding: four AC and four DC tables.
- Interleaved and noninterleaved scans.

3. Lossless Process

- Coding: Predictive, sequential.
- Resolution: From two bits per pixel to 16 bits per pixel.
- Huffman coding or arithmetic coding: four DC tables.
- Interleaved and noninterleaved scans.

4. Hierarchical Process

- Coding: DCT-based or lossless process.
- Multiple frames (nondifferential and differential).
- Interleaved and noninterleaved scans.

JPEG has been developed for the compression of still-images; however, the proliferation of low-cost hardware for JPEG has led to the development of an additional mode of operation for video sequences: motion-JPEG. Under motion-JPEG, each frame of a video stream is compressed independently using the baseline JPEG algorithm. However, there is no standard syntax for motion-JPEG coded streams, and encoded data may not be able to be decoded across different platforms.

5.7 IMPLEMENTATION ISSUES

The specific implementation of the JPEG standard depends on the application requirements and whether a software or a hardware implementation is adopted. In this section, we will address several issues relevant to the implementation of the *baseline* JPEG standard.

5.7.1 Hardware versus Software Implementation

Using a scaled DCT, such as the one described in Chapter 3, JPEG compression of a 640×480 RGB image requires close to 23 million operations. Thus, a typical general-purpose processor can compress such an image in less than a second.

Computation of the DCT requires close to 45 percent of the overall time. In Chapter 3 we described several algorithms for computing the 8×8 DCT and IDCT. In a software implementation, the scaled DCT method of Winograd and Feig, which requires 80 multiplications and 464 additions, is a good choice. A scaled DCT method is particularly advantageous for JPEG, since the quantization tables are fixed for the entire image. In a hardware implementation, a DCT method that has a more regular dataflow, such as the Chen DCT, is preferable.

5.7.2 IDCT Complexity

An 8×8 IDCT requires the same number of operations as the DCT. However, for most JPEG compressed images, the 8×8 block that is input to the IDCT is quite sparse. In a software implementation, the sparseness can be exploited to realize a more efficient evaluation of the IDCT. For instance, in a smooth region of an image, one would find that the corresponding DCT blocks have all AC coefficients set to zero. For such blocks, the IDCT is simply a scaled version of the DC coefficient that is copied to all the 64 locations in the block. Similarly, in regions of low activity, one may only need to compute simpler 2 × 2 or 4 × 4 IDCTs. The appropriate IDCT can easily be determined from the output of the Huffman decoder.

5.7.3 Arithmetic Precision Requirements

For eight bits per pixel data, the output of the 8 × 8 DCT will yield DCT coefficients that have a dynamic range between -1,023 and 1,023. Determining the requirements for arithmetic precision in the DCT requires a careful study of the flowgraph associated with the specific DCT method. In general, using 12 bits of precision for the constants in the DCT flowgraph and 16 bits of precision for all arithmetic operations yields the same output quality as 32-bit floating-point arithmetic. In Chapter 7, we provide a standard testing procedure for determining the accuracy of the IDCT, assuming that the DCT output is not quantized.

5.7.4 JPEG Coding Tables

JPEG specifies no default coding tables; however, most implementations use the example tables described in the Annex K of the ISO document. Designing custom Huffman tables is relatively easy. One can use either the techniques described in Chapter 2 or a similar method described in Annex K of the standard. However, the design of custom quantization tables is not trivial. In a previous section, we described a design method based on bit-rate control before the entropy coder. Recently, Ratnakar and Feig have proposed the design of quantization tables that are optimum in a rate-distortion sense; that is, if the user specifies the desired rate for a specific image, say, 0.5 bits per pixel, then this design method will provide the quantization tables that achieve this rate, provided that a custom Huffman table is used. Note that, JPEG does not have an explicit rate-control; hence, this design method is a useful tool for providing rate-control within JPEG. However, this design method is image specific.

For a typical grayscale image, Figure 5.14 shows the effects of custom coding tables on the output bit rate of a JPEG coder. Several observations can be

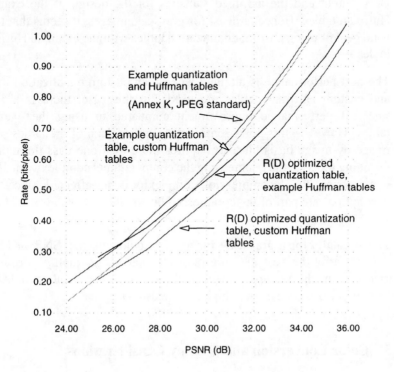

Figure 5.14 JPEG coder performance for example and custom coding tables.

made from these plots.

- Using the example quantization tables at compression ratios around 40:1 (0.20 bits per pixel for the 8 bits per pixel grayscale image), custom Huffman tables yield nearly 1 dB performance improvement over the example Huffman tables. Performance gains greater than 0.5 dB usually yield visible differences between the corresponding reconstructed images. At compression ratios around 10:1, custom Huffman tables are not beneficial.

- Using the example Huffman tables, custom quantization tables yield better performance only at low compression ratios. At high compression ratios, there is significant mismatch between the actual statistics of the

DCT output and the assumed statistics for the design of the example Huffman tables. Hence, at least for grayscale images, it seems that there is no benefit using custom quantization tables with the example Huffman tables.

- The best performance is attained when both custom quantization tables and custom Huffman tables are used. In this case, there is a 1 dB to 2 dB performance improvement compared to using the example tables. However, this approach requires the redesign of the tables on an image-by-image basis. It also requires more than one pass through the encoding process, thus, increasing the computational complexity of JPEG compression; the decompression complexity is not affected, since table descriptions are part of the compressed bit stream.

The above data is for a grayscale image. For color images, SNR or PSNR is not necessarily the best measure of perceptual image quality. From our experience, using the example Huffman tables, custom quantization tables can yield better performance even at high compression ratios.

5.7.5   Color Conversion and Display Considerations

For image compression of RGB images, the RGB to YCbCr color transformation may represent up to 39 percent of the overall compression process. Similarly, in many applications, there is a need to decompress a YCbCr JPEG image and display it as an RGB image. Furthermore, the Cb and Cr components are usually half the resolution of the Y component; hence, during the color conversion process, the Cb and Cr components have to be upsampled to facilitate the conversion to RGB. Upsampling, color converting, and displaying of an image may represent 33 percent of the computational complexity of the JPEG decompression process. Color conversion using lookup tables can be used to speed up the process. In recent years, there is a trend towards implementing the color conversion and upsampling functions in the display system hardware, and thus eliminating the overhead associated with the display.

5.7.6   The JPEG Compressed Data File Format

The JPEG standard specifies an interchange format so that compressed images can be reconstructed reliably. However, the interchange format only provides information pertaining to the image dimensions, resolution of each of the components that make up the image, and the quantization and Huffman code tables that are used in the compression process. JPEG does not explicitly specify the color space that is used in compression. In recent years, a de facto standard, referred to as the JPEG file interchange format (JFIF), has evolved and is supported by many image processing applications. This format is a superset of the JPEG interchange format and supports either grayscale or YCbCr images. The bit-stream syntax of JFIF also provides for inclusion of a thumbnail RGB representation of the image; this can be used for browsing and cataloging purposes.

5.8   JPEG EXTENSIONS AND APPLICATIONS

Part 3 of the JPEG standard (ITU-T Recommendation T.84 or ISO/IEC 10918-3) specifies requirements and guidelines for encoding and decoding extensions to the processes defined in Part 1. There are four major extensions defined in this part: variable quantization, selective refinement, tiling, and a new image-interchange format.

5.8.1   Variable Quantization

In lossy JPEG, the amount of loss or compression is controlled by the elements of the quantization tables. JPEG allows quantization tables to be redefined prior to the start of a new scan, but not within a scan. The variable quantization extension introduces a quantizer scale factor that may be used to scale the original quantization matrix on a block-by-block basis. This scale factor may be coded in the compressed bit stream at the start of any 8×8 block. This extension does not apply to the baseline JPEG process.

This extension allows for greater control on image quality and compression ratios. For example, in an image with both text and graphics, one may use different scale factors for blocks characterized as *image* and different scale factors for blocks characterized as *text*. Since text regions and image regions tend to have different statistical properties, the different scale factors in these areas should allow for more efficient coding of the image.

5.8.2   Selective Refinement

This extension allows for the selective refinement of parts of an image. The three different types of selective refinement are hierarchical, progressive, and component.

Hierarchical selective refinement is used in the hierarchical mode of operation. It allows the encoder to specify which areas of the image are coded in greater detail than the rest of the image. In the decoder, the differential frames are then added only to those specified areas.

Progressive selective refinement is used in the progressive mode of operation. As in hierarchical refinement, the encoder may supply additional DCT data only for specific regions in the image.

Component selective refinement may be used in all modes of operation. It allows the encoder to specify that a certain region of the image uses fewer color components than the total number of components defined in the frame header. This extension could be used for the more efficient representation of images that combine both grayscale and full color regions.

5.8.3   Tiling

The tiling extension allows an image to be subdivided into subimages, also called tiles. It can be used to represent images larger than 65,535 pixels on a side, for random access within parts of an image, or for more efficient compression. In *simple tiling*, the image is divided into an array of non-overlapping rectangular tiles. All simple tiles have the same size, except maybe for the tiles at the right and the bottom sides of the image. These tiles must have the same component identifiers, sampling factors, and entropy coders; however, each tile may be coded with a different quantization table. *Pyramidal tiling* allows multiple resolution versions of the same image to be stored in the same compressed bit stream. This feature allows access to lower resolution versions of a larger image. *Composite tiling* is the most complex tiling method. Complex tiles may overlap and have different sizes; however, they still need to have the same component identifiers. It is expected that most of the future applications will only support simple tiling.

5.8.4   The Color Facsimile Standard

In November of 1994, ITU formally approved extensions to the Group 3 and Group 4 facsimile standards for the transmission of continuous-tone images. The new color facsimile standard uses the CIE (1976) $L^*a^*b^*$ (CIELAB) color space for the representation of color images and the baseline JPEG algorithm for data compression.

In the baseline implementation of the color facsimile pipeline, a scanned 24-bit RGB image (scanned at 200 dpi) is first translated into the CIELAB color space using the standard CIE D_{50} illuminant. The gamut range is restricted to

$$L^* = [0, 100], \tag{5.11}$$
$$a^* = [-85, 85], \tag{5.12}$$
$$b^* = [-75, 125]. \tag{5.13}$$

This color space was chosen because it provides for a more flexible, relatively uniform, and device-independent color specification. It also enables the receiver to reproduce a hard copy of the original image without ambiguity.

After color transformation, the chroma components (a^* and b^*) of the image are subsampled by a factor of two in both the horizontal and the vertical directions. The subsampled image is coded using the JPEG algorithm (lossy or lossless), and the compressed bit-stream is transmitted using the standard Group 3 or Group 4 facsimile protocol. At the receiver, after Group 3 or Group 4 decoding, the image is decompressed, color transformed, and printed on a color printer.

Optional features of the standard allow for higher spatial resolutions, no color subsampling, higher precision (12 bits per pixel), and custom color gamut range.

5.8.5   The Still-Picture Interchange File Format (SPIFF)

SPIFF is yet another image file format developed for the efficient interchange of files containing image data. It is more comprehensive than the JFIF format, but still not as complex as some other image file formats, such as TIFF. It is backward-compatible with JFIF, but in addition to grayscale and color JPEG images, it also supports bilevel images and many of the facsimile compression algorithms.

5.8.6   The FlashPix Image File Format

FlashPix is an image file format that has been developed recently by Eastman Kodak, in collaboration with Hewlett-Packard Company, Live Picture, Inc., and Microsoft. FlashPix is not an international standard, but it has the potential to become the next industry de facto standard for storing and manipulating digital images in the internet and on the desktop. A FlashPix image file includes a hierarchical digital representation of an image, a 96×96 pixel thumbnail, and viewing parameters. In addition, it may also include result images, file links, and application-specific data.

In FlashPix, image data are stored hierarchically at multiple and independent resolutions. This feature allows imaging applications to select the best resolution possible for a specific platform. For example, an internet-based image viewer may elect to display a low-resolution view of an image, but a printer driver may use the image at its full resolution.

At the bottom of the image hierarchy is the image at full resolution. This resolution is determined by the resolution of the capturing device. At the top of the image hierarchy is a 64×64 pixels representation of the image. Each of the levels in the image hierarchy is created by repeatedly decimating the image by a factor of two, both vertically and horizontally. Each level is also divided into a set of 64×64 nonoverlapping tiles. The tiling format allows applications to quickly access any part of the image at any given resolution. Tiles can be stored either uncompressed or compressed using the baseline JPEG standard. For compressed tiles, JPEG coding tables can either be shared among tiles or they can be part of the compressed bit stream of each tile.

If the original image is edited, FlashPix allows applications to save those edits as viewing parameters in an edit-script. Supported operations include spatial transformations, cropping information, color correction, contrast adjustment, and filtering. Thus, applications can minimize the computational load by applying the specified edit-script only to the specified tiles and at the most appropriate resolution. Furthermore, with file links to the original data, FlashPix requires minimal extra storage for multiple representations of the same image data. Result images, that is, fully preprocessed images, can also be part of a FlashPix file. These prestored images may be useful to applications that do not have adequate computing power to process data from the original image.

5.9   TO PROBE FURTHER

The JPEG standard [108] is published in two parts. Part 1 sets requirements and guidelines for coding and decoding and describes the interchange format. Part 2 describes compliance tests for the processes described in Part 1. A new part (ISO/IEC 10918-3 or ITU-T Rec. T.84) provides extensions to the coding processes defined in Part 1 and describes in detail the new SPIFF file format.

A good source for the JPEG standard is the book by Pennebaker and Mitchell [204]. The book includes a complete copy of the ISO document and provides additional implementation details. For a concise overview of the standard, we also recommend an excellent tutorial overview by Wallace [255].

Experiments for determining visibility thresholds for the DCT basis functions using the YCbCr color space are described by Lohscheller and Franke in [158] and [159]. Similar experiments for the RGB color space are presented in [205]. The optimal bit allocation for quantization tables based on rate-control theory is described in the textbook by Jayant and Noll [121]. An adaptive coding scheme based on the same principles is also described by Chen and Smith [39]. In [46], de Queiroz and Rao weight the image statistics with a matrix that contains spatial information about the human visual system (HVS). Recently, quantizer design methods optimal in a rate-distortion sense have been developed in [218] and [219]. The techniques developed here can provide a fixed-rate implementation of JPEG even though the JPEG standard as developed does not include explicit rate-control. In many cases, a compressed image will be decompressed, halftoned, and printed. Recently, Vander Kam et al. [124] were able to improve the compression of grayscale images that will be printed by taking into consideration both the sensitivity of the HVS and the frequency characteristics of the printer's halftoning algorithm.

Many imaging applications currently support the JPEG standard. However, to facilitate the interchange of JPEG compressed data across applications, there is a need to develop a file format. The TIFF format is widely used in the imaging community and has provisions for describing a JPEG bit stream. For the simpler JFIF format, details can be found in [88]. The FlashPix image format is described in [1].

The section on the JPEG extensions is based on a recent article by Lee [141]. The details of the color facsimile standard are defined in new Annexes to the facsimile standards documents. For Group 3, Annex E of ITU-T T.30 [115] specifies new entries in the DIS and DCS frames of the facsimile communication protocol and Annex I of ITU-T T.4 [116] describes the appropriate JPEG syntax and JPEG application markers. A general overview of the standard is also given in [21].

There are numerous software and hardware implementations of the baseline JPEG standard. Sites with public domain JPEG coders are given in Appendix C. Hardware implementations of the JPEG standard are discussed in Chapter 13.

6

THE MPEG VIDEO STANDARDS

6.1 INTRODUCTION

Advances in digital video technology and storage have made it possible to use digital video into a number of multimedia applications. In 1988, in response to a growing need for a common format for coding and storing digital video, ISO established the Moving Pictures Expert group (MPEG) with the mission to develop standards for the coded representation of moving pictures and associated audio information on digital storage media. MPEG, formally known as group ISO-IEC/JTC1 SC29/WG11, completed the first phase of its work in 1991 with the development of the ISO standard 11172, *Coding of moving pictures and associated audio - for digital storage media at up to about 1.5 Mbit/s*. This standard is also known as MPEG-1.

In 1990, MPEG started the second phase of its work, namely, to develop extensions to MPEG-1 that would allow for greater input-format flexibility, higher data rates (as needed by high-definition TV), and better error resilience. That work led to the ISO standard 13818 (or ITU-T Recommendation H.262), *Generic coding of moving pictures and associated audio*. This standard is also known as MPEG-2. In 1994, MPEG initiated a new standardization effort, commonly referred to as MPEG-4. Unlike MPEG-1 and MPEG-2, wherein the emphasis was primarily on coding efficiency, the objective of MPEG-4 is to provide a standard that supports new ways of communicating, accessing, and manipulating digital audiovisual data. MPEG-4 plans to offer a common technical solution to various communication paradigms - telecommunications, broadcast, and interactive - among which the borders are disappearing. In contrast to the frame-based paradigm of MPEG-1 and MPEG-2, MPEG-4 proposes an object-based paradigm for scene representation. The release of the MPEG-4 international standard is targeted for November of 1998. All MPEG standards are *generic*; that is, application independent. They do not specify the operations of the encoder. Instead, they specify the syntax of the coded bit stream and the decoding process. Thus, they provide enough flexibility in the specifications so that different vendors can include specific optimization elements.

The MPEG standards are published in several parts; MPEG-1 comprises five parts, whereas MPEG-2 is comprised of nine parts. The major parts of the MPEG-1 and MPEG-2 standards are: systems, video, audio, and compliance testing. The systems part specifies the system coding layer of the standards and defines how data, audio, and video streams can be multiplexed. For example, in MPEG-1, the systems layer provides sufficient information for synchronization of multiple video streams, random access to the data, and buffer management. In MPEG terminology, the video, audio, and data streams are generically referred to as elementary streams. Note that, in a system layer stream, more than one audio, video and data stream can be included. A detailed discussion of the system stream is beyond the scope of this book.

The video part specifies the coded representation of video data and a decoding process for reconstructing MPEG coded pictures. Audio is an integral part of the MPEG standards. The audio part specifies the coded representation of audio coded data and a decoding process. The compliance part of the standards specifies procedures to determine the characteristics of coded bit streams and to test compliance of bit streams and decoders.

In this chapter, we cover only the video part of the MPEG standards. The MPEG audio standards are covered in a later chapter, together with other algorithms for audio coding and compression.

6.2 THE MPEG-1 VIDEO STANDARD

6.2.1 Preliminaries

The MPEG-1 video coding algorithm is a lossy compression scheme that can be applied to a wide range of input formats and applications; however, it has been optimized for applications that support a continuous transfer bit rate of about 1.5 Mbits/s (such as a CD-ROM).

MPEG-1 deliberately uses the term *picture* and not *frame*, because it does not recognize interlaced sources. In interlaced video, each frame is comprised of fields, namely, a top field and a bottom field. Within a frame of interlaced video, scanlines from the two fields are interleaved (Figure 6.1a). Note that, spatially adjacent scanlines are not temporally adjacent. For example, at 30 fps, the adjacent scanlines are $\frac{1}{60}$-th of a second apart. In progressive (or noninterlaced) video there is no notion of a field. Unlike interlaced video, spatially adjacent scanlines are also temporally adjacent. A frame begins from the top-left corner and continues through successive lines to the bottom of the frame (Figure 6.1b). In MPEG-1, interlaced video, such as that generated by a TV camera, may have to be converted to a noninterlaced coding format before coding. This process of scan conversion is not specified within the MPEG-1 standard.

There were two key requirements during the development of the MPEG-1 standard: need for high compression and need for random-access capability. Intraframe coding alone is best suited for random-access but cannot meet the requirements for high compression. To meet these conflicting requirements, MPEG resorted to a combination of intraframe and interframe coding techniques. Furthermore, to improve the compression ratio, it proposed using both predictive and interpolative coding schemes.

Compression functions within MPEG include the following:

1. sample rate reduction in the spatial and temporal domains of both the luminance and chrominance components,

2. block-based DCT for the intraframes and interframes,

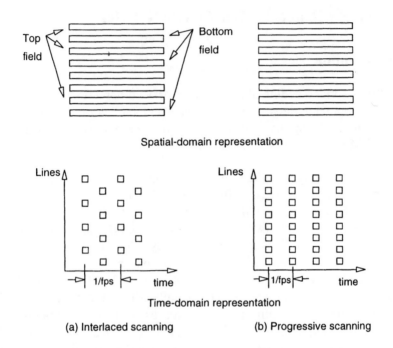

Spatial-domain representation

Time-domain representation

(a) Interlaced scanning   (b) Progressive scanning

Figure 6.1   Interlaced and progressive scanning schemes.

3. block-based motion compensation for predictive and interpolative inter-frames, and

4. Huffman coding for the lossless compression of motion vectors and the quantized DCT coefficients.

For a picture size of 352 × 240 pixels, at 12 bits per pixel and a picture rate of 30 pictures per second, MPEG-1 compresses the 30.4 Mbits/s bit stream down to 1.15 Mbits/s. For this compression ratio of 26:1, the picture quality can be better than that provided with the analog VHS representation. We will describe the compression method in detail later in this chapter.

The source input format

The MPEG-1 bit-stream syntax allows for picture sizes of up to 4,095 × 4,095 pixels; however, many of the applications using MPEG-1 compressed video have been optimized for the SIF (source input format) video format, which is a simple derivative of the CCIR 601 format for video frames in digital television. According to CCIR 601, a color video source has three color components: a luminance component (Y) and two color-difference or chrominance components (Cb and Cr). There are two options for the CCIR 601 spatial resolutions. The first option (for NTSC television systems) uses 525 lines per frame at 60 frames per second. The luminance frame has 720 × 480 active pixels, and each of the chrominance frames has active 360 × 480 pixels. This is referred to as the 4:2:2 subsampling format. The second option (for PAL television systems) uses 625 lines per frame at 50 frames per second. Here, the luminance frame has 720 × 576 active pixels, and each of the chrominance components has 360 × 576 active pixels. To accommodate compressed rates as low as 1.5 Mbits/s, MPEG-1 defines (but does not require) the source input format. SIF sequences have a luminance resolution of 360 × 240 pixels per picture at 30 pictures per second or 360 × 288 pixels per picture at 25 pictures per second. In both cases, the resolution of the chroma components is half of the luminance resolution in both the horizontal and the vertical dimensions. This is referred to as the 4:2:0 subsampling format. SIF pictures can easily be derived from a CCIR 601 frame using filtering and subsampling.

In MPEG-1, the YCbCr color components are always interleaved. A macroblock is defined as the minimum coded unit and consists of four 8 × 8 blocks of luminance, one 8 × 8 block of Cb, and one 8 × 8 block of Cr (Figure 6.2). The maximum dimension of a macroblock is 16 pixels. Each picture is divided into a series of macroblocks, from left to right and from top to bottom. This requires that the horizontal and vertical resolutions of each picture be multiples of 16. If they are not, then the coder adds padding pixels to the right or at the bottom of each picture that are later discarded by the decoder. Since the horizontal resolution of 360 pixels in a SIF picture is not divisible by 16, one can duplicate the last pixel in each line eight times so that the total number of pixels per line is 368. In an alternative implementation, one can discard the leftmost four pixels and the rightmost four pixels from each line; then the remaining picture has 352 pixels per line and is called the *significant*

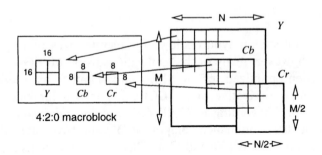

Figure 6.2 Definition of a macroblock in MPEG-1.

pixel area. Table 6.1 summarizes the picture characteristics for the CCIR 601, the SIF, and the significant pixel area formats.

Picture Rate (Hz)	30	25
CCIR 601		
Y	720 × 480	720 × 576
Cb, Cr	360 × 480	360 × 576
SIF		
Y	360 × 240	360 × 288
Cb, Cr	180 × 120	180 × 144
Significant Pixel Area for SIF		
Y	352 × 240	352 × 288
Cb, Cr	176 × 120	176 × 144

Table 6.1 Picture sizes for the CCIR 601, SIF, and significant pixel area formats.

MPEG-1 provides a great degree of flexibility; however, it would be unrealistic to expect that every MPEG decoder could support all coding options. Thus, MPEG-1 defines the *constrained parameters bit stream*, that is, the bit stream that every MPEG-1 compatible decoder should be able to support. Table 6.2 shows the parameters of a constrained bit stream; however, conforming to these sets of parameters is still not a requirement of the standard. From Table

Coding Parameter	Maximum Value
Horizontal picture size	768 pixels
Vertical picture size	576 lines
Macroblocks	396
Pixel rate	396 × 25 macroblocks/s
Picture rate	30 pictures/s
Range of motion vectors	+/- 64 pixels (half-pixel resolution)
Size of input buffer	327,680 bits
Bit rate	1,856 kbits/s

Table 6.2 Coding constraints for a constrained parameters bit stream.

6.2, the constraint on the pixel rate (396 × 25 macroblocks/s) restricts the picture size to approximately 352 × 288 pixels, which is the significant area of a SIF image at 25 Hz.

6.2.2 Picture Types

Some of the common operations on video streams include video editing, random access, and searches. Support of these operations often conflicts with the requirements for high compression. MPEG-1 defines four different types of pictures that offer the flexibility to trade off between coding efficiency and random access.

Intra-pictures (I-pictures) are compressed using intraframe coding; that is, they do not reference any other pictures in the coded bit stream. They provide for fast random access, but offer only moderate compression. This situation is similar to applying lossy JPEG on individual pictures. Predicted pictures (P-pictures) are coded using motion-compensated prediction from past I-pictures or P-pictures. The compression for P-pictures is better than for I-pictures, and P-pictures can be used as reference points for additional motion compensation. Bidirectionally predicted pictures (B-pictures) provide the highest degree of compression. They are coded using motion-compensated prediction from either past and/or *future* I-pictures or P-pictures. Since B-pictures are not used in the prediction of other B or P-pictures, such pictures can accommodate more distortion and hence yield more compression than I- or P-pictures. DC-coded pictures (D-pictures) are similar to I-pictures; however, only the DC coefficients from the DCT output are present. D-pictures cannot be mixed with other picture types and are used mainly for fast searches. D-pictures are not used in MPEG-2.

Figure 6.3 shows the relationship among the three main picture types in a video sequence with eight pictures. Pictures p_1 and p_8 are I-pictures, pictures

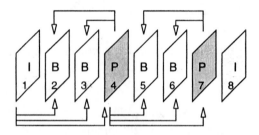

Figure 6.3 Example of inter-dependence among I-, P-, and B-pictures in a video sequence.

p_4 and p_7 are P-pictures, and the remaining are B-pictures. In this example, p_3 is coded using motion-compensated prediction from p_1 and p_4. There is no requirement to use either P-pictures or B-pictures in the MPEG bit stream. The size of I-, P-, and B-pictures for a typical video sequence coded at SIF resolution and at 1.15 Mbits/s is shown in Figure 6.4. Note that B-pictures require significantly fewer bits than either I-pictures or P-pictures. Increasing the number of B-pictures between an I- and a P-picture may not lead to better compression due to a drop off in temporal correlation as the distance between a B-picture and the corresponding I- and P-pictures increases. In the next section, we provide a brief overview of the encoding process.

6.2.3   Video Encoder

The MPEG standards *do not* define an encoding process. They only specify the syntax of the coded bit stream and the decoding process. Based on these

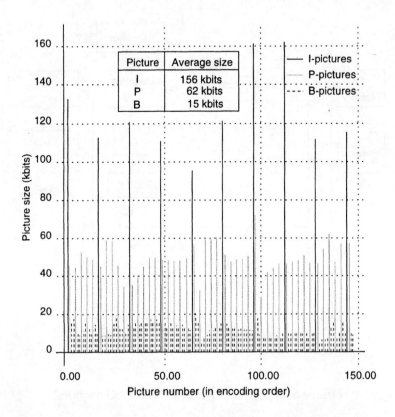

Figure 6.4 Example of bit distribution among I-, P-, and B-pictures in an MPEG-1 coded bit stream. Pictures are coded in IPBBPBBPBBPBBPBB sequence.

requirements, Figure 6.5 shows the functions that need to be executed by a typical MPEG encoder.

Preprocessing. The encoding process usually begins with some preprocessing. This may include color conversion to YCbCr, format translation (interlaced to progressive), prefiltering, and subsampling. None of these operations is specified in the standards.

Motion estimation and compensation. After preprocessing, the encoder selects the coding type for the input picture. I-pictures require no motion

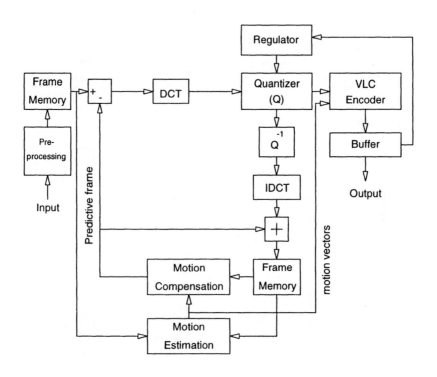

Figure 6.5   Block diagram of an MPEG encoder.

estimation or compensation. Each macroblock is DCT coded, and the DCT coefficients are quantized, coded using a variable-length coder (VLC), and stored in the output buffer. Within each macroblock, processing is performed on 8×8 blocks. In applications requiring constant bit rate, a buffer regulator may adjust the quantization matrix so that the rate of the compressed bit stream is relatively constant. For variable bit-rate applications, the regulator function may not be needed.

The quantized blocks are also inverse quantized (Q^{-1}) and transformed into the spatial domain by an inverse DCT. This operation duplicates the behavior of the decoder and yields a copy of the encoded picture as it will be seen by the decoder. That copy is then stored in local memory and will be used for future predictive coding. In effect, this operation allows the encoder to monitor the quality of the transmitted image so that it does not diverge from the original signal. Since the VLC operation is lossless, there is no need to include the VLC unit in the feedback path.

If the input is coded as a P-picture or a B-picture, then the encoder does not code the picture macroblocks directly. Instead it codes the prediction errors. In MPEG terminology, this is referred to as *interframe predictive coding*. For P-pictures, this process is explained in Figure 6.6. For each macroblock in the

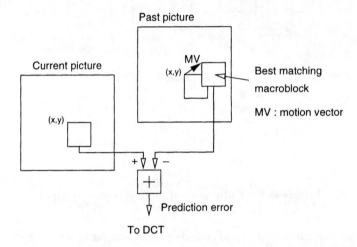

Figure 6.6 Forward motion compensation.

current picture, the motion estimator yields the coordinates (motion vector) of the macroblock that best matches its characteristics in a search area in the past picture. The two macroblocks are subtracted, and their difference is then DCT coded. Any of the block-based motion estimation methods described in Chapter 4 can be used.

For B-pictures, the motion estimation process is performed twice, once for a past picture and once for a future picture, so as to yield two motion vectors. The encoder can form a prediction error macroblock from either of the two candidate macroblocks or from their average. This process is shown in detail in Figure 6.7. In MPEG terminology, this is referred to as *interframe interpolative coding*. The prediction error is then coded using the block-based DCT. The quantized DCT coefficients of the prediction errors, together with the motion

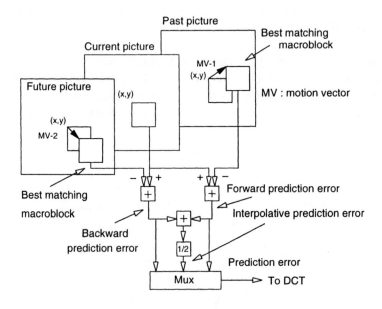

Figure 6.7   Bidirectional motion compensation.

vectors are then multiplexed and coded using the variable-length coder. Unlike JPEG, MPEG uses only a Huffman coder for variable-length coding.

Because of the bidirectional prediction, some reordering of the input pictures is needed at the encoder so that they are delivered in the correct order to the decoder. For example, consider the sequence of pictures in Figure 6.3. Since picture p_2 is a B-picture that depends on pictures p_1 and p_4, it can be encoded only after the input of picture p_4. After reordering of the input pictures, the correct encoding sequence is $p_1, p_4, p_2, p_3, p_7, p_5, p_6$, and p_8. This bit stream provides the decoder with all the information needed to decode each received picture; however, the decoder will need to reorder these pictures so that they can be displayed in the correct order.

6.2.4   Video Decoder

Figure 6.8 shows a block diagram of an MPEG video decoder. As expected,

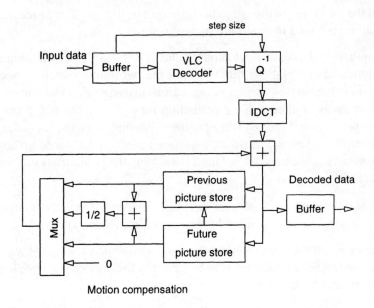

Figure 6.8 Block diagram of an MPEG decoder.

the circuitry of the video decoder is very similar to the circuitry in the feedback loop of the video encoder. Input pictures are first Huffman decoded and their type determined from the header information. For each macroblock, the coded coefficients are dequantized and translated into the spatial domain by an IDCT. Consider the encoded sequence $p_1, p_4, p_2, p_3, p_7, p_5, p_6$, and p_8. For the purposes of this discussion, we will assume that all macroblocks in a P-picture are coded as P-macroblocks and all macroblocks in a B-picture are coded as B-macroblocks. MPEG-1 offers more flexibility in macroblock coding, and we will discuss this in a later section in this chapter. The decoding process is as follows:

1. Input p_1 (I-picture). No motion compensation is performed (the input to the multiplexor in Figure 6.8 is set to zero). Compute the IDCT and save the output in a display buffer and in the previous picture store.

2. Input p_4 (P-picture). For each macroblock, compute the IDCT and perform motion compensation; that is, add the area pointed to by the motion vector

in the previous picture store to the output of the IDCT. The reconstructed picture is saved in the future picture store.

3. Input p_2 (B-picture). Compute the IDCT and perform bidirectional motion compensation; that is, using the two motion vectors, access the pixel values in the corresponding regions in the previous and future picture store areas, and compute a prediction for p_2. Sum the IDCT output and the prediction. Since p_2 is the picture following p_1 in display order, p_2 can be displayed now. Note that, B-pictures need not be stored in the picture store area since they are not used in forming the predictions of subsequent pictures.

4. Input p_3 (B-picture). Repeat the steps performed for decoding p_2. After displaying p_3, p_4 (decoded in Step 2) is displayed next.

5. Input p_7 (P-picture). Repeat the steps performed for p_4. However, the reconstructed picture needs to be saved in the previous picture store; this overwrites p_1 in the previous picture store.

6. Input p_5 and p_6 (B-pictures): Repeat the steps outlined for p_2. Note that, predictions for p_5 and p_6 use p_4 and p_7. After p_5 and p_6 are decoded, we first display p_4 and then the pictures p_5, p_6, and p_7.

This completes the decoding of one group of pictures. For the next group, starting from p_8, these decoding steps are repeated. Since at most three pictures have to be kept in store at any given instant, an MPEG-1 decoder needs at least a half Mbyte of memory for SIF resolution pictures. Note that, the order of display for the pictures is defined in the presentation time stamp that is contained in the system layer.

6.2.5   Structure of the Coded Video Bit stream

The MPEG-1 syntax for the coded video bit stream has a hierarchical representation with six layers: (1) a sequence layer, (2) a group-of-pictures layer, (3) a picture layer, (4) a slice layer, (5) a macroblock layer, and (6) a block layer.

The sequence layer is the top coding layer. It includes a sequence header and is followed by one or more groups of pictures. It ends with a *sequence-end-code*.

Among the information included in the sequence header are the vertical and the horizontal size of each picture, the pixel aspect ratio, the picture rate in pictures per second, the bit rate in units of 400 bits/s, and the minimum buffer size needed by the decoder (in units of 2,048 bytes). A single bit also specifies whether the coded stream meets all the requirements of a constrained parameter bit stream or not. In addition, the header may also include the DCT quantization matrices for intra and nonintra-pictures and optional user data. The default quantization matrix for intra-pictures is the one specified by the JPEG standard for luminance components. For nonintra-pictures, all values of the default quantization matrix are equal to 16. The rationale for uniform quantization might be that nonintra-pictures provide error information; hence, there is no reason to favor any particular set of frequency components. In contrast, in I-pictures, the frequency components are directly related to the activity within the picture.

A group of pictures (GOP) is a set of pictures in contiguous display order. It has to contain at least one I-picture. A GOP can begin with either a B-picture or an I-picture and must end with either an I-picture or a P-picture. If the first picture is an I-picture or a B-picture that does not depend on the pictures of the prior GOP, then this GOP can be coded and displayed independently from any other group. This is defined as a *closed GOP*. The header of a GOP includes timing information, a closed-GOP flag, extension and user data. Figure 6.3 shows a closed GOP with a GOP size of seven.

The picture layer defines the coding information for each picture. Because of the possibility of reordering P-pictures and B-pictures, the header provides a temporal reference number that can be used to define the display order of a picture. Additional header data provide information about the picture type, synchronization, and the resolution and range of the motion vectors.

Each picture is divided into slices. Slices can be as big as the whole picture and as small as a single macroblock. In case of data corruption, the information in the slice headers allows for a smoother recovery by the decoder. For example, in the case of transmission errors, a decoder can drop a slice and not the whole picture. A slice header contains information for its position within a picture and a quantizer scale factor, between one and 31, that can be used by the decoder to dequantize the coded DCT coefficients. Specification of quantizer scale factor allows the encoder to perform rate regulation at the slice level.

A slice is divided into macroblocks. The header of the macroblock defines the macroblock type, positional information, codes for the horizontal and vertical motion vectors, and which blocks within a macroblock are actually coded and transmitted. Optionally, the encoder may include *macroblock stuffing*. The pattern "0000 0001 111" is a stuffing code and can be inserted into the bit stream wherever the encoder detects the possibility of a buffer underflow. This stream is ignored by the decoder. Macroblock stuffing is not permitted in MPEG-2.

Figure 6.9 summarizes the syntax layers in the coded MPEG bit stream.

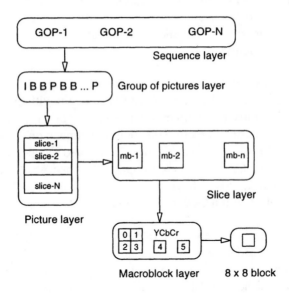

Figure 6.9   Syntax layers in MPEG-1 video coding.

6.2.6   Macroblock Coding

As mentioned before, there are three main picture types in MPEG: I-, P-, and B-pictures. However, even within a single I-, P- or B-picture, macroblocks can be coded differently. The decision trees for coding macroblocks is shown in Figure 6.10. This section summarizes the main options in macroblock coding.

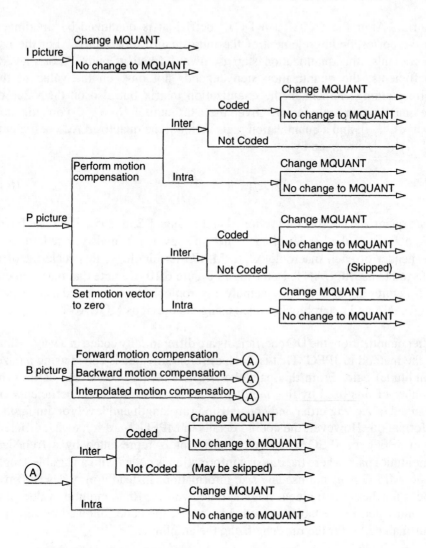

Figure 6.10 Decision trees for coding macroblocks in I-, P-, and B-pictures.

I-pictures

I-pictures are coded without motion compensation; however, the MPEG syntax allows each macroblock to be coded using a different effective quantization matrix. After the DCT, each DCT coefficient is quantized by dividing it by the corresponding element of the intra quantization matrix. For the DC coefficients, the quantization step is always fixed at eight. For the AC coefficients, the quantization step depends not only on the value of the corresponding element in the quantization matrix but also on the value of the quantization scale factor. Given the DCT output $Dct[i, j]$, a quantization matrix $Q[i, j]$, and a quantization scale factor q, the quantized AC coefficients ($QDct[i, j]$) are given by

$$QDct[i, j] = \frac{8Dct[i, j]}{qQ[i, j]}. \qquad (6.1)$$

Each quantized coefficient is clipped in the range [-255, 255]. The scale factor q is defined in the header of each slice; however, it can also be redefined in the header of each macroblock. In MPEG terminology, this scale factor is referred to as MQUANT. Referring to Figure 6.10, we note that macroblocks in I-pictures are of two types, namely, macroblocks that are coded using a new MQUANT and macroblocks coded using the previous MQUANT.

After quantization, the DC coefficients are differentially coded in a way similar to the method in JPEG. That is, each differential value is coded using a (size, amplitude) pair. From this pair, the *size* is Huffman coded and defines the number of bits used by the *amplitude*. As in JPEG, the AC coefficients are scanned in zig-zag order and coded using run-length and level (or amplitude) information. However, the coding details in MPEG-1 are somewhat different from JPEG. In JPEG, each AC coefficient is represented by a (run/size, amplitude) pair, where the run/size information is coded with a variable-length code. MPEG does not use any size information. Instead it provides Huffman codes for the most frequent run/amplitude values. Run/amplitude values that are not listed in the table are coded by an escape code followed by separate Huffman codes for the run-length and the amplitude.

Example. Consider the case of coding the run-length/amplitude values of 2/2 and 2/25. Since amplitude = 2 can be represented by two bits (10) and amplitude = 25 can be represented by five bits (11001), in JPEG we would have to code the pairs (2/2, 10) and (2/5, 11001). In MPEG, 2/2 can be coded directly as 00001000. The symbol 2/25 is not listed in the coding tables; hence, one has to use the escape code (000001) followed by the Huffman codes for run-length = 2 (000010) and amplitude = 25 (0001 1001).

Coding of P-pictures

P-pictures are also divided into slices and macroblocks; however, because of the motion-compensation process, the encoder has far more coding choices for each macroblock. The coding choices are depicted in Figure 6.10. The decisions are summarized as follows.

- Decide if you want to set the motion vector to zero. In many cases, the prediction error using a nonzero motion vector may be close enough to the prediction error for this macroblock, assuming a zero-valued motion vector. Since nonzero motion vectors require additional coding bits, it is more efficient to code this macroblock using the zero-valued motion vector. A simple rule of thumb is to use nonzero motion vectors only if the minimum prediction error is at least 1.1 times smaller than the error calculated for the zero motion vector.

- Decide if you want to code the macroblock as an intra- or inter-type macroblock. In many cases, it may require fewer bits to code a macroblock as an intra-macroblock (I-macroblock) even though it belongs to a P-picture. This may happen if motion estimation fails due to a high level of temporal activity.

- Decide if a macroblock needs to be coded or not. If, after quantization, all coefficients in the macroblock are zero, then that macroblock is not coded. Such a macroblock is referred to as a skipped macroblock. Decoding of such macroblocks requires only a simple copy of corresponding macroblocks in the previous picture. In all cases, when a macroblock is coded, not every block within the macroblock needs to be coded. There may be cases, where after quantization, all coefficients within one or more blocks in a macroblock will be zero. In such cases, a six-bit pattern, referred to as the coded block pattern, will indicate to the decoder which of the six blocks within the macroblock have been coded. Note that, I-pictures cannot have skipped macroblocks, and all blocks within an I-macroblock have to be coded.

- Decide if MQUANT needs to be changed or not. The encoder may decide to change MQUANT if it estimates possible buffer overflows or underflows.

Coding of B-pictures

The selection of the macroblock types in B-pictures is very similar to the selection of macroblock types in P-pictures. From Figure 6.10, the decisions can be summarized as follows.

- Decide if you want to use forward, backward, or interpolated motion compensation.

- Decide if you want to code the macroblock as an intra-type or as an inter-type macroblock.

- Decide if a macroblock needs to be coded or not. The entire macroblock can be skipped only if the previous macroblock was an inter-type macroblock and its motion compensation is good enough.

- Decide if the quantizer scale (MQUANT) needs to be changed or not.

As in the P-picture case, even when a macroblock is coded, some of the blocks within it may be skipped if all the quantized DCT values within these blocks are zero. The coded block pattern will indicate which of the blocks within a macroblock are coded. The ISO document provides several guidelines for the efficient selection of macroblock types; however, none of these is mandatory.

In Table 6.3, we show a distribution of I-, P- and B-macroblocks for 150 pictures of an MPEG-1 coded video sequence. The MPEG-1 encoder uses a GOP of 15 pictures with two B-pictures for every P-picture, and the coded bit rate is 1.15 Mbits/s. Zero MV refers to macroblocks that are coded using a zero motion vector. Note that, for B-pictures, there is a considerable number of macroblocks that are coded using predictive (P) macroblocks. This usually occurs when there is a scene change or when objects present (absent) in a P-picture before a B-picture disappear (appear) in the P-picture following that B-picture.

As mentioned before, MPEG-1 allows for skipped blocks within a macroblock. For example, for the data in Table 6.3, Table 6.4 shows the number of the 8 × 8 blocks that are actually coded. If all blocks were coded, then we should have a block count that is six times the macroblock count shown in Table 6.3.

Picture	Macroblock Type				
Type	I	P	B	Zero MV	Skipped
I	3,300				
P	897	8,587		5,128	568
B	60	7,356	22,845		429

Table 6.3 Example of the distribution of different macroblock types in a video sequence.

Picture	Macroblock Type			
Type	I	P	B	Zero MV
I	19,800			
P	5,382	30,730		18,146
B	360	8,176	18,853	

Table 6.4 Example of the number of 8 × 8 blocks that are actually coded in a video sequence.

6.3 MPEG-1 IMPLEMENTATION ISSUES

Complexity estimates for the MPEG-1 decoding of SIF resolution pictures, suggest that software-based decoding is well within the realm of general-purpose processors. Table 6.5 shows a typical breakdown of the computational load among the decoding functions in MPEG-1. Note that, the data here are based on simulations of SIF resolution pictures coded at 1.15 Mbits/s and having two B-pictures for every P-picture.

There is a general belief that the IDCT is the most compute-intensive task in MPEG-1 decoding; however, this is not always true. If the encoder performs motion estimation and rate control effectively, then, as shown in Table 6.4, IDCT is not required for many of the blocks. In such cases, as shown in Table 6.5, motion compensation may be the most compute-intensive task. Note also that YCbCr to RGB color transformation and upsampling is as expensive as the IDCT.

Decoding Function	Load (%)
Bit stream header parsing	0.44
Huffman decoding and inverse quantization	19.00
Inverse 8 × 8 DCT	22.10
Motion compensation	38.64
Color transformation and display	19.82

Table 6.5   Example of the distribution of the computational load in MPEG-1 decoding.

Scaled DCTs could be used in MPEG coders; however, frequent changes of the quantization scale will decrease their overall computational efficiency. In MPEG, unlike JPEG, pixel values in one or more pictures are used in the reconstruction of subsequent pictures; thus, careful attention should be paid to the accuracy of the IDCT, so that noise due to arithmetic precision does not influence the picture quality.

The MPEG-1 constrained parameter set does not deliver video quality that is acceptable for broadcast applications. Towards this end, the MPEG-2 standard has been developed using many of the principles of MPEG-1 coding. In the next section, we provide a brief overview of the MPEG-2 video standard.

6.4   THE MPEG-2 VIDEO STANDARD

MPEG-2 is the outcome of the second phase of work by MPEG. The original goal of MPEG-2 was to define a generic standard that could be applied to as wide a class of applications as possible and to support compressed bit streams at rates close to 5 Mbits/s for NTSC/PAL quality or close to 10 Mbits/s for near studio video quality. Among the original requirements were (1) compatibility with MPEG-1, (2) good picture quality, (3) flexibility of input format, (4) random access capability, (5) fast forward, reverse play, and slow motion capability, (6) bit stream scalability, (7) low delay for two-way communication, and (8) resilience to bit errors. Bit stream scalability is defined as the ability to discard portions of the bit stream, but still render the encoded bit stream at reasonable quality.

MPEG soon realized the following: (1) There was no reason to restrict the maximum coded bit rate to 10 Mbits/s. MPEG could successfully support higher bit rates; for example, 80 to 100 Mbits/s for HDTV applications. (2) It would be impossible to define a single standard that could satisfy all requirements. (3) Most applications would use only a small subset of the features offered by the standard. Hence, MPEG decided to adopt a toolkit-like approach; that is, MPEG-2 is a collection of tools defined in such a way as to satisfy the requirements of specific major applications.

The range of coding support provided by MPEG is divided into *profiles* and *levels*. For each profile/level, MPEG-2 provides the syntax for the coded bit stream and the decoding requirements.

A profile is a defined subset of the entire bit stream syntax specified by MPEG-2. The profiles are Simple, Main, 4:2:2, SNR, Spatial, High, and Multiview. Within a profile, a level is defined as a set of constraints imposed on the parameters of the bit stream, such as picture resolution or maximum bit rate. For each profile, the four levels are Low (for SIF resolution pictures), Main (for CCIR 601 resolution pictures), High-1440, and High (for HDTV resolution pictures). Before we discuss the specific constraints for each profile-level pair, we need to describe some of the coding extensions in MPEG-2.

The MPEG-2 syntax has two categories: (1) nonscalable syntax which is a superset of the coding syntax for MPEG-1, with additional extensions that support the coding of interlaced signals, and, (2) scalable syntax, which allows for a layered coding of the video stream. Decoders can either decode the basic stand-alone layer for a signal of lower quality or use the additional layers to increase the quality of the decoded signal.

Like MPEG-1, MPEG-2 is a lossy video compression scheme, based on DCT coding, block-based motion compensation, predictive and interpolative interframe coding, and Huffman coding. Many of these functions are similar to the ones in MPEG-1, and we refer the reader to the previous section for the coding details. In the following sections, we highlight the functions that are either new to MPEG-2 or are extensions of those found in MPEG-1.

MPEG-2 is backward compatible with MPEG-1; that is, an MPEG-2 decoder should be able to decode an MPEG-1 coded stream. Using the scalable syntax, the lower layer of an MPEG-2 signal could also be decoded by an

MPEG-1 decoder, but this is not required. Here, we will primarily focus on the nonscalable syntax. Some of the new features supported by MPEG-2 are discussed in the following sections.

6.4.1   Interlaced Pictures

MPEG-2 supports both interlaced and noninterlaced pictures. Fields in an interlaced picture can be coded separately (field pictures), or they can be interleaved and coded as one picture (frame pictures). Like MPEG-1, all input pictures can be coded as I-, P- or B-pictures. If the first picture of a coded frame is an I-field picture, then the second picture can be either an I-field picture or a P-field picture; that is, the first field can be used as a predictor for the second field. If the first picture is a P-field or a B-field picture, then the second field-picture has to be of the same type as the first field picture.

6.4.2   Color Subsampling

As in MPEG-1, input pictures are coded in the YCbCr color space; however, in addition to the 4:2:0 format used in MPEG-1, MPEG-2 also supports the 4:2:2 and 4:4:4 color subsampling formats. In the 4:2:2 format (also known as CCIR 601 format), the chrominance components have the same vertical resolution as the luminance component, but the horizontal resolution is halved. In the 4:4:4 format, all components have identical horizontal and vertical resolutions. Table 6.6 shows picture sizes for a 60-Hz CCIR 601 signal using three different subsampling formats. From Table 6.6, a CCIR 601 input signal has 720 active

Color	Pixels per line × lines		
Component	4:2:0	4:2:2	4:4:4
Y	720×480	720×480	720×480
Cb	360×240	360×480	720×480
Cr	360×240	360×480	720×480

Table 6.6   Active pixels for a 60-Hz CCIR 601 signal using various subsampling formats.

pixels per line. In practice, a preprocessor to the coder removes the eight leftmost pixels and the eight rightmost pixels to generate a picture with 704 pixels per line so that an integral number of macroblocks are obtained for each row. The informative section of the MPEG documents includes various filtering and picture conversion techniques.

Figure 6.11 shows the macroblocks used in the 4:2:2 and 4:4:4 formats. Macroblocks in the 4:2:0 format are identical to that of MPEG-1. A 4:2:2

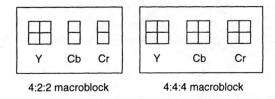

4:2:2 macroblock   4:4:4 macroblock

Figure 6.11 The 4:2:2 and 4:4:4 macroblocks in MPEG-2.

macroblock has four blocks of Y, two blocks of Cb, and two blocks of Cr. A 4:4:4 macroblock has four blocks of Y, four blocks of Cb, and four blocks of Cr. When the coded bit stream has no scalable extensions, then all blocks within a macroblock are 8×8. With scalable extensions, macroblocks may contain *scaled blocks* with lower resolutions, such as 1×1, 2×2, or 4×4. The structure of the luminance macroblocks is different for frame DCT coding and field DCT coding, as shown in Figure 6.12. In frame DCT coding, all blocks contain data from both the top field and the bottom field. In field DCT coding, the top two blocks have data from the top field, and the bottom two blocks have data from the bottom field. For chrominance blocks, all blocks are interlaced (as in frame DCT coding).

6.4.3 Prediction Modes and Motion Compensation

In MPEG-2, the picture sequence can be either a collection of frame pictures or a collection of field pictures. Two classes of prediction are supported, namely, frame prediction and field prediction.

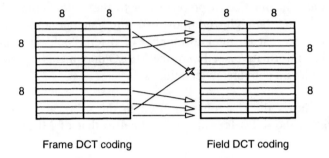

Frame DCT coding   Field DCT coding

Figure 6.12   Structure of luminance macroblocks in frame and field DCT coding.

Field and frame prediction

For field pictures, predictions are made independently for each field from reference fields. This is similar to what is done in MPEG-1, with each field considered an independent picture.

In frame prediction, predictions are formed from reference frames, which may have been coded using frame or field prediction. If a frame picture is assumed to have originated from two fields, then both fields must have the same coding type, except for the case when the first field is an I-picture, in which case the second field of this frame can be an I- or P-picture.

Figure 6.13 shows examples of the different prediction modes for frame pictures. Frame pictures may use both field and frame predictions. In frame prediction (Figure 6.13a), the whole (usually interlaced) frame is considered a single picture. In field prediction, each frame is treated as two separate fields. A motion vector can point either to a field in another reference frame or to a field in the current frame. For example, in Figure 6.13b, the first field (which can be either the top or the bottom field) can be predicted either from the top or the bottom fields in a reference frame. In Figure 6.13c, the second field (the bottom field in this example) can be predicted either from the bottom field of another frame or from the top field of the current frame. The above examples can easily be extended to other cases. Field prediction modes are beneficial in coding interlaced scan video. For progressive scan video, frame prediction is sufficient.

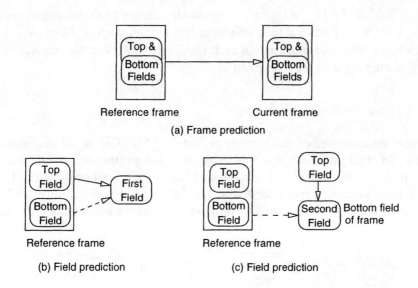

Figure 6.13 Examples of prediction modes for frame pictures.

In MPEG-2, unlike MPEG-1, all motion vectors are specified at half-pixel resolution. Let (u, v) denote the motion vector of the luminance components in a macroblock. Due to subsampling of the chroma components, their motion vectors are given by $(\frac{u}{2}, \frac{v}{2})$ for the 4:2:0 format and by $(\frac{u}{2}, v)$ for the 4:2:2 format. For the 4:4:4 sampling format, the chroma components use the motion vectors of the luminance component without any scaling.

Since MPEG-2 video may be used in harsh broadcast environments, some immunity to transmission errors is needed. In a compressed video bit stream, erroneous reception of motion vectors can be catastrophic, unlike, say, erroneous reception of some of the high-frequency DCT coefficients within an 8×8 block. Thus, the MPEG-2 syntax has provisions for including specially coded motion vectors, referred to as concealment motion vectors, within intra-macroblocks. The assumption here is that the transmission of intra-macroblocks is done in a more reliable manner.

To improve prediction efficiency, MPEG-2 provides two additional motion compensation modes. The *16 × 8 motion-compensation* mode allows a 16 × 16 macroblock to be treated as an upper 16 × 8 region and a lower 16 × 8 region. Each of the 16×8 regions is then independently motion-compensated. Thus, a 16×16 P-macroblock will have two motion vectors; likewise, a B-macroblock will have four motion vectors. The 16×8 motion-compensation mode is only allowed for field pictures.

Dual-prime motion compensation

Another motion-compensation mode provided in MPEG-2 is the *dual-prime* mode. The dual-prime mode is applicable only to P-pictures and to GOPs that have no B-pictures between these P-pictures and the reference pictures. This motion-compensation mode for frame and field type pictures is illustrated in Figure 6.14. For frame pictures, two motion vectors are associated with each

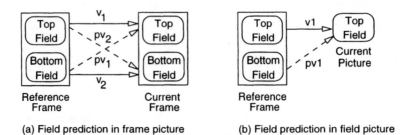

(a) Field prediction in frame picture   (b) Field prediction in field picture

Figure 6.14 Dual-prime prediction mode in frame and field pictures.

field. For instance, in Figure 6.14a, motion vectors v_1 and pv_1 are associated with the top field, and motion vectors v_2 and pv_2 are associated with the bottom field of the current frame. Let $v_1(x)$, $pv_1(x)$, $v_2(x)$, and $pv_2(x)$ denote the horizontal components of the motion vectors v_1, pv_1, v_2, and pv_2. The vertical (y) components are denoted in a similar manner. In dual-prime prediction, motion vectors pv_1 and pv_2 are derived from v_1 and v_2 as follows.

Consider the derivation of pv_1. Since pv_1 is a motion vector from an opposite parity field, namely, from the bottom field in the reference frame to the top field in the current frame, the existing motion vector v_1 is scaled to reflect the temporal distance between the fields. First, a correction is made to the vertical component of the motion vector to reflect the vertical shift between lines of the top and bottom fields. Second, a small differential motion vector δ is added.

The appropriate choice of δ is determined by the encoder. In a similar manner, we can derive pv_2. From the MPEG-2 document, the relevant equations for frame picture prediction, as depicted in Figure 6.14, are given by

$$pv_1(x) = \frac{v_1(x)}{2} + \delta(x), \tag{6.2}$$

$$pv_1(y) = \frac{v_1(y)}{2} - 1 + \delta(y), \tag{6.3}$$

$$pv_2(x) = \frac{3 \times v_2(x)}{2} + \delta(x), \tag{6.4}$$

$$pv_2(y) = \frac{3 \times v_2(y)}{2} + 1 + \delta(y), \tag{6.5}$$

where $\delta(x)$ and $\delta(y)$ can take the values in -1, 0 or 1. Note that, pv_1 and pv_2 are not included in the encoded bit stream; instead, along with v_1 and v_2, the corresponding δ is included in the bit stream so that the decoder can derive pv_1 and pv_2.

For the field picture shown in Figure 6.14b, the relevant equations for pv_1 are

$$pv_1(x) = \frac{v_1(x)}{2} + \delta(x), \tag{6.6}$$

$$pv_1(y) = \frac{v_1(y)}{2} - 1 + \delta(y). \tag{6.7}$$

6.4.4 DCT and Quantization

MPEG-1 uses only two quantization matrices for the DCT coefficients: one for intra-blocks and one for nonintra-blocks. In MPEG-1, the quantization matrices can be changed only at the sequence level of the hierarchy, whereas, in MPEG-2, the matrices can be changed at the picture level. For the 4:2:2 and the 4:4:4 formats, MPEG-2 allows for more efficient coding by supporting the use of different quantization matrices for the luminance and chrominance components. Thus, one can use two matrices for luminance blocks (intra and nonintra) and two matrices for chrominance blocks (intra and nonintra). Quantization of the AC coefficients is similar to the process in (6.1). In MPEG-1 and MPEG-2, the quantizer scale factor can be changed on a macroblock basis in order to achieve a constant bit rate at the encoder output. The method to change the quantizer scale factor is not part of the MPEG-1 and MPEG-2 standard.

6.4.5 Scalable Bit streams

A new feature of MPEG-2 is bit stream scalability, which allows for a layered representation of the coded bit stream. The MPEG-2 syntax allows for four basic modes of bit-stream scalability: data partitioning, SNR scalability, spatial scalability, and temporal scalability. These basic scalability schemes can be combined to form a hybrid scalability scheme.

Data partitioning

With data partitioning, the bit stream is split into two layers, called partitions. For example, the first partition may include all critical header information (such as headers and motion vectors), and the second partition may include the remaining bit stream. This mode is intended for use in applications that can allocate two channels for a single bit stream.

SNR scalability

SNR scalability can be used in applications that support video transmission at multiple quality levels. All layers have the same spatial resolution but different video qualities. The lower layer is coded by itself and provides the basic video quality. The enhancement layers are coded as to enhance the basic quality by providing refinement data for the DCT coefficients of the lower layer. SNR scalability provides for better resilience to transmission errors. For example, the enhancement layers can be sent over a channel with poor performance, and the lower layer can be sent over a channel with better error performance.

Spatial scalability

With spatial scalability, the bit stream is divided into layers of different spatial resolution (this is similar to the hierarchical coding in JPEG). The lower layer is coded by itself to provide the basic spatial resolution. The enhancement layers can use the spatially interpolated lower layer to provide a video stream at the full spatial resolution. Layers may have different frame sizes, frame rates, or chrominance formats. In a typical application, the lower layer may be MPEG-1 compliant at SIF resolution, whereas the enhancement layer might be used to generate CCIR 601 resolution video.

✓ *Temporal scalability*

Temporal scalability allows the migration from systems with low temporal resolution to systems with higher temporal resolution. In this case, the lower layer is coded by itself and provides the basic temporal rate, and the enhancement layers are coded with temporal prediction with respect to the lower layer. The lower layer and the enhancement layers can be combined to generate a stream at full temporal resolution. All layers have the same frame size and chrominance formats but different frame rates.

6.4.6 Profiles and Levels

Table 6.7 summarizes the key characteristics of the profiles and levels defined for nonscalable MPEG-2 bit streams. The profiles and levels for the scalable syntax of MPEG-2 are shown in Table 6.8. In these tables, the upper bounds for the picture resolution, frame rate, and bit rates are provided. Picture resolution is given as pixels per line multiplied by the number of lines. The *Bitrate* data refer to the maximum compressed bit rate supported by the input buffers of a decoder. Areas with "N/A" indicate that there are no conformance restrictions for these variables.

It is expected that early implementations of the standard will only support the Main Profile at the Main Level. The Main Profile provides the basic functionality of MPEG-2 and is expected to be supported by most of the initial applications. The Simple Profile is a low-cost version of the Main Profile. Coding is similar as in the Main, but without any bidirectional prediction. Levels are related to picture resolution. The upper bound on the Main Level resolutions corresponds to the CCIR 601 picture format. The High-1440 and High Level resolutions correspond to picture resolutions for HDTV. The Low Level resolutions correspond to the SIF format.

These profiles and levels have a hierarchical relationship. Hence, the syntax supported by a *higher* profile/level includes all the syntactical elements of *lower* profiles/levels. A decoder that can decode bit streams at a certain profile/level should be able to decode also all bit streams of lower profiles/levels than its own. The only exception is that decoders of the Simple Profile at Main Level are also required to decode Main Profile at Low Level bit streams, that is, MPEG-1 coded streams.

Levels		Profiles (Nonscalable)		
		Simple 4:2:0 (I,P)	Main 4:2:0 (I,B,P)	4:2:2 (I),(I,B,P)
High	Max resolution/ rate (Hz)	N/A	1920 x 1152/60	N/A
	Bitrate (Mbits/s)	N/A	80	N/A
High-1440	Max resolution/ rate (Hz)	N/A	1440 x 1152/60	N/A
	Bitrate (Mbits/s)	N/A	60	N/A
Main	Max resolution/ rate (Hz)	720 x 576/30	720 x 576/30	720 x 608/30
	Bitrate (Mbits/s)	15	15	50
Low	Max resolution/ rate (Hz)	N/A	352 x 288/30	N/A
	Bitrate (Mbits/s)	N/A	4	N/A

Table 6.7   Profiles and levels in MPEG-2 video coding, nonscalable modes.

The MPEG-2 4:2:2 Profile

The MPEG-2 Main Profile at Main Level definition addresses the needs of applications targeting the consumer market. There is a growing interest in the use of digital video in the studio and post-production environment. Digital Betacam is widely used in the studios and provides high quality compressed video at 90 Mbits/s. However, it does not offer lower rates and it is a proprietary standard. Given the success of MPEG-2 in video distribution products, there was a need to use MPEG-2 in the studio and post production environments, and to have a unified compression process from video production through video distribution. The distribution format of MPEG-2 Main Profile at Main Level, with a chrominance subsampling of 4:2:0 and a maximum bit rate of 15 Mbits/s, is found to be unsatisfactory for the video production environment, wherein the video stream may have to be processed across several generations;

Levels		Profiles (Scalable)			
		SNR 4:2:0	Spatial 4:2:0	High 4:2:0, 4:2:2	Multiview 4:2:0
High	Enhancement (Auxiliary)	N/A	N/A	1920 x 1152/60	1920 x 1152/60
	Lower (Base)	N/A	N/A	960 x 576/30	1920 x 1152/60
	Bitrate (Mbits/s)	N/A	N/A	100 (all layers) 80 (base + mid) 25 (base layer)	130 (all layers) 50 (auxiliary) 80 (base layer)
High-1440	Enhancement (Auxiliary)	N/A	1440 x 1152/60	1440 x 1152/60	1920 x 1152/60
	Lower (Base)	N/A	720 x 576/30	720 x 576/30	1920 x 1152/60
	Bitrate (Mbits/s)	N/A	60 (all layers) 40 (base + mid) 15 (base layer)	80 (all layers) 60 (base + mid) 20 (base layer)	100 (all layers) 40 (auxiliary) 60 (base layer)
Main	Enhancement (Auxiliary)	720 x 576/30	N/A	720 x 576/30	720 x 576/30
	Lower (Base)	-	N/A	352 x 288/30	720 x 576/30
	Bitrate (Mbits/s)	15 (all layers) 10 (base layer)	N/A	20 (all layers) 15 (base + mid) 4 (base layer)	25 (all layers) 10 (auxiliary) 15 (base layer)
Low	Enhancement (Auxiliary)	352 x 288/30	N/A	N/A	352 x 288/30
	Lower (Base)	-	N/A	N/A	352 x 288/30
	Bitrate (Mbits/s)	4 (all layers) 3 (base layer)	N/A	N/A	8 (all layers) 4 (auxiliary) 4 (base layer)

Table 6.8 Profiles and levels in MPEG-2 video coding, scalable modes.

that is, multiple compression and decompression cycles. For example, the 4:2:0 representation can introduce chrominance bleeding in color edges and fading of colors, which in turn can worsen after multigeneration processing. The High Profile at High Level or the High Profile at High-1440 could be used in the studios; however, the scalability requirements and high picture resolutions would bring with it high cost and complexity. To promote the use of MPEG-2 within the studio environment, the 4:2:2 profile was developed and was added to ISO standard 13818-2 as Amendment No. 2.

Coding systems based on the 4:2:2 profile have the following features:

- The upper bound for bit rate is 50 Mbits/s (lower than Digital Betacam's bit rate, but higher than the bit rate of MPEG-2 Main Profile at Main Level).

- Main Profile at Main Level compatibility.

- No scalability mode. This avoids additional cost and lower efficiency.

- The DC precision of intracoded blocks can be 8, 9, 10 or 11 bits. The use of 11 bits is beneficial when very fine quantization parameters are used for pictures coded at higher data rates, since the improved DC precision compensates for the loss due to the rounding and integer arithmetic.

- The maximum number of bits in a macroblock is unconstrained. On the other hand, the 4:2:0 format constrains the maximum bits to 4,608.

- Excellent multigeneration performance (typically up to eight generations).

- GOP selection is flexible and permits all encoded pictures to be I-pictures.

Motion-JPEG has been used in digital studios, because as an intra-only coding method it is more amenable to frame-based editing. For the same bit rate across several generations, the MPEG-2 4:2:2 profile provides better image quality than Motion-JPEG. Typically, this performance gap can be at least 2 dB in PSNR. Furthermore, MPEG-2 has better coding efficiency since it can incorporate temporal prediction through the use of P- and B-frames.

Currently, the video storage technology of choice in the professional studio is component analog video tape recording. The MPEG-2 4:2:2 profile has been proposed to address the needs of this market. Video storage systems employing the MPEG-2 4:2:2 profile outperform current analog solutions in terms of preserving the image quality across several generations. In the analog approach, the image quality rapidly degrades after the second generation, whereas, in a digital storage system, employing the 4:2:2 profile, the loss in quality stabilizes and no further losses occur after a few generations.

The MPEG-2 Multiview Profile

Recent advances in MPEG-2 compression standards, as well as the computing capabilities of desktop PCs, have made it plausible to support multiple viewpoint digital video in many applications. For instance, the stereo view is a simple scheme to portray depth on a two-dimensional display. The stereo view can enhance the viewing experience considerably in both television systems and in video games. In surgery and microscopy, the perception of depth could also be very useful. There are other applications, such as a virtual walk-through a building, wherein there is a need to process several views in order to generate a view at an arbitrary angle. The MPEG-2 *multiview profile* (MVP), shown in Table 6.8, was developed to facilitate these applications. In September of 1996, MVP was added to the ISO standard 13818-2 as Amendment No. 3.

The key feature of this profile is that it is a two-layer representation, namely, a base layer and an enhancement layer. The SNR, Spatial, and High profiles in Table 6.8 do not use the notion of temporal scalability; on the other hand, MVP uses a coding structure wherein the enhancement layer uses the temporal scalability tool within MPEG.

A generalized stereoscopic video codec conforming to the MVP is shown in Figure 6.15. MVP supports stereoscopic pictures for all picture resolutions, namely, Low, Main, High-1440, and High. The chroma format is restricted to 4:2:0, and each layer of MVP has the same number of samples per line, lines per frame, and frames per second. The base layer of MVP is assigned to a left-view and its compatibility relationship to other profiles and levels is the same as the one in Main Profile at the same level. Thus, the base layer is coded as an MPEG-2 Main Profile bit stream. The base layer coding format ensures backward compatibility with Main Profile-conforming decoders so as to generate mono-view scenes of the same program.

In a dual-view scene acquisition system, there is a high degree of correlation between the video frames of the two views. The correlation exists not only among video frames within one view but also among video frames in the two views. The use of the latter form of correlation can lead to efficient compression and the process of removing redundancy due to this cross-correlation is sometimes referred to as disparity prediction. For a given block of pixels in the current right-view frame, disparity prediction requires finding a corresponding block of matching pixels in the left-view frame. The

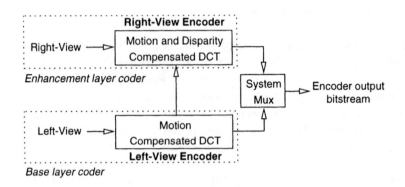

Figure 6.15 A generalized video codec for MVP.

formulation of a prediction residual using disparity prediction is referred to as disparity compensation. In addition to disparity prediction, the correlation amongst frames within a single view can also be exploited to form a motion-compensated residual.   Thus, in the MVP codec of Figure 6.15, for the enhancement layer, a motion and disparity-compensated signal is coded using the DCT.

A typical disparity and motion-prediction structure for the base and enhancement layers of MVP is shown in Figure 6.16. In this figure, the base layer is coded using a typical IBBP.... configuration for the GOP. For the enhancement layer, unlike the base layer, P-picture prediction can be either the most recently decoded enhancement layer picture, even if this were a B-picture or a base layer picture. Furthermore, while B-picture prediction in the base layer is as in the MPEG-2 Main Profile, in the enhancement layer, B-picture predictions can be from I-, P-, or B-pictures in the base layer and past P- or B-pictures from the enhancement layer. In the base layer, if B-pictures are used, the bit stream arriving at the decoder will not be in display order, whereas, for the enhancement layer, the bit stream will always be in presentation order. This has some implications in synchronizing the two views, since these views are two separate elementary bit streams.

When forming the disparity prediction for the enhancement layer, studies within the MPEG group revealed that using the *most recent decoded enhancement picture* and/or the *most recent base layer picture in display order* provides

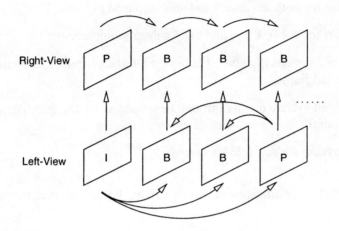

Figure 6.16 A prediction configuration for an MVP codec.

a superior configuration in terms of system complexity and performance over other more complex schemes.

The demand for stereoscopic video may not be high enough to require coding of base and enhancement layers at the same quality. Subjective tests have shown that reduced resolution in one of the views does not degrade perceived stereoscopic quality. Hence, in the MVP, the bit stream for the enhancement layer is often allocated fewer bits than that for the left view; the upper bounds on these bit rates are given in Table 6.8. Note that, in order to support stereoscopic video, one does not need to use the MVP approach. An alternate approach might be to code both views independently using, say, the parameters of Main Profile at Main Level, and then perform some post-processing to regenerate the stereo view. Experiments indicate that the MVP approach outperforms the latter scheme by nearly 2-3 dB, for bit rates around 2-4 Mbits/s. Further coding gains can be obtained using either brightness balancing of two views for disparity and motion compensated prediction or joint rate-control of both views.

In summary, MPEG-2 is a superset of MPEG-1. Some of the features in MPEG-2 that are not found in MPEG-1 are given below:

- Support for both interlaced and noninterlaced pictures.

- Support for 4:2:0, 4:2:2, and 4:4:4 subsampling schemes.

- Motion compensation based either on interlaced frames or noninterlaced fields and frames.

- Improved picture quality through new options for the quantization of the DCT coefficients and alternate zig-zag ordering.

- New syntax for scalable bit streams.

The MPEG-2 standardization process is still evolving. The work items include:

- Modes to support extensions that may be developed by ITU-T.

- Digital Storage Media Command and Control (DSM-CC) is the specification of a set of protocols for managing MPEG-1 and MPEG-2 bit streams. These protocols may be used to support applications in both stand-alone and heterogeneous network environments. In the DSM-CC model, a stream is generated by a server and delivered to a client. Both the server and the client are considered to be users of the DSM-CC network. The DSM-CC conformance testing specifications part for the MPEG-2 standards is expected to be completed by April of 1998.

The current phase of the MPEG standardization effort is the MPEG-4 suite of standards and this is discussed next.

6.5   THE MPEG-4 CODING STANDARD

The MPEG-1 and MPEG-2 set of standards are now widely used in commercial products such as CD-interactive, digital video cameras, and digital audio and video broadcasting. MPEG-1 and MPEG-2 deal with *frame-based video and audio* and in many applications they have enabled the solution providers to offer a digital system replacement for the analog systems that existed before. The most important goal of these standards has been to make storage and transmission very efficient. As digital media becomes widely used, there is a blurring of borders between three distinct service models: *communications*, *interactivity*, and *broadcasting*, and their corresponding industry sectors, namely, telecommunications, computers, and TV/film. While the convergence among these sectors may take a long time, the distinction between the service models is disappearing. For instance, in recent years, interactivity is being added to broadcast services and many communication/interactive applications are appearing on the Internet. In anticipation of this trend, in July of 1993, the MPEG group initiated a new standardization phase, referred to as MPEG-4. The objective of MPEG-4 was to standardize algorithms for audiovisual coding in multimedia applications, allowing for interactivity, high compression, scalability of video and audio content, and support for natural and synthetic audio and video content. The MPEG-4 coding standard has the ISO designation 14496 and is expected to be approved as an international standard by November of 1998.

6.5.1 MPEG-4 Architecture

A general model of an MPEG-4 system is depicted in Figure 6.17. MPEG-4

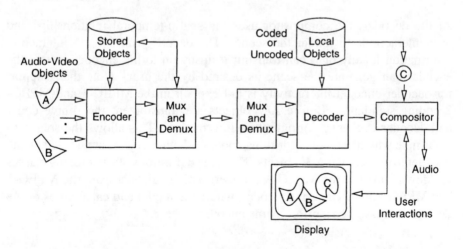

Figure 6.17 Schematic overview of an MPEG-4 system.

defines an audiovisual scene as a coded representation of *audiovisual objects* (A/V) that have certain relationships in space and time. This is different from MPEG-1 and MPEG-2, wherein the audiovisual scene is thought of as *video frames with associated audio.* In the MPEG-4 notion of A/V objects, the object could be a video object within a scene or it could be the complete background. It could also be an audio object, such as one instrument in an orchestra or a single voice. This new approach to information representation offers much more flexibility for versatile reuse of data, intelligent schemes to manage bandwidth, processing resources, and error protection. The object notion allows for mixing of natural and synthetic (computer-generated) objects as well as other data types such as text overlays and graphics.

As shown in Figure 6.17, at the encoder, the A/V objects and their spatio-temporal relationships are used by the encoder to generate the encoded bit streams. These bit streams, after optional error-protection, are multiplexed with stored objects and then transmitted downstream. The bit streams could be transmitted across multiple channels, where each channel offers a different quality of service. This permits different objects to be reconstructed at the decoder at different qualities. The multiplexor in the MPEG-4 system combines the elementary data streams into one output data stream. The demultiplexor also provides functions needed to recover the system clock, synchronize multiple streams, interleave multiple streams for use by the compositor, etc.

At the decoder, the compositor uses the spatio-temporal relationships and user interactions to render the scene. The decoder can use the interaction information locally or it can transmit it upstream to the encoder so that the encoder can generate the scene as desired by the user. Note that, support for decoder-encoder interactivity is not explicit in the MPEG-1 and MPEG-2 coding standards. Before A/V objects are transmitted, the source coder and decoder exchange configuration information. This allows the source to determine which class of algorithms, tools, and other objects are needed by the decoder to process the A/V objects. Then, the definitions of any missing classes are downloaded to the MPEG-4 decoder. This is distinct from the MPEG-1 and MPEG-2 hardwired push model, where the model and capabilities of the decoder are assumed a priori by the encoder.

6.5.2   MPEG-4 Video Coding

In MPEG-4, like MPEG-1 and MPEG-2, there is a hierarchical representation of the information. Each video frame is segmented into a number of arbitrary shaped image regions, called video object planes (VOP). The process of segmentation is outside the scope of the MPEG-4 standard, and since this is an encoder function, like MPEG-1 and MPEG-2, MPEG-4 will specify only the minimum set of functions that are needed for interoperability. Successive VOPs belonging to the same physical object in a scene are referred to as video objects (VO). A VO can be viewed as the MPEG-4 equivalent of a GOP in the MPEG-1 and MPEG-2 standards. The shape, motion and texture information of the VOPs belonging to the same VO is encoded into a separate video object layer (VOL). In addition, relevant information needed to identify each of the VOLs and how various VOLs are composed is also encoded. This allows for selective decoding of VOPs and also provides object-level scalability at the decoder.

Video object plane (VOP)

The notion of VOPs and their use in video coding in MPEG-4 is illustrated in Figure 6.18. The example in Figure 6.18a consists of a person in the foreground and a stationary weather map in the background. The entire frame comprising the background and foreground can be classified as a single VOP. VOP coding is then a straightforward application of MPEG coding techniques. An alternate segmentation is to decompose the scene into two VOPs, say, VOP_1 for the background object and VOP_2 for the foreground object. A binary alpha-plane-image sequence, as depicted in Figure 6.18b, is coded in this example to indicate to the decoder the shape of the foreground object VOP_2 and its location with respect to VOP_1. In general, MPEG-4 may support the coding of grayscale alpha planes to allow the decoder to compose the VOPs with various levels of transparency. For the purposes of coding, VOP_1 is as shown in Figure 6.18c and VOP_2 is as shown in Figure 6.18d. Note that, the two regions covered by VOP_1 and VOP_2 are nonoverlapping and that the sum of the pixels covered by the two VOPs is identical to the pixels contained in the original sequence shown in Figure 6.18a. Since each VOP is coded separately, based on the decoded information from the alpha channel, the decoder can decode and display each VOP separately or reconstruct the original sequence by decoding and compositing both of them.

MPEG-4 supports the overlapping configuration for VOPs as well. For instance, if the entire background in Figure 6.18a was known a priori at the encoder, VOP_1 could be the entire background and VOP_2 could be as in Figure 6.18d. If the background is stationary, only one frame needs to be coded for

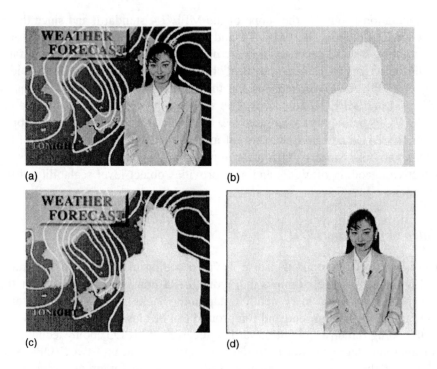

Figure 6.18 Description of VOP. (a) One frame from a scene
- before scene segmentation, (b) segmentation mask specifying
location of foreground VOP, (c) image content of VOP_1 as input
to encoder, and (d) image content of VOP_2 as input to encoder.

the background. Thus the foreground and the background can have different
display rates at the decoder.

The VOP coding process for this example is summarized in Figure 6.19.
MPEG-4 supports content-based scalability. Thus, at the decoder, the compositor can choose to decode only, say, VOP_2 and replace VOP_1 by a local
object before it produces the final image. For example, in Figure 6.19, VOP_1
is being replaced by a local weather map.

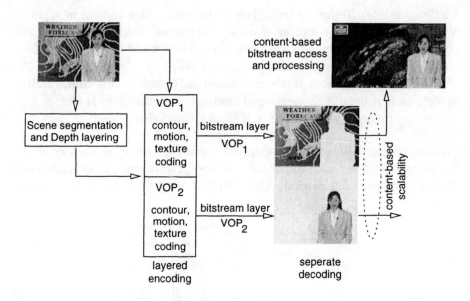

Figure 6.19   Example of a VOP coding process.

Coding of information for each VOP

For each VO, the shape, motion, and texture information of VOPs comprising this VO are coded. There will be many ways in which the shape, motion, and texture information will be coded. We will restrict discussion to the baseline scheme as adopted by the MPEG-4 verification model.

Shape Coding: The shape information is referred to as *alpha planes*. The techniques to be adopted by the standard will provide lossless coding of alpha planes and lossy coding of shapes and transparency information; thus, allowing for tradeoffs between bit rate and accuracy of shape representation. Furthermore, intra- and inter-shape coding functionalities employing motion-compensated shape prediction is envisioned so as to allow both efficient random access operations as well as efficient compression of shape and transparency information for diverse applications. After coding the VOP shape information, each VOP image in a VO is partitioned into non overlapping macroblocks.

Motion Coding: Temporal redundancies between video content in separate VOPs within a VO are exploited using block-based motion estimation and compensation. In general, these techniques can be viewed as extensions of the standard block-matching techniques used in MPEG-1, MPEG-2, H.261, and H.263 to image sequences of arbitrary shape. Advanced motion compensation modes, such as the use of overlapped motion compensation in H.263, as well as coding of motion vectors for 8×8 blocks, could also be used.

To perform block-based motion estimation and compensation between VOPs of varying location, size, and shape, a shape-adaptive macroblock approach shown in Figure 6.20 is used. The reference window is the original image

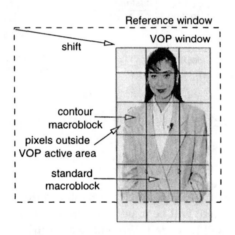

Figure 6.20   Macroblock grid for coding foreground VOP$_1$.

border. A shift parameter is coded to indicate the location of VOP$_1$ with respect to the borders of the reference window. A VOP window surrounds the foreground video object. The VOP window contains an integer number of macroblocks, horizontally and vertically, and the size of each macroblock is 16 $\times$ 16 pixels. Macroblocks can be either *standard* or *contour* macroblocks. For a standard macroblock, where all of its pixels are inside the active VOP area (see Figure 6.20), any of the motion estimation and compensation techniques described in Chapter 4 can be used. For a contour macroblock, where some of its pixels may be outside the active VOP area (see Figure 6.20), prior to motion estimation and compensation of the current VOP in frame t, a simple image padding technique is used to reference the VOP of frame $t - 1$. The VOP padding method extrapolates pixels outside of the VOP based on pixels inside the VOP. After padding the reference VOP, the motion-estimation and compensation process is the same as in the case of standard macroblocks. However, during block matching only pixels belonging to the active area of the VOP are used in the motion estimation process.

Texture Coding: The intra VOPs, as well as residual errors after motion compensated prediction, are coded using DCT on 8×8 blocks, in a manner similar to that employed in MPEG-1, MPEG-2, H.261, and H.263. For each macroblock, there could be four 8×8 luminance blocks and two 8×8 chrominance blocks. As in the motion-estimation step, 8×8 blocks well within the VOP active area can be coded in a straight forward manner. For the coding of motion-compensated prediction error blocks (P-VOPs) that straddle the VOP boundary, pixels outside the active area are set to a value of 128 prior to DCT coding. After computing the DCT, zig-zag scanning, quantization, and run-length coding of DC and AC coefficients are used as in the previous MPEG standards. Efficient prediction of DC and AC coefficients for intra and inter-coded blocks can also be employed (this approach is not available in MPEG-1 and MPEG-2).

Multiplexing Shape, Motion, and Texture Data: The compressed alpha plane, motion vectors, and DCT coded information are multiplexed into a VOL bit stream by coding the shape information followed by motion and texture coded data. Motion vectors and DCT coefficients can be coded either jointly, as in H.263, or separately. This will allow new and more efficient motion and texture-coding techniques to be developed. In a typical coding scenario, 75% of the available coding-bit budget is used in the coding of texture information, 14% of the bits is needed to code the contour information, 4.2% of the bits is needed to code the motion information, and the remaining is used to signal decision modes employed at the encoder.

Scalability modes

MPEG-4 allows for *content-based* access or transmission of arbitrarily shaped VOPs at various temporal and spatial resolutions. This is in contrast to the frame-based scalability modes developed for MPEG-2. Thus, MPEG-4 supports object scalability as well as quality scalability. Decoders not capable of reconstructing at full-resolution all VOPs can decode subsets of the layered bit stream to display selected VOPs - this is an example of object scalability. Furthermore, quality scalability can be employed on an object-by-object basis to decode objects at lower spatial or temporal resolution. Temporal scalability within MPEG-4 allows the decoder to provide different display rates to different VOLs; that is, a foreground object of interest may be displayed at a higher rate than the remaining background or other objects.

6.5.3   MPEG-4 Tools, Profiles, and Decoder Flexibility

MPEG-1 and MPEG-2 define complete algorithms for audio, video, and systems. In contrast, MPEG-4 will attempt to standardize video tools, where a video tool could be a fully defined algorithm, a shape coding module, a motion compensation module, a texture coding module, or related techniques. The *glue* that will bind these tools together is the MPEG-4 systems description language (MSDL) which will have several components, including: (1) definitions for the interfaces between the coding tools, (2) a mechanism to combine coding tools and to construct algorithms and profiles, and (3) a mechanism to download new tools. The MSDL will transmit to the decoder the bit stream and the manner in which the tools have to be used at the decoder to reconstruct the audio and video. At a more advanced level, MSDL will allow the downloading of tools not available at the decoder. Thus, the MPEG-4 MSDL will provide a very flexible framework by allowing a wealth of algorithms to be supported by the standard.

The MSDL toolbox approach is depicted in Figure 6.21. Note that the MSDL approach covers tools for both natural and synthetic data. A set of tools such as DCT and motion-compensation could be combined to form an algorithm for, say, motion-compensated prediction in DCT domain. Algorithms may themselves be composed of tools and/or algorithms. Profiles define subsets of tools and/or algorithms suitable for specific application requirements, such as low delay coding and low-complexity decoding. Levels in MPEG-4 refer to a specification of the constraints and performance criteria needed to satisfy one or more applications.

A decoder with all the features of MPEG-4 will be impractical. MPEG-4 defines three levels of decoder programmability that define flexibility and extensibility.

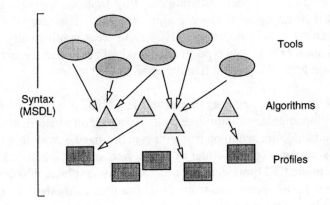

Figure 6.21 MPEG-4 tools, algorithms and profiles.

Flex_0 : A nonprogrammable decoder which contains a standardized set of audio, video, and systems tools.

Flex_1 : A flexible decoder with a finite set of audio, video, and system tools and their standardized interfaces, which may be flexibly configured into arbitrary algorithms. Flexible decoders use locally optimized decompression tools, but they can download descriptions of new classes of data (A/V objects) with associated decompression methods. Such flexible terminals will enable content providers to adapt the decompression processes to the specific data they are dealing with, by reconfiguring the algorithms with standardized tools. Depending on technology and market needs, these decoders may be extended to support the downloading of tools.

Flex_2 : An extensible decoder which will support a standardized mechanism to describe arbitrary algorithms made of arbitrary tools. Currently, no Flex_2 specification is being attempted within the MPEG-4 standardization process.

6.5.4 MPEG-4: Synthetic-Natural Hybrid Coding

Work on the basic concepts of the MPEG-4 video coding part has been completed. However, several extensions are being developed. One of the key extensions is the ability to handle natural as well as synthetic content.

Presently, graphics has been communicated by first converting it to a video object and then applying a video coding scheme. The aim of MPEG-4's synthetic-natural hybrid coding (SNHC) is to treat synthetically generated contents as another new data type, and to standardize, from a communications viewpoint, on how to represent it and compress it.

Let us look at some specific examples where SNHC might be used. A CD-ROM encyclopedia shows audio/video clips to a user on a historical site, where the audio and video are accompanied with synchronized, scrolling, explanatory notes, an overlay marker or pointer by a historian, a scrolling map, a 2-D inset showing animated 3-D models, and other overlay graphics, which come and go in reaction to user selections from a side-bar menu. Another example could be a 3-D video game for networked players, which utilizes clever blends of audio/video, animated, 2-D/3-D sprites, and frugal 3-D graphics to challenge users on low-cost mixed-media PCs. Applications such as these, requiring a rich mix of audio, video, and graphics, are expected to benefit from the MPEG-4 SNHC standardization process. Much of the work on SNHC will be quite basic in nature. Other needs will be considered in future phases of MPEG-4 or other standards.

A bottom-up approach has been chosen for MPEG-4 SNHC, focusing in the early phases on two particular applications: audio/video with 2-D graphics and interpersonal communication. These ultimately will involve interactions with audio, video, and 2-D/3-D objects, as well as human representations within 2-D/3-D environments. Three levels of extensions are defined. The first level of extension is concerned with the coding of facial and human body animation combined with text-to-speech synthesis. Here, an initial area of focus is to extend the models available in the virtual-reality modeling language (VRML). VRML can be used to create models of simple objects like tables, chairs, etc. However, it is not effective in creating a good model for the human face and body. A second level provides media integration of audio/video objects with 2-D graphics. The third level integrates 3-D synthetic objects, behaviors and interaction in a multiuser 3-D virtual environment. At each level of extension, the application focus, requirements, and expected standards are specified. Aspects of the first and second level are currently being pursued to support facial/body animation with speech and graphics overlay for audio/video. The decoder could include a 3-D articulated facial model to reproduce facial expressions animated by remote parameter streams. The size of the parameter set after compression can be less than 1 kbits/s, allowing for animation at extremely low rates. The synchronization delay of audio or speech text with the facial deformations needs to be around 10-20 ms, since audio/visual correlation aids speech intelligibility.

6.6 FUTURE DIRECTIONS: MPEG-7

MPEG-1 and MPEG-2 have addressed primarily the storage and transmission aspects of audio and video. MPEG-4 will extend the functionality of the underlying data representations and will also maintain some backward compatibility with MPEG-1 and MPEG-2. These standards will help promote the widespread use of audiovisual information. However, as such information becomes widely available, finding the desired information will become increasingly difficult. Currently, efficient solutions exist for retrieving textual information. Multimedia databases on the market today allow searching for pictures using characteristics like color, texture, and information about shape of objects in pictures.

MPEG proposes to specify a new standard, called *Multimedia Content Description Interface* and referred to as MPEG-7, that will extend the limited search capabilities of today to include more information types. This description will be associated with the content itself, to allow fast and efficient searching for material that a user might be interested in. The types of information will include: still pictures, graphics, image sequences, composition information as well as some special cases such as facial expressions.

The MPEG-7 description may not depend on the coded representation of the material; however, the standard in a way may build on the MPEG-4 standard. The standardized description of information within MPEG-7 may exist at a number of semantic levels. For instance, in the case of visual material, a lower abstraction level could be a description of shape, size, texture, and color. At a higher level, a semantic description such as *this is a scene with a dog* could be coded in an efficient form.

In addition to having a description of the content, the MPEG-7 description may include other forms of information about multimedia data, such as:

- Coding scheme used, for example, JPEG. This information helps in determining whether the material can be read by the user.

- Conditions for accessing the material.

- Links to other relevant material.

To fully exploit the possibilities of such a description, as a preprocessor to an MPEG-7 system, there might be a feature extractor; however, this feature extractor and the search engines themselves will be outside the scope of the standard.

There are many applications and application domains that will benefit from the MPEG-7 standard. Applications include, digital libraries (image catalog, musical dictionary), multimedia directory services (such as yellow pages) broadcast media selection (radio and TV channel), and multimedia editing (personalized electronic news service, media authoring). The potential applications are spread over the following domains: education, journalism (for example, searching speeches of a certain politician), tourist information, entertainment, investigation services (human characteristics recognition, forensics), geographical information systems, medical applications, architecture, real estate, and interior design. It is envisioned that the MPEG-7 will become an international standard in November of 2000.

6.7   TO PROBE FURTHER

The MPEG-1 and MPEG-2 standards are described in the ISO documents [109] and [111]. Good overviews of these standards are also given in [169] (MPEG-1) and [92] (MPEG-2). In addition to the systems, video, audio, and conformance testing parts, MPEG-2 will also include a fifth part (13818-5 software) that will provide software-based coding models for the first three parts in the standard. The video portion of MPEG-2 has also the nomenclature Recommendation ITU-T H.262. The standardization process itself is described by Chiariglione [41]. Early descriptions of MPEG-1 can be found in [68] and [175]. A good overview of the systems part in MPEG-1 is given by MacInnis [163].

For MPEG-2, an early overview of the standardization process and the coding requirements is given by Wells [258]. MPEG-1 and MPEG-2 support constant bit-rate and variable bit-rate encoding. The MPEG-2 4:2:2 profile is described in the MPEG-2 related standards document [112]. A detailed description, including extensive performance data for this profile, is given by Horne [97]. The MPEG-2 standards document [113] provides details on the multiview profile of Table 6.8. An overview of the multiview profile and its coding performance can be found in [40]. Bit-rate control is primarily achieved by changing the quantizer scale factor. However, the specific mechanism by which the quantizer scale factor is changed is not specified in the MPEG-1 and MPEG-2 standards. The MPEG-2 test model document [114] provides one approach for bit-rate control. A brief description of the test model can be found in [53]. Bit-rate control is an area of active research. Various approaches suitable for MPEG-1 and MPEG-2 can be found in [127], [50], [36], [210], [61], and [81]. Most of these techniques are based either on concepts from control system theory or on image classification. Various researchers have also attempted to develop functional models that relate bit rate to the quantizer scale factor, much like the rate-distortion functions we described in previous chapters. Recently, MPEG-2 was chosen for the video coder of the Grand Alliance HDTV system for North America. A general overview of this system is given in [96] and [215].

Software implementations of MPEG coders have been reported in [25], [52], and [23]. Sites with public-domain MPEG encoders and decoders are given in Appendix C. Hardware implementations of MPEG coders are described later in this book.

An extensive overview of MPEG-4 can be found in [42]. Additional papers describing the activities of the MPEG-4 standardization process can be found in related papers appearing in the same issue as [42]. Additional information can be found in [212] and in papers appearing in the same issue as [212]. MPEG-4 defined a video verification model in January 1996. The intent of this model is to provide a fully defined core algorithm platform for the development of the MPEG-4 standard. The structure of the verification model can give an indication of the tools and algorithms that will be provided by the MPEG-4 standard. The verification model is described in [232]. A brief overview of the proposed standardization activities related to MPEG-7 can be found at the official MPEG Web site listed in Appendix C.

Video generation, preprocessing, post processing, and display are not defined by the standards, but are integral parts of any video coding system. There are many fine textbooks on color and digital television, including [103] and [224]. The CCIR 601 standard for digital TV is described in [3].

7

VIDEO TELECONFERENCING
STANDARDS

7.1 INTRODUCTION

In recent years, there have been significant developments in videoconferencing and audiovisual services. First, data rates in digital telecommunication services have been constantly improving. For example, integrated services digital networks (ISDNs) may provide switched transmission services at data rates close to 2 Mbits/s. Second, advancements in video compression technology allows videoconferencing signals to be compressed below 64 kbits/s.

In the late 1970s, the telecommunications industry realized that continuous growth of audiovisual services was possible only through international standards. Customers expected and demanded video terminal equipment to have the same level of compatibility as other telecommunication equipment, such as modems and facsimile machines. In the early 1980s, the CCITT Recommendations H.120 and H.130 were the first international standards for videoconferencing. The two standards provided guidelines for transmitting PAL or NTSC television signals at data rates close to 2 Mbits/s. The conversion between PAL and NTSC was integrated into the coding process.

However, these standards found acceptance only in Europe. Manufacturers in the United States and Japan continued to develop and market proprietary video coding techniques that offered better signal quality at data rates lower than 2 Mbits/s. In the late 1980s, a new collaboration among telecommunication operators and manufacturers of videoconferencing equipment led to the development of the H.320 videoconferencing standards for videoconferencing over circuit-switched media like ISDN and switched-56 connections. H.261 is the video coding component of this standard. It is also known as the P × 64 standard since it describes video coding and decoding methods at the rates p × 64 kbits/s, where p is an integer from 1 to 30. H.320 was ratified in Geneva in December of 1990.

In September of 1993, ITU established a program to develop an international standard for videoconferencing over the public switched telephone network (PSTN). This standardization effort is referred to as H.324 and it is expected to find extensive global use since it is intended to use the most pervasive communications medium, ordinary telephone lines. It is anticipated that the H.324 standards suite will facilitate interoperability in two principal applications: (1) stand-alone, low-cost, videophones, and (2) desktop videoconferencing and document sharing. H.324 was ratified by the ITU in November of 1995.

7.2   H.320 VIDEO TELECONFERENCING

A video teleconferencing system processes not only video but also audio and ancillary digital data. Thus, additional considerations have to be made for audio compression, data multiplexing, and overall system control. Recommendation H.320 is a complete family of standards, and defines the technical requirements for narrow-band visual telephone systems and terminal equipment.

Figure 7.1 shows the block diagram of a generic, H.320 compliant, visual telephone system and which Recommendations are relevant for each one of the blocks. Video equipment includes cameras, monitors, and video processing units. Audio equipment includes microphones, speakers, and audio processing units. Telematic equipment includes data terminals and visual aids, such as electronic blackboards or a still-picture transceiver. The system control unit performs end-to-network signaling and end-to-end control to establish a common mode of operation among terminals. The video and the audio codecs perform video and audio compression and decompression. A delay in the audio path compensates for the video coding delay and allows the system to maintain audio and video synchronization. The mux/dmux unit multiplexes audio, video, and data into a single bit stream during encoding and demultiplexes them during decoding. Finally, the network interface

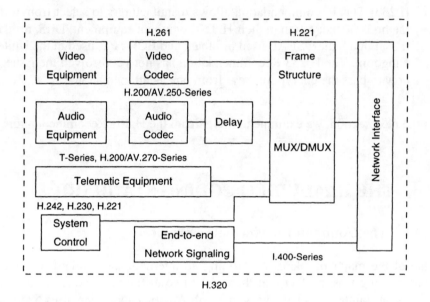

Figure 7.1 Block diagram of a generic visual telephone system, and associated Recommendations.

provides the interface between the networks and the terminal as defined in the I.400-series Recommendations.

H.320 refers to a number of Recommendations, including:

- H.261: A video coding algorithm for compressing signals at data rates from 64 kbits/s to 1,920 kbits/s.

- G.722, G.726, and G.728: A series of algorithms for the compression of audio signals at data rates from 16 kbits/s to 64 kbits/s. These Recommendations are examined in more detail in Chapter 16.

- H.221: This Recommendation specifies the frame structure for multiplexing video, audio, and data into a single bit stream.

- H.230 and H.242: These Recommendations specify the handshaking protocols between H.320 compliant equipment.

■ H.233: This Recommendation allows manufacturers to select from three methods of encryption in their H.320 compliant equipment: DES, used in the United States; SEAL used in Japan; and BCRYPT, used in the United Kingdom. The H.233 Recommendation has not yet resolved the issue of how to pass the encryption keys from one location to another.

In the next section, we examine in more detail the H.261 Recommendation.

7.3   THE H.261 VIDEO CODING STANDARD

7.3.1   The Common Intermediate Format

One of the major problems in defining an international standard for video-conferencing was the fact that there exist two different line and picture-rate television standards. NTSC, used in North America and Japan, uses 525 lines per interlaced picture at 30 pictures per second. On the other hand, most of the other countries use 625 lines per interlaced picture at 25 pictures per second. To eliminate the problem of interoperability among systems with different formats, a new common intermediate format (CIF) was adopted. Both the 625 and the 525 line systems need to include pre- and postprocessing modules to convert to and from CIF.

CIF is a noninterlaced format. It is based on 352 pixels per line, 288 noninterlaced lines per picture at 30 pictures per second. These values represent half the active lines of a 625/25 television signal and the picture/rate of a 525/30 (NTSC) signal. Therefore, 625/25 systems need only to perform a picture-rate conversion and NTSC systems need to perform only a line-number conversion.

Color pictures are coded using one luminance and two color-difference components (YCbCr) as specified by the CCIR 601 standard. The Cb and Cr components are subsampled by a factor of two on both the horizontal and vertical directions and have 176 pixels per line and 144 lines per frame. The picture area covered by these numbers of pixels and lines has an aspect ratio of 4:3. Table 7.1 summarizes the characteristics of a CIF frame.

Color component	Image size (pixels × lines)
Y	352 × 288
Cb	176 × 144
Cr	176 × 144

Table 7.1 Picture characteristics of the common intermediate format (CIF).

For low bit-rate applications, in addition to CIF, video coders may also use a quarter-CIF (QCIF) format, which has half the number of pixels and lines required for CIF. Support for CIF coding and decoding is optional; however, all coders must be able to operate using QCIF.

7.3.2 H.261 Encoding

Like MPEG, the H.261 encoding algorithm uses a combination of DCT coding and differential coding. Figure 7.2 shows a block diagram of an H.261 video encoder. The main elements are frame prediction, DCT transformation, quantization (Q), and variable length coding (VLC).

The DCT coding path is similar to the one used in JPEG and MPEG. A video frame is first translated into a CIF frame and then stored into frame memory. Noise filtering or other signal preprocessing can also be performed at this stage. Similarly to the operations in JPEG, the DCT operates on 8 × 8 picture blocks. Four luminance (Y) blocks and one Cb and one Cr color difference block are combined to form a macroblock.

Note that frame prediction is done in a manner similar to that in MPEG-1, with the exception that only I-pictures and P-pictures are used. The macroblock organization and the classification of a macroblock as an intra-type or inter-type follow the approach in MPEG-1. Differential coding allows the DCT coder to operate on either input macroblocks (INTRA mode) or the differential macroblocks between the current frame and the prior frame (INTER mode). Not all macroblocks need to be coded and transmitted. For example, at low bit rates, macroblocks (and up to three full frames) can be skipped. The criteria

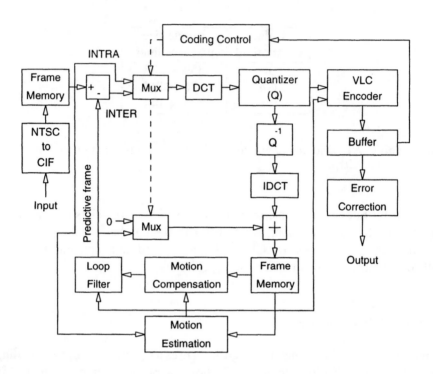

Figure 7.2   Block diagram of an H.261 video encoder.

for choosing either to transmit or skip a macroblock or the control mechanism for intra- or interframe coding are not part of the standard and may vary dynamically, depending on the complexity of the input signal and the output data rate constraints.

Following the DCT, the output DCT coefficients are quantized, coded using a variable length coder (a Huffman coder in this case), and stored into an output buffer. By monitoring the capacity of the buffer, the coder can perform data rate control dynamically. The rate-control strategy is not specified in the H.261 standard; here, approaches similar to that adopted for MPEG-1 can be used. A proposal for rate control is included in the informative part of the H.261 document.

Since H.261 is a predictive coder and since transmission errors may cause significant image quality problems, a BCH(511,493) coder is used to add parity bits for error detection and correction. However, the use of error detection and correction is optional for an H.261 decoder.

Because of the use of predictive coding, the encoder needs to monitor the quality of the transmitted image so that it does not diverge from the original signal. After an inverse quantization (Q^{-1}) and an inverse DCT (IDCT), the encoder is able to reconstruct a frame as it will be seen by the decoder. The reconstructed frame is stored into frame memory. Note that, since the VLC operation is lossless, there is no need to include the VLC unit in the feedback path that generates the prediction frames.

As shown in Figure 7.2, the prediction path may also include optional circuitry for motion estimation and compensation, and a spatial loop filter. The role of the loop filter is to minimize the prediction error by smoothing the pixels in the previous frame. The loop filter is a separable 2-D filter that operates on 8 × 8 blocks. The corresponding 1-D filter is a three-tap finite impulse response filter. At block edges, the filter coefficients are 0, 1, and 0. Otherwise, they are $\frac{1}{4}$, $\frac{1}{2}$, and $\frac{1}{4}$. Recent studies suggest that tap weights of 0.1248, 0.7495, and 0.1248 yield nearly a 1 dB improvement in signal quality when an accurate motion-estimation technique is used. On the other hand, for a less accurate motion-estimation method, the H.261 recommended tap weights may be more appropriate.

The compressed data stream is arranged hierarchically in four layers: picture, group of blocks (GOB), macroblock (MB), and block.

- A picture is the top layer.

- Each picture is divided into groups of blocks (GOBs). A GOB is either one-twelfth of a CIF picture or one-third of a QCIF picture.

- Each GOB is divided into 33 macroblocks.

- Each macroblock consists of six 8 × 8 blocks, that is, four blocks of luminance (Y), one block of Cb, and one block of Cr.

7.3.3   The H.261 Video Decoder

Figure 7.3 shows a block diagram of the H.261 video decoder. After optional

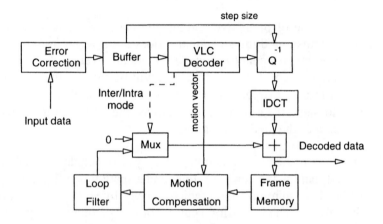

Figure 7.3   Block diagram of an H.261 video decoder.

error correction, the compressed input is buffered and processed by the variable length decoder. The decoded data are parsed and then processed by an inverse quantizer and an inverse DCT. Depending on the transmission mode (INTRA or INTER), macroblocks from a prior frame may also be added to the current data to form the reconstructed data. The decoded CIF data may then be converted to the appropriate display format.

7.4   IMPLEMENTATION ISSUES

Not all implementation details are specified by H.261. This gives manu-facturers the opportunity to tailor their implementations to suit their target applications and to provide for product differentiation. Among the details that are open to variations are rate-control mechanisms, motion-estimation algorithms, and preprocessing and postprocessing algorithms.

In early field trials of H.261 codecs, it was noted that small variations in the implementation of the inverse DCT between two coders may yield noticeable picture artifacts. To resolve this problem, the H.261 Recommendation provides guidelines for measuring the accuracy of a particular implementation of the inverse DCT. The procedure to test the accuracy of an IDCT is outlined in Annex A of the H.261 standard and consists of the following steps: (1) generate random data in a specified range, and form 8×8 data blocks, (2) perform a DCT on each block, (3) perform an IDCT, and (4) measure peak, mean, and mean square error between the original data and the output of the IDCT. To be H.261 compliant, the following conditions must be met:

- For any pixel, the peak error should be less than 1.0, the average error (over 10,000 blocks) should be less than 0.015, and the mean square error should be less than 0.06.

- Overall, the average error should be less than 0.0015, and the mean square error should be less than 0.02.

- All zeros in must produce all zeros out.

In addition to the above conditions, the standard requires that there be at least one intraframe coded macroblock for every 132 interframe coded macroblocks. This number corresponds to four GOBs or to one-third of a CIF picture. Intraframe coded macroblocks flush out any errors that defeat the error-correction circuitry and allow for the errors in the frame prediction to be reset. For better coding efficiency, most implementations of H.261 prefer not to perform intraframe coding on all the macroblocks of a picture. Instead, they perform intraframe coding on a few macroblocks in every picture using a rotational scheme.

H.261 shares much functionality with the MPEG video coding standards. However, even though the key coding algorithms are the same, the two standards target different applications with different requirements in data rates, picture quality, and end-to-end coding delay. Table 7.2 shows the main differences between H.261 and MPEG.

Presently, there is a great deal of interest in developing software implementations for H.261 on general-purpose processors. Due to the motion estimation process, encoding is significantly more complex than decoding. In fact, motion estimation may be close to 60 percent of the overall computational load. One way to reduce the computational load is by reducing the search-range in motion estimation.

MPEG	H.261
Uses CIF, SIF, or higher spatial resolutions.	Uses QCIF or CIF spatial resolutions.
Variable image aspect ratio (defined in the header).	Fixed 4:3 aspect ratio.
Uses groups of pictures.	No notion of GOPs.
I, P, and B macroblocks.	No B macroblocks.
Typical bit rates are around 1.1 Mbits/s.	Typical bit rates are around 384 kbits/s. Max. bit rate is 2 Mbits/s.
No restrictions on skipped pictures.	Only 1, 2, or 3 skipped pictures allowed.
Sub-pixel accurate motion vectors.	Pixel accurate motion vectors.
Typical motion vector range is +/- 15 pixels.	Typical motion vector range is +/- 7 pixels.
The end-to-end coding delay is not critical.	Used mostly in interactive applications. End-to-end delay is very critical.

Table 7.2 Main differences between the MPEG and the H.261 standards.

In H.261, the maximum range for a motion estimation vector is [-15, 15] pixels. Using the maximum range in all H.261 applications will not necessarily improve the quality of the compressed signal. H.261 applications can operate at various bit rates, ranging from 64 kbits/s to 1,984 kbits/s. At high frame rates, the temporal distance between frames is smaller; thus, one can afford to have smaller search ranges. At low frame rates, the situation is reversed, and one needs larger search ranges for better image quality.

Furthermore, videoconferencing primarily deals with scenes where most of the movement is in the head and the shoulders. Experiments show that one can use a smaller, diamond-shaped, search region, instead of the conventional rectangular-search region, with no noticeable loss in video quality. Several search regions are depicted in Figure 7.4a. From Figure 7.4b, at very low bit rates (64 kbits/s), and frame rates below 10 fps, motion estimation using

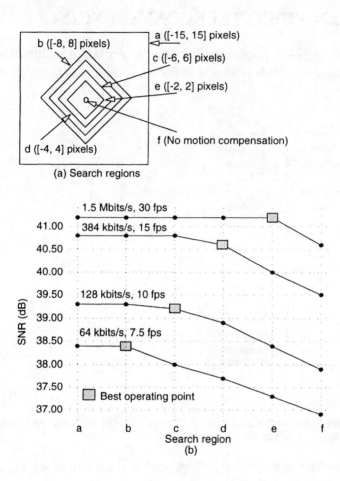

(a) Search regions

(b)

Figure 7.4 Search regions and video quality for typical H.261 encoding. (a) Diamond-shaped search regions; (b) video quality for the corresponding search regions.

the diamond-shaped search region b ([-8, 8] pixels) yields virtually the same video quality as motion estimation using the square-search region a ([-15, 15] pixels). At high bit rates (1.5 Mbits/s), for typical videoconferencing scenes, there appears to be no benefit in using a large search range. The shaded squares in Figure 7.4b, show the best motion-estimation schemes for various cases of practical interest. For these search ranges, the computational complexity of motion estimation can be reduced by at least eighty percent.

7.5   H.324 VIDEO TELECONFERENCING

Like H.320, H.324 defines several standards. An H.324 multimedia system based on these standards is shown in Figure 7.5. From Figure 7.5, an H.324

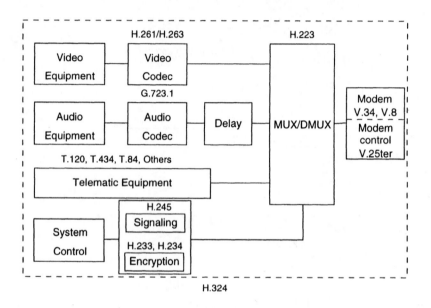

Figure 7.5   Block diagram of a generic H.324 multimedia system, and associated Recommendations.

system is similar to the H.320 system described earlier. However, four new companion Recommendations are included:

- H.263: A video coding scheme for low bit-rate communications.

- G.723.1: A dual-rate audio and speech coder for multimedia communications at 5.3 and 6.3 kbits/s.

- H.223: A multiplexing protocol.

- H.245: A control protocol that can be used to specify the coding and decoding abilities of the receiver and the transmitter. This feature can be used to enable the various coding modes in the video encoder. The procedures in H.245 are also planned to be used in videoconferencing standards for ATM networks (Recommendation H.310) or local-area networks (LAN's) with non-guaranteed bandwidth (Recommendation H.323). This will enable interworking of H.324 terminals with ATM-based and LAN-based videoconferencing systems.

For modem-based communication, H.324 specifies the use of the V.8 (V.8bis) procedure to start and stop data transmission, and the V.34 modem. The V.34 standard supports data rates of up to 28.8 kbits/s. For data applications, H.324 specifies the use of the T.120 protocol as one possible means of data exchange. For still-image transmission, the T.84 Recommendation (JPEG) could be used.

The speech coding part of the H.324 standard is described briefly in Chapter 16. In the next section, we examine in more detail Recommendation H.263, the video coding part of H.324. Furthermore, we restrict our discussion to the near-term H.263 coding standard. A long-term Recommendation is still evolving and will be briefly discussed later in this chapter.

7.6 THE H.263 VIDEO CODING STANDARD

The H.263 video coding algorithm employs many of the coding features found in H.261 and MPEG-1. For example, both H.261 and H.263 share the same basic encoder and decoder structures shown in Figures 7.2 and 7.3. Because of the many similarities between H.261 and H.263, in this section, we focus only on the key differences between these two video coding schemes.

7.6.1 H.263 versus H.261

Bit rate

The H.263 Recommendation puts no constraints on the bit rate; however, its target bit rate is up to 64 kbits/s. For applications on general switched telephone networks (GSTN) via a 28.8 kbits/s modem, the maximum video bit rate will probably be around 20 kbits/s, since the remaining bandwidth must be reserved for speech, data, and control. Note that, for H.261, the target bit rate is $p \times 64$ kbits/s, where p is an integer between 1 and 30.

Picture Format

Table 7.3 shows picture sizes that are supported by H.263. All H.263 decoders

Format	Image Size					
	Y			Cb and Cr		
sub-QCIF	128	×	96	64	×	48
QCIF	176	×	144	88	×	72
CIF	352	×	288	176	×	144
4CIF	704	×	576	352	×	288
16CIF	1408	×	1152	704	×	576

Table 7.3   Picture formats for H.263.

must be able to operate using sub-QCIF and QCIF bit streams. An encoder should be able to operate with at least one of the sub-QCIF or QCIF formats. The encoder determines the format to be used and is not obliged to operate with both. For display, decoders may upsample the sub-QCIF pictures or downsample the QCIF pictures. These constraints ensure that an H.263 codec can be realized at a low cost and yet permit interworking with H.261.

Group of blocks structure

H.261 and H.263 use a hierarchical syntax in which the video sequence is decomposed into pictures, group of blocks (GOB), macroblocks (MB), and blocks. In H.263, for error resilience, each GOB contains only one macroblock row. Thus, for QCIF, each GOB has 11 MBs, compared to the $11 \times 3 = 33$ MBs used in H.261. Furthermore, in H.263, optional header information can be inserted in the GOB layer. This allows the coder to insert extra synchronization codewords for improved error resilience.

Error correction

Many H.263 video codec applications will operate over GSTN or mobile channels. In these channels, the noise manifests as burst errors, and the BCH single-error correcting code adopted by H.261 is not appropriate. No error correction or detection is specified in H.263. An optional error correction scheme is described in Annex H of the H.263 Recommendation.

Precision of motion compensation

Motion vectors for H.263 have half-pixel accuracy compared to the integer pixel accuracy for motion vectors used in H.261. Half pixel values are found using bilinear interpolation. The improvement due to the use of half-pixel accuracy is well established, and half-pixel accurate motion estimation has been used in other coding standards, such as MPEG-2. For example, in comparison to using integer-pixel accurate motion vectors, there is a 1 to 3 dB performance gain in PSNR when half-pixel accurate motion vectors are used. If only integer-pel accuracy is used for the motion vectors, then the H.263 coder performs worse than an H.261 coder. This is attributable to the loss of the filtering effect that results from bilinear interpolation.

Loop filter

H.261 employs a spatial-domain loop filter in the coding loop to reduce the block effects due to block-based motion estimation. H.263 does not employ such a filter since the bilinear interpolation used in H.263 for half-pixel motion compensation introduces some low-pass filtering as a side-effect.

Motion vector coding

As in H.261 and MPEG, the motion vectors are differentially encoded. That is, if MV is the true value of the motion vector, then it is differentially encoded as

$$MV_d = MV - MV_p, \tag{7.1}$$

where MV_p is the predictor. In H.261, the predictor is the motion vector from the previous macroblock. However, in H.263, the predictor is the median of the motion vectors from three previously coded blocks, as shown in Figure 7.6. If $MV1$, $MV2$, and $MV3$ denote three such motion vectors,

$$MV_p = \text{median}(MV1, MV2, MV3). \tag{7.2}$$

This results in more efficient prediction of the motion vectors and fewer bits

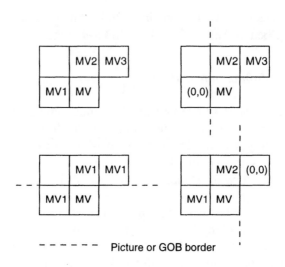

Figure 7.6   Prediction of motion vectors in H.263.

for coding motion vectors. Note that, as shown in Figure 7.6, special rules apply at the borders of the current GOP or picture.

Other differences

- In H.261, the quantizer can be changed on a per macroblock basis. In H.263, within a GOB, quantizer transitions from one macroblock to the next are restricted to the two nearest finer or coarser quantizers.

- The quantized DCT coefficients are coded with a 3-D VLC (run, level, last coefficient) instead of a 2-D VLC (run, level) and the end-of-block marker. This improves the compression ratio.

- In H.261, the macroblock address is used to indicate how many macroblocks have been skipped, whereas, in H.263, a single bit is transmitted for every skipped macroblock.

- Unlike H.261, there is no still-picture mode defined for H.263. Instead the JPEG standards (T.81, T.84) could be used for transmitting still-pictures.

Most of the differences with respect to H.261 are intended to reduce the overhead in the bit stream and make H.263 suitable for very low bit-rate applications. The H.263 video coding standard specifies optional coding techniques that can further improve the performance. These options are discussed next.

7.6.2 H.263 Advanced Coding Options

Depending on the capabilities of the encoder and the decoder, the H.263 codec may employ several advanced coding modes to improve the performance; that is, reduce the bit rate and/or improve the image quality. The use of a particular option has to be negotiated with the decoder by external means (for example, Recommendation H.245). Even without the advanced coding options, at the same bit rate, an H.263 coder outperforms an H.261 coder by 2 dB. When all of the advanced coding options are used, a 1 dB additional gain is obtained. Equivalently, for the same image quality, an H.263 coder with the advanced coding options may operate at 50 percent lower bit rate. In a software implementation, an H.263 decoder employing all of the advanced coding options is three times more complex than a basic H.263 decoder. A software-only H.263 encoder using the advanced coding options is 40 percent more complex than a basic H.263 encoder. Next, we briefly describe the four advanced coding options of H.263.

Unrestricted motion vectors mode

In standards such as H.261, MPEG-1, and MPEG-2 (H.262), the motion vectors are restricted to point to the pixels inside a previous frame. Thus, many of the macroblocks at the border of the picture will have suboptimal prediction. In H.263, if the unrestricted motion vector (UMV) mode is enabled, the motion vector can point to a position outside the actual frame region. In this mode, when a pixel referenced by the motion vector lies outside the frame, then the closest edge pixel is used. Note that, the motion vector range in this mode is extended from the default [-16, 15.5] to [-31.5, 31.5]. The UMV mode is useful when moving objects are entering or moving around the frame border.

Advanced prediction mode

This option includes *overlapped block motion compensation* and the possibility of *four motion vectors for a macroblock* instead of the original single motion vector. In this mode, the UMV mode is automatically enabled. Therefore, the maximum range of the motion vectors is also [-31.5, 31.5]. This mode leads to significant improvements in image quality.

Four motion vectors per macroblock: This coding mode can be enabled on a per macroblock basis and the one/four vectors decision is indicated in the macroblock header. When four motion vectors are used, each of the motion vectors is used by all pixels in any one of the four luminance blocks within the macroblock. The motion vectors for the Cb and Cr blocks are derived by calculating the sum of the four luminance motion vectors and dividing the result by 8; this result has to be rounded to the nearest half-pixel position.

In the basic H.263 scheme (see Figure 7.6), the motion vector for the current macroblock is predicted from the median of three neighboring motion vectors. In the four motion vectors per macroblock case, the neighborhood for motion vector prediction has to be modified for each 8×8 block within a macroblock. The four different cases are illustrated in Figure 7.7. For example, for the top-left block in the current macroblock, the modified neighborhood is shown in Figure 7.7a. The decision as to when to use one or four motion vectors is not specified within the H.263 standard.

Overlapped block motion compensation (OBMC): In this mode, motion compensation is performed on an 8×8 block basis. Prediction of each pixel within the 8×8 block is a weighted sum of three prediction values given by

$$\hat{p}(i,j) = (q(i,j)H_0(i,j) + r(i,j)H_1(i,j) + s(i,j)H_2(i,j) + 4)/8, \quad (7.3)$$

where $\hat{p}(i,j)$ is the final pixel value and $q(i,j)$, $r(i,j)$, and $s(i,j)$ are pixels from the referenced picture obtained from three motion vectors, namely, the motion vector of the current block and two out of its four *neighbor* motion vectors, as shown in Figure 7.8. The matrices $H_0(i,j)$, $H_1(i,j)$, and $H_2(i,j)$ are shown in Figures 7.8b, 7.8c, and 7.8d, respectively. The remote motion vectors need not belong to the same GOB. OBMC produces a smoother motion field while yielding similar or lower prediction error than conventional block matching prediction. Annex F of the H.263 standard provides additional details

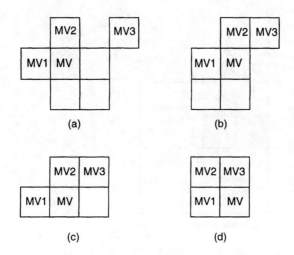

Figure 7.7 Prediction neighborhood for motion-vector prediction. Here, MV indicates the relative position of the current 8×8 block in the macroblock.

for this option. Using only OBMC or four motion vectors per macroblock results in approximately 1 dB of coding gain.

Syntax-based arithmetic coding mode

In this mode, the variable length coding/decoding processes employ arithmetic coding. Since arithmetic coding utilizes fractional parts of a binary representation, the coding restriction of one bit per symbol is removed. This improves overall coding efficiency. For inter-coded macroblocks, arithmetic coding yields a three to four percent reduction in bit rate. For intracoded macroblocks, there is a typical ten percent reduction in bit rate.

PB-frames mode

In H.261, there are no B-frames; however, H.263 specifies the optional use of PB-frames. In this mode, two frames are coded as one unit. The first frame in this unit is a P-frame which is predicted from a previously decoded P frame. The second frame is a B-frame which is predicted from the P-frame in this

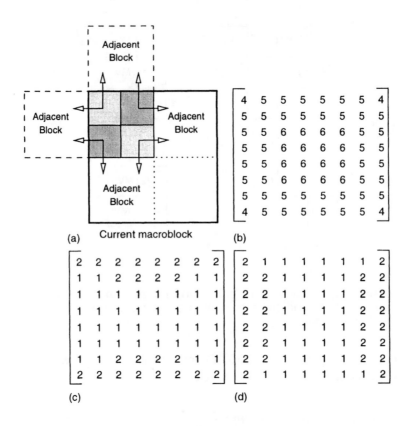

Figure 7.8 The OBMC scheme in H.263. (a) Remote motion vector selection for OBMC, (b) weighting matrix $H_0(i,j)$, (c) weighting matrix $H_1(i,j)$, and (d) weighting matrix $H_2(i,j)$.

unit and a previously decoded P-frame. The PB-frame structure is illustrated in Figure 7.9a. In a PB-frame, there will be 12 blocks per macroblock. For each macroblock, the six blocks belonging to the P-frame are transmitted first, followed by the six blocks of the B-frame. Unlike MPEG, as shown in Figure 7.9b, only parts of a B-block are bidirectionally predicted. For pixels where the backward motion vector points to the inside of the current P-macroblock, bidirectional prediction is employed. The remaining pixels in the B-block are predicted using the forward motion vector.

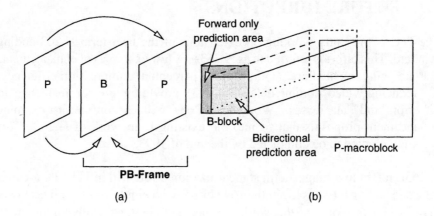

Figure 7.9 PB-frame. (a) PB-frame prediction, (b) B-block prediction in PB-frame.

For relatively simple sequences, with this mode, the frame rate can be doubled without increasing the bit rate by too much. For sequences with a lot of motion, PB-frames do not work as well as B-pictures in MPEG. This is because there are no separate bidirectional vectors in H.263. The forward and backward vectors are derived from the P-frame motion vector and a delta vector. The advantage over MPEG is that this scheme requires less overhead for the B-picture part. This is really useful for the low bit rates and relatively simple sequences that are most often generated by videophones.

In a PB-frame, 15-20 percent of the bits are typically used for the B-frame. From an encoding point of view, for the same frame rate, the use of PB-frames can result in lower encode time since motion estimation is done only for every other frame. Furthermore, for B-frames, a smaller search range is used. On the other hand, the use of PB-frames can result in coding delays, since two frames have to be stored and coded. Compared to a basic H.263 decoder software implementation, the PB-frame mode requires 50 percent more CPU cycles.

7.7   FUTURE DIRECTIONS

The ITU is presently working on extensions of the near-term H.263 coding standard. This effort is referred to as H.263+. A few examples of enhancements that are being pursued are: (1) improving perceptual compression efficiency, (2) reducing video coding delay, and (3) providing greater resilience to bit errors and data losses. Work on H.263+ will be confined to examine enhancement proposals that fit into the existing framework of H.263. This work is expected to be completed by the end of 1997.

H.263L refers to a longer-term standardization effort within ITU. Its focus is on video coding bit rates lower than 64 kbits/s, with emphasis on bit rates lower than 24 kbits/s. Some of this work is expected to be technically aligned with the MPEG-4 effort in low bit-rate coding. H.263L related work is expected to be completed by November of 1998.

7.8   TO PROBE FURTHER

The H.320 and H.261 Recommendations are described in the ITU documents [86] and [84]. In [32], Carr provides additional details on the early efforts for the development and testing of a video teleconferencing standard. The methodology adopted for the development of H.261 and the evolution of the standard is discussed in [189]. Brief descriptions of the H.261 standard and comparisons with JPEG and MPEG are also presented in [175] and [8]. The role of the loop filter in H.261 is investigated in detail in [196].

The H.324 and H.263 Recommendations are in ITU documents [87] and [85]. The status of H.324 is summarized in [226]. Some of the implementation issues regarding H.263 can be found in [136]. In [78], Girod provides an extensive comparison of the H.261 and H.263 coding standards. The concept of overlapped block motion compensation is described in [16], while implementation issues and the underlying theory are developed in [192] and [239]. Motion estimation techniques associated with overlapped block motion compensation are presented in [214]. Documents describing the current activities related to the near-term and long-term extensions to H.263 can be found at the H.324 documents site listed in Appendix C.

8

PROCESSING REQUIREMENTS

8.1 INTRODUCTION

In Chapters 1 to 7, we examined the core algorithms of image and video compression standards without regard to their hardware implementation. However, ease of implementation and low processing requirements are among the key factors that have influenced the final recommendations of the standards committees. In the past, the computational needs of intensive image-processing tasks were satisfied with expensive custom multiprocessor systems. However, the standards committees believed that it would be easier for the industry to accept the new standards if low-cost implementations were possible.

The latest RISC (reduced instruction set computers) architectures have enough compute power for the implementation of some of the video standards. For example, workstations from DEC, Sun Microsystems, and Hewlett-Packard can perform MPEG-1 and MPEG-2 decompression at 30 frames/s (fps) now. For video teleconferencing using the H.261 standard, general-purpose processors can already encode and decode 8 to 10 QCIF frames per second. However, even the latest designs cannot adequately address the processing requirements of some of the standards, such as real-time MPEG encoding. In such cases, custom hardware may be necessary. In the next eight chapters, emphasis will be on the hardware implementation of the image and video compression standards. Specifically, in this chapter we examine their processing requirements. In later chapters we discuss design tradeoffs, the most common hardware designs for the core algorithms (like the DCT, motion estimation, and entropy coding), and IC designs of various video processors.

8.2   MEASURING COMPLEXITY

There are many different measures of algorithmic complexity. For example, in digital signal processing, one measure is the number of multiplications and additions. For the video compression standards, a good measure of complexity is the number of required RISC-like operations. For example, the evaluation of $r1 = a + b$, where a and b are data stored in external memory, and $r1$ is a processor register, requires two data loads and one addition, for a total of three operations. For signal and video processing algorithms, the total number of operations is usually expressed in MOPS (million operations per second) or in GOPS (giga operations per second).

Example. Consider the MOPS requirements for computing a two-dimensional (2-D) DCT on a QCIF frame. Using row-column decompositions on 8×8 blocks, a 2-D DCT can be computed by performing 1-D DCTs on the rows of the input data matrix followed by 1-D DCTs on the columns. An eight-point 1-D DCT is given by

$$ y_i = \sum_{k=1}^{8} c_{ik} x_k, i = 1, 2, ..., 8, \qquad (8.1) $$

where y_i denotes the output elements, x_k denotes the input data, and c_{ik} denotes the DCT coefficients. From the above, this implementation requires at least: eight image data loads, eight coefficient data loads, eight multiply-accumulate operations, and one data store operation, for a total of 25 operations per pixel. Because of the row-column decomposition, an 8×8 DCT will require $2 \times 25 \times 64 = 3,200$ operations. In a YCbCr QCIF frame, with a 4:2:0 sampling ratio, Y has 176×144 pixels, and Cb and Cr have 88×72 pixels each. A complete frame has 594 8×8 blocks. At 15 fps, the total number of operations is $15 \times 594 \times 3,200 = 28.5$ MOPS. For a CIF frame, a similar implementation would require 114 MOPS.

Following the same approach, one can estimate the processing requirements of the various compression algorithms for different frame sizes and data rates. Table 8.1 shows the MOPS requirements for compression and decompression using H.261 on CIF resolution sequences at 30 fps. These estimates were computed assuming fast implementations of the DCT or IDCT algorithms and logarithmic (instead of exhaustive) searches for motion estimation.

Compression	MOPS
RGB to YCbCr	27
Motion estimation (25 searches in a 16 × 16 region)	608
Inter-/Intraframe coding	40
Loop filtering	55
Pixel prediction	18
2-D DCT	60
Quantization, zig-zag scanning	44
Entropy coding	17
Frame reconstruction	99
Total	**968**
Decompression	
Entropy decoder	17
Inverse quantization	9
Inverse DCT	60
Loop filter	55
Prediction	30
YCbCr to RGB	27
Total	**198**

Table 8.1 MOPS requirements for H.261 compression and decompression (CIF at 30 fps).

From Table 8.1, decompression requires approximately 200 MOPS, and that is now easily achievable by several general-purpose RISC or DSP (digital signal processing) processors. However, an encoder requires more than 1,000 MOPS of processing power, which is beyond the capabilities of a general-purpose processor at this time.

Table 8.2 shows MOPS estimates for the baseline implementation of MPEG, for different frame sizes, and different percentages of frames computed with bidirectional motion estimation (B-frames). These estimates are for 30 fps, assuming no preprocessing or postprocessing (that is, color transformations), no audio, and no other system-related operations. In all cases, we assume a 4:2:0 color encoding. For HDTV, we assume that each frame has 1440 × 1152

Compression	SIF	CCIR 601	HDTV
No B-frames	738	3,020	14,498
20% B-frames	847	3,467	16,645
50% B-frames	1,011	4,138	19,865
70% B-frames	1,120	4,585	22,012
Decompression			
No B-frames	96	395	1,898
20% B-frames	101	415	1,996
50% B-frames	108	446	2,143
70% B-frames	113	466	2,241

Table 8.2   MOPS estimates for MPEG compression and decompression at 30 fps.

pixels (MPEG-2, Main Profile at High-1440 Level). For comparison, Figure 8.1 shows the trends in computing power for general-purpose DSPs, RISC processors, and programmable image processors. From the data in Figure 8.1,

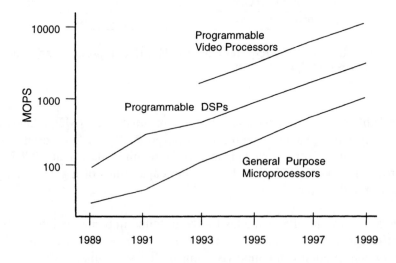

Figure 8.1   Trends in the performance of programmable processors.

it is clear that general-purpose processors will need to integrate additional features before they can be used for real-time video encoding and decoding.

8.3 DISTRIBUTING THE LOAD

In Figure 8.2, we show the main processing flow in JPEG, MPEG, H.261, and H.263 encoding. Note that, there is considerable functionality that is common among all these compression standards. They all include three key functions:

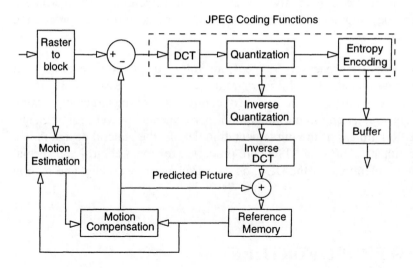

Figure 8.2 Main processing flow in JPEG, MPEG, H.261, and H.263 encoding.

(1) the computation of the DCT, (2) quantization of the DCT coefficients, and (3) entropy coding. MPEG, H.261 and H.263 share additional functions, namely, motion estimation and motion compensation. In addition to these key operations, any implementation must also allocate significant computational power for data preprocessing and postprocessing, audio processing, data I/O, and host and display interface. For example, data preprocessing may include noise filtering, deinterlacing, color transformations (RGB to YCbCr), and subsampling. Postprocessing may include raster-to-interlaced transformations,

color transformations, and audio-video synchronization. Since no single programmable processor has enough processing power for all these operations, parallel processing architectures are not uncommon for video processors.

Until recently, there were few commercially available video processing ICs. Furthermore, their functionality was limited to low-level operations, such as filtering. This was due to the limitations in circuit integration and to the fact that only low-level imaging functions had a large enough range of applications that could justify the development of special hardware. The emergence of standards for image and video compression opened new market areas, such as multimedia computing, video set-top units for interactive television, and HDTV. It is now commercially viable for IC manufacturers to invest in the design and development of a new generation of image and video processing ICs.

Designs of video processors range from fully custom architectures with minimal programmability to fully programmable multiprocessor architectures. Their architecture depends on the speed requirements of the target application and the constraints on circuit integration, performance, power requirements, and cost. Regardless of the implementation details, the general design theme is to use either a DSP or a RISC core processor for main control and special hardware accelerators for the DCT, quantization, entropy coding, and motion estimation.

8.4   TO PROBE FURTHER

The main scope of this chapter was to present some estimates of processing complexity for the image and video compression standards. In related work, Fujiwara et al. [62] and Guttag et al. [83] present similar results for the real-time implementation of the H.261 standard. Guttag et al. estimate the total number of operations to be close to 1,200 MOPS, very close to our estimate. Fujiwara et al. use a more conservative RISC model and thus provide much higher MOPS estimates. For example, they estimate that a single multiply-accumulate operation will require five (instead of three in our model) instructions. Without using fast algorithms for the DCT, and by assuming full search (instead of logarithmic search) for motion estimation, their total estimated number of operations is approximately 3,400 MOPS. However, in relative numbers, there is general agreement that motion estimation is the most compute-intensive operation, requiring nearly 60 percent of the processing power of an H.261 or MPEG encoder.

In [269], Zhou et al. provide a detailed analysis of the MOPS requirements for MPEG video decompression on a general purpose RISC-like processor. For a 4 Mbits/s, CCIR 601 input stream, they estimate that real-time MPEG-2 decoding requires close to 400 MOPS. The authors count one multiplication as four generic operations and assume that each group of pictures (GOP) has one I-frame, four P-frames, and ten B-frames. The above estimate does not include the computational requirements for the final color conversion from YCbCr to RGB. Lee and Kim [149] also examine the computational requirements of MPEG-2 decoding on a programmable processor by analyzing the core tight loops and mapping them to a sequential RISC processor. For a 10 Mbits/s input stream, they estimate that real-time MPEG-2 decoding of a CCIR 601 stream, with no B-frames, requires at least 564 MOPS. They estimate that a VLIW RISC processor with three processing units and running at a modest 190 MHz can decode such a bit stream in real-time.

9

RISC AND DSP CORES

9.1 INTRODUCTION

As noted in the previous chapter, most video compression ICs utilize either a RISC or a DSP core as their central controller and processing unit. By RISC or DSP core we imply a standard programmable processor, without the I/O pads, buffers, and associated circuitry. These parts are replaced by custom circuits that interface with the rest of the architecture. The main goal in using a processing core is to take advantage of existing hardware and software resources (such as arithmetic processing units, memory, operating system support, and language compilers) and to minimize the design time of a more complex and dedicated IC. In this chapter we describe the basic architectures of a RISC core and a DSP core and some of their similarities and differences.

9.2 THE RISC CORE

9.2.1 Architecture

The fundamental principle of reduced-instruction-set computers is to achieve high performance with the minimum amount of complexity. As a result, instructions are very simple and perform mostly register-based operations. Register-to-register operations allow most of the instructions to take place inside the chip. This allows for faster implementation (shorter clock cycles) and simpler control. A RISC core decodes and executes load, store, arithmetic,

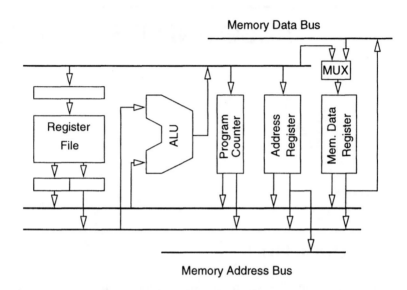

Memory Data Bus

Memory Address Bus

Figure 9.1   Block diagram of a RISC core.

and control transfer operations. The only instructions that access external data are the load and store operations. During arithmetic operations, the processor will read two registers, perform an operation, and store back the result into the register file.

Figure 9.1 shows a typical data path of a RISC architecture. It includes a register file with source and destination latches, an ALU (arithmetic and logic unit) and a program counter. A RISC processor may also have additional registers for data and instruction addressing or other control related functions. Most RISC designs use the same ALU to compute both algebraic operations and memory addresses for load and store operations. The justification for such a design is that, because during load and store operations the ALU is not busy, such an implementation does not cause any performance penalty.

Table 9.1 shows three typical RISC instructions from the Berkeley RISC processor: an integer add, a memory read, and a conditional jump. The corresponding instruction formats are shown in Table 9.2. For register-to-register instructions, an operation (specified by OPCODE) is performed on the registers specified by SRC1 and SRC2, and the result is stored in the

Instruction	Operands	Operation	Comments
ADD	$Rs, S2, Rd$	$Rd = Rs + S2$	Integer addition
LDL	$(Rx)S2, Rd$	$Rd = M[Rx + S2]$	Memory load
JMP	$CON, S2(Rx)$	$PC = Rx + S2$	Conditional jump

Table 9.1   Examples of RISC instructions.

OPCODE	DEST	SRC1	SRC2
OPCODE	DEST	SRC1	IMMEDIATE OPERAND
OPCODE	DEST	IMMEDIATE OPERAND	

Table 9.2   Examples of RISC instruction formats.

register specified by DEST. In some instructions, the last field may denote an immediate operand instead of a register source. For memory accesses, SRC1 specifies the index register and SRC2 specifies the offset. Data is exchanged by the register specified in DEST. For program control instructions, like branches, multiple operand fields may be combined to generate a PC-relative address.

9.2.2   Memory Interfaces

RISC architectures were developed as general-purpose compute engines. As such, they provide support for virtual memory addressing. Because of their high-speed internal clock, it is not cost-effective to connect external memory directly to a RISC core. Instead, a RISC core communicates with external memory via data and instruction caches. Figure 9.2 shows a typical configuration of a RISC-core based architecture. It includes the core processor, a floating-point coprocessor, an instruction cache, a data cache, and a memory-management unit. The memory interface unit usually includes a translation look-aside buffer (TLB) that translates the virtual addresses to physical memory space.

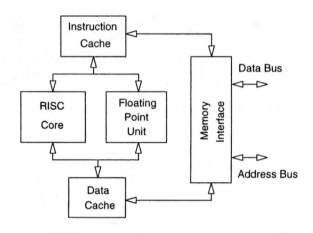

Figure 9.2   Typical RISC-core-based architecture.

9.2.3   Superscalar and VLIW Processors

The original RISC machines had a very simple ALU and no on-chip multiplier. Multiplication was performed using shift-and-add instructions or external math coprocessors. Such a RISC processor would execute on the average one instruction per cycle. Now, many designs include multiple fix-point ALUs, load and store function units, a special branch unit, and on-chip floating-point units, all operating in parallel. These RISC processors attempt to exploit parallelism at the machine-instruction level and use either a superscalar or a VLIW (very long instruction word) architecture. Both of these architectures can execute multiple instructions in parallel, however, they apply completely different methodologies to extract parallelism.

Superscalar architectures rely on an on-chip hardware dispatch unit that uses dynamic scheduling to evaluate data dependencies and issue instructions to the processing units at run time. In contrast, VLIW architectures rely on the compiler to statically schedule the instructions at compile time. VLIW architectures take their name from the fact that at every cycle they initiate multiple operations within a very long instruction word. For example, Figure 9.3 shows a VLIW instruction that initiates four instructions: one load, one store, one integer add, and one floating-point multiplication. Examples of superscalar architectures include the UltraSPARC processor from Sun

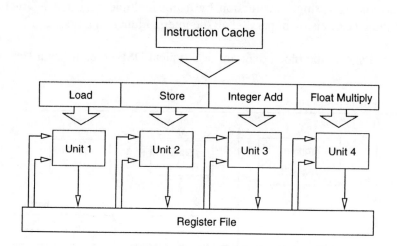

Figure 9.3 Example of instruction-level parallelism in a VLIW architecture.

Microsystems and the RISC family of processors from Hewlett-Packard. Examples of VLIW processors include the Trimedia processor from Philips and the Mpact architecture from Chromatic Research. These two VLIW processors are described in more detail in Chapter 14.

9.3 THE DSP CORE

9.3.1 Architecture

In contrast to RISC architectures that were developed for general-purpose computing, DSP designs are optimized for the efficient implementation of digital signal-processing applications. The core operation of most DSP algorithms is the multiply-accumulate operation $r = b + ax$. Hence, it is not surprising that DSP architectures are optimized for this operation. For example, to execute the operation $r = r + ax$, where r denotes an accumulator register, a typical general-purpose processor would require two memory loads (for reading a and x into local registers), one multiplication, and one addition, for a total of four instructions. A programmable DSP can execute the same operation using a single instruction by using multiple data and instruction buses and by exercising in parallel all the internal function units.

Figure 9.4 shows the block diagram of a typical DSP core. It has a Harvard

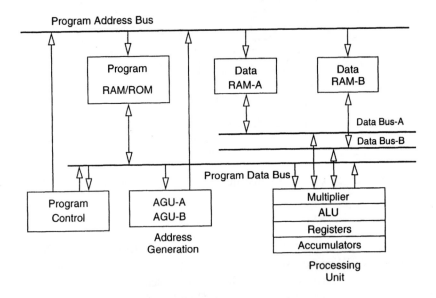

Figure 9.4   Block diagram of a typical DSP core.

architecture, that is, separate data and instruction buses and separate instruction and data memories, a processing unit, two data memories with their own address generation units (AGUs), a program controller, and program memory. The processing unit includes a parallel multiplier, an ALU, accumulators, and registers.

9.3.2   Instruction Format

Most DSPs have a long instruction word (LIW) format. All instructions are of the same size and execute in a single cycle. As in RISC VLIW processors, each field of an instruction controls a separate unit of the architecture. For some instructions, multiple fields may be combined or shared for immediate

ALU OPCODE	AGU-1 OPCODE	AGU-2 OPCODE	PC OPCODE	IMMEDIATE OPERANDS

Figure 9.5 Typical DSP instruction format.

Instruction	Operands	Operation	Comments
RPTB	S	$RE = S$ $RS = PC + 1$	Repeat a block of instructions
MPYF3 ‖ ADDF3	$S1, S2, D1$ $S3, S4, D2$	$D1 = S1 \times S2$ $D2 = S3 + S4$	Multiply- Accumulate

Table 9.3 Examples of DSP instructions.

operands (Figure 9.5). In a typical cycle, a DSP will fetch the next instruction, do two simultaneous data reads and one data write, perform a multiply-accumulate and an ALU operation, and update all data address generation units.

To demonstrate the special nature of DSP instructions, Table 9.3 shows two instructions from the TMS320C30 processor. Other DSPs have similar instructions. The first entry (RPTB) is a repeat-block instruction. It allows a block of instructions to be executed repeatedly without any loop penalty. RE is the repeat-end register and is loaded with the address of the end of the block (S). RS is the repeat-start address register and is loaded with the next value of the program counter. The number of times the loop will be repeated is determined by the value of a separate repeat-counter register.

The second entry in the table shows a multiply-accumulate instruction. The two vertical lines next to ADDF3 show that addition will be executed in parallel with the multiplication. The source operands ($S1 - S4$) can be either register values or pointers to memory locations. If they are address pointers, then addressing operations can be performed in parallel with the multiply-accumulate operation. For example, the value of an address pointer could be incremented or decremented by a specified value.

Efficient data-address generation is very important for DSPs. Hence DSPs are the only processors to incorporate modulo-like arithmetic in data-address generation or hardwired support for bit-reverse address generation. These features are essential for the efficient implementation of filtering or FFT algorithms.

Since most of the DSP algorithms have short processing loops, most of the time-critical data, and usually all the instructions, fit on-chip. Thus, most DSPs have simple memory interface circuitry and no data or instruction caches. For I/O intensive operations they rely on internal or external DMA (direct memory access) units. Architectures with limited internal instruction memory and data paths may also have a loop-instruction cache for storing instructions for a loop operation.

9.4 RISC VERSUS DSP

Both RISC and DSP processors share the same philosophy: simpler is faster. Hence, both have simple addressing schemes and a limited number of instructions. Both rely on fast access to the on-chip data, either in local memory (DSPs) or registers and caches (RISC). In general, DSP cores can provide better performance than RISC cores for signal-processing applications. However, RISC architectures have better support in operating systems and compiler design, and they can be used in a larger range of applications. The VLIW structure of a typical DSP makes it very difficult to program it and to fully utilize its processing power. DSPs have more efficient loop control (zero overhead loops) and data addressing but limited data-space addressing power. In general, companies with long history in DSP design prefer to use and expand their DSP cores for the design of video processors. On the other hand, new developers of video ICs tend to use RISC cores, most probably because they provide better programming flexibility and faster time to market.

9.5 TO PROBE FURTHER

Both RISC and DSP architectures are relatively new. In the mid-1970s, G. Radin led the development of the first RISC-like processor, the IBM 801 minicomputer [213]. However, RISC designs did not find their way into low-cost workstations until the late 1980s, after the success of the RISC and MIPS projects at the University of California at Berkeley [228] and Stanford University [93]. The book on computer architecture by Patterson and Hennesy [201], the leaders of the Berkeley and Stanford projects, covers in depth all the hardware and software design issues in the development of a modern RISC processor. For a detailed description and evaluation of specific RISC architectures, a book edited by Slater [235] has articles on most of the commercially available RISC processors. An early description of the design tradeoffs in VLIW architectures is given by Fisher [59].

In related work, Patel and Douglas [200] describe the architectural features of the Intel i860, and Shacham et al. [229] present the series 32000/EP processor from National Semiconductors. Both of these processors merge DSP-like functionality into a RISC core. The 32000/EP is a 100 MIPS/50 MFLOPS superscalar RISC processor with two integer units, a floating-point unit, and a DSP unit with a 16×16-bit fixed-point multiplier. Special instructions allow the efficient implementation of complex-arithmetic operations, a feature very useful in DSP algorithms like the FFT.

Programmable digital signal processors became very popular after the success-ful introduction of the first TMS320 processor (the TMS32010) from Texas Instruments (TI) in 1982 Since then, TI continues to lead in the DSP market with a complete family of fix-point and floating-point processors [242]. Other major developers of DSPs are AT&T, Motorola, Analog Devices, LSI Logic, Zoran, and SGS-Thomson. Many of them are expanding now in the video processing market.

A very good tutorial article by Lee [142] discusses in detail both the history and the main design features of DSP architectures. A detailed description of the design of a typical DSP processor is also given by Fellman et al. [57]. A recent book edited by Bayoumi and Swartzlander [19] covers the current developments in the design of programmable DSPs, DSP cores, and image processors. For a more detailed description of specific designs, Michalina [168] presents a very good overview of the DSP cores available from SGS-Thomson. The Motorola and TI DSP cores are described in [143] and [131]. For a short description of most commercially available DSPs, *EDN*, a trade magazine published by Cahners Publishing, publishes annually a DSP-chip directory [151].

10

ARCHITECTURES FOR THE DCT

10.1 INTRODUCTION

Due to the importance of the two-dimensional (2-D) DCT in digital image processing, particularly in video compression, various algorithms and architectures have been proposed for its implementation. These methods can be divided into two main categories. The first category includes techniques from linear matrix analysis and decomposition. Lee's fast DCT is a typical example of this category. The second category includes algorithms from polynomial and number theory. For example, Winograd's algorithm fits into this class. Both categories can be further partitioned into algorithms that either use or do not use row-column decompositions. For example, with a row-column decomposition, the 2-D DCT can be computed using one-dimensional DCTs, allowing for simpler implementations. Algorithms that do not use row-column decompositions are more computationally efficient but require complex hardware.

In general, a basis of comparison for the various DCT algorithms is the number of multiplications and additions they require. However, for a VLSI implementation other factors are also important. These factors include complexity of control logic, requirements for memory, size, and power, complexity of interconnect, and efficient implementation of the inverse transform using the same hardware. Polynomial transforms require fewer multiplications but have irregular structure and complex interconnection schemes among processing elements. Furthermore, they are not as computationally efficient for small-size transforms. Hence, these algorithms may be used for a software implementation on a general purpose processor, but they are seldom the basis for dedicated DCT processors in video codecs.

Single-chip video codecs need to integrate the DCT-IDCT hardware with a core processor, an entropy coder, and other circuitry. For such strict size requirements for the implementation of the DCT, two techniques seem to be widely used: (1) vector processing using four parallel multipliers and (2) parallel processing using distributed arithmetic. These designs have a regular structure, have simple control and interconnect, and achieve a good balance between performance and complexity of implementation.

10.2   VECTOR PROCESSING

Using matrix notation, let F denote the output of an $N \times N$ DCT for input f. Then

$$F = CfC^t, \tag{10.1}$$

where C is a matrix with the cosine basis functions, and C^t is the transpose of C. Using a row-column decomposition, F can be computed using 1-D DCT transforms as

$$Y = Cf^t, \tag{10.2a}$$

$$F = CY^t = CfC^t, \tag{10.2b}$$

where Y is an intermediate product matrix. (The columns of Y correspond to the output of the 1-D DCTs of the rows of f.) A 2-D transform can thus be computed by applying 1-D transforms, first on each row of f and then on the transpose of Y. Figure 10.1 shows a block diagram of such an architecture.

For $N{=}8$, (10.2) can be written as

$$Y_{k,l} = \sum_{m=1}^{8} c_{m,l} f_{k,m}, \quad l,k = 1,2,...,8. \tag{10.3}$$

From (10.3), the evaluation of each row of Y requires eight eight-point inner products. However, due to the special properties of the DCT coefficient matrix, it can be shown that

$$Y_{k,l} = \sum_{m=1}^{4} c_{m,l} u_{k,m}, \quad l = 1,3,5,7, \tag{10.4a}$$

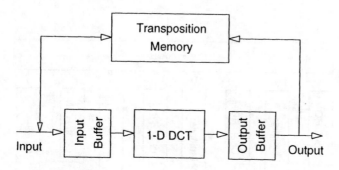

Figure 10.1 Architecture for two-dimensional DCT using row-column transformation.

$$Y_{k,l} = \sum_{m=1}^{4} c_{m,l} v_{k,m}, \quad l = 2, 4, 6, 8, \tag{10.4b}$$

where $u_{k,m} = f_{k,m} + f_{k,8-m+1}$ and $v_{k,m} = f_{k,m} - f_{k,8-m+1}$, for $m = 1$ to 4. The above equations imply that an eight-point DCT can be computed using two four-point DCTs, provided that one preprocess the original data to generate the u and v sequences. (This is usually the first stage in most of the butterfly-based fast DCT algorithms). Using the above formulation, the number of multiplications is reduced by a factor of two and each row of Y can now be computed using eight four-point inner products.

Figure 10.2 shows a block diagram for the implementation of the 1-D DCT using an array of four multipliers. The preprocessor computes the u and v data. Each column-multiplier is a parallel multiplier based on the modified Booth algorithm. The four products from the multipliers, after appropriate scaling, are summed into the output accumulator. A Wallace tree-adder is usually used at this stage. In pipelined mode, this architecture allows a four-input inner product to be computed each cycle, and an eight-point DCT to be computed in eight cycles.

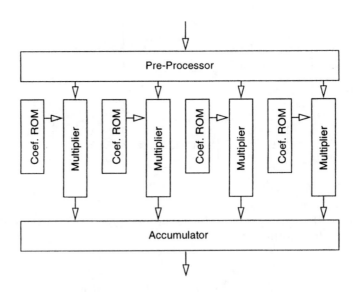

Figure 10.2   Eight-point DCT using four parallel multipliers.

10.3   DISTRIBUTED ARITHMETIC FOR THE DCT

Distributed arithmetic is a technique that allows the hardware implementation of a sum of products without using multipliers. By storing first a finite number of intermediate results, a sum of products can be obtained through repeated additions and shifting operations and without the use of any multiplications. The technique allows the design of signal processors with a reduced gate count and a very regular structure. Hence, it is ideally suited for integrating a DCT processor into a video codec.

10.3.1   Technical Overview

Consider the evaluation of the following sum of products:

$$y = \sum_{m=1}^{N} a_m x_m, \qquad (10.5)$$

where x_m denotes the input data and a_m are fixed coefficients. Assuming (mostly for convenience and without loss of generality) that each x_m is a two's complement binary number with B-bits of precision and that $|x_m| < 1$, then x_m can be expressed as

$$x_m = -x_m^{(0)} + \sum_{j=1}^{B-1} x_m^{(j)} 2^{-j}, \qquad (10.6)$$

where $x_m^{(j)}$ is the jth bit of x_m and has a value of either 0 or 1. (The above formulation implies a decimal point next to the most significant bit ($x_m^{(0)}$)). For example, $1001 = -1 + \frac{1}{8} = -0.875$.

Substituting (10.6) into (10.5) yields

$$y = \sum_{m=1}^{N} a_m \left[-x_m^{(0)} + \sum_{j=1}^{B-1} x_m^{(j)} 2^{-j} \right]. \qquad (10.7)$$

After an interchange of the order of summations,

$$y = \sum_{j=1}^{B-1} \left[\sum_{m=1}^{N} a_m x_m^{(j)} \right] 2^{-j} - \sum_{m=1}^{N} a_m x_m^{(0)}. \qquad (10.8)$$

Let

$$F_N(a, x^{(j)}) = \sum_{m=1}^{N} a_m x_m^{(j)}. \qquad (10.9)$$

Since $x_m^{(j)}$ may only be either 0 or 1, $F_N(a, x^{(j)})$ can take only 2^N possible values. These values can be precomputed and stored in a lookup-table (or ROM). Substituting (10.9) into (10.8), yields

$$y = \sum_{j=1}^{B-1} F_N(a, x^{(j)}) 2^{-j} - F_N(a, x^{(0)}). \qquad (10.10)$$

This is the key expression for the implementation of a sum of products using distributed arithmetic. Multiplication by 2^{-1} corresponds to a right shift by one bit, hence, y in (10.10) can be computed using repeated table look-ups, additions, and shifts.

Figure 10.3 shows a block diagram for the implementation of (10.10). Data is processed bit-serially, the least significant bit first. After each cycle, the output of the accumulator is shifted by one bit. The final sum is computed in B cycles. This circuit is commonly referred to as a ROM-accumulator (RAC).

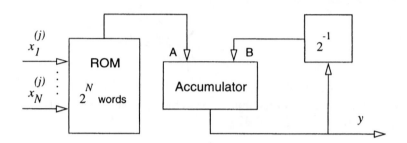

Figure 10.3 Block diagram of a ROM-accumulator unit for the implementation of sum of products using distributed arithmetic.

As an example, let $N = 2$, $B = 4$, $a_1 = 0.125$, $a_2 = -0.75$, $x_1 = -0.625$ (1011), and $x_2 = 0.25$ (0010). Table 10.1 shows the values of $F_2(a, x^{(j)}) = a_1 x_1^{(j)} + a_2 x_2^{(j)}$ for $j = 0, 1, 2$, and 3. For convenience, the decimal point is shown explicitly. From (10.10) and $B = 4$, the summation can be expanded as

$x_1^{(j)}$	$x_2^{(j)}$	$F_2(a, x^{(j)})$
0	0	0
0	1	$a_2 = 1.010$
1	0	$a_1 = 0.001$
1	1	$a_1 + a_2 = 1.011$

Table 10.1 Lookup table data - Example.

$$y = \left[\left[\left[F_2(a, x^{(3)})\right] 2^{-1} + F_2(a, x^{(2)})\right] 2^{-1} + F_2(a, x^{(1)})\right] 2^{-1} - F_2(a, x^{(0)}).$$
(10.11)

Given (10.11), Table 10.2 shows the input and output data flow in the ROM-accumulator unit for this example. After four cycles, the final sum is $y = -0.265625$ (1.101111).

Bit	Input Data		Adder		
j	$x_1^{(j)}$	$x_2^{(j)}$	Input-A	Input-B	Output
3	1	0	0.001	0.000	0.001
2	1	1	1.011	0.0001	1.0111
1	0	0	0.000	1.10111	1.10111
0	1	0	-0.001	1.110111	1.101111

Table 10.2 Example of data flow using distributed arithmetic.

10.3.2 Implementation of the DCT

The design methodology of the previous section can now be directly applied to the hardware implementation of the DCT using distributed arithmetic. For B bits of data precision, let

$$f_{k,m} = -f_{k,m}^{(0)} + \sum_{j=1}^{B-1} f_{k,m}^{(j)} 2^{-j}, \qquad (10.12)$$

and

$$F_8(c_l, f_k^{(j)}) = \sum_{m=1}^{8} c_{m,l} f_{k,m}^{(j)}, \qquad (10.13)$$

where $f_{k,m}^{(j)}$ denotes the jth bit of $f_{k,m}$. From (10.3), using distributed arithmetic, an eight-point DCT can be computed as

$$Y_{k,l} = \sum_{j=1}^{B-1} F_8(c_l, f_k^{(j)}) 2^{-j} - F_8(c_l, f_k^{(0)}), \quad k, l = 1, 2, \cdots, 8. \qquad (10.14)$$

As before, each $Y_{k,l}$ can be computed using the ROM-accumulator unit shown in Figure 10.3, where the size of the ROM is 256 words. The size of the ROM can be reduced to only 16 words if we take into consideration that an eight-point DCT can be computed using two four-point inner-products (see equation (10.4)). In either case, an output will be available after B cycles. However, since each row of Y ($Y_{k,1}, Y_{k,2}, \cdots, Y_{k,8}$) utilizes the same $f_{k,m}$ data, all row elements can be computed in parallel.

Figure 10.4 shows a parallel implementation of the eight-point DCT-IDCT using distributed arithmetic. For each row, first, the f_k data vector is loaded

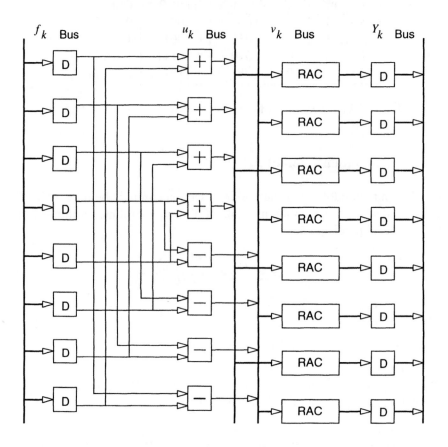

Figure 10.4   Parallel implementation of an eight-point DCT using distributed arithmetic.

into the input registers (denoted by D). Then, data are processed bit-serially by the bit-serial adders or subtractors to generate the u and v data. Two data buses distribute the bits into eight ROM-accumulator units. Each RAC has 16 words of memory for the forward DCT and 16 words for the inverse DCT. After B cycles, the outputs from the RAC units are loaded into the output registers and then transferred sequentially to memory.

10.3.3 Doubling the Speed of Distributed Arithmetic

In most video applications, the bit-length of the input data is usually eight bits; however, to comply with the accuracy requirements specified by the standards, DCT architectures employ a higher internal dynamic range. For 16 bits of precision, an array of eight RAC units can complete an eight-point DCT in 16 cycles. This is half of the input data rate into the DCT unit. The output rate can be increased by either doubling the clock of the DCT unit (not always possible or desirable) or by increasing the processing rate of a RAC unit.

To increase the processing rate in distributed arithmetic, the RAC unit can be modified to process two bits at a time from each operand. For example, (10.11) can be rewritten as

$$y = [F_2(a, x^{(3)})2^{-1} + F_2(a, x^{(2)})]2^{-2} + [F_2(a, x^{(1)})2^{-1} - F_2(a, x^{(0)})].$$
(10.15)

If we can access two $F(\ )$ values simultaneously, then operations within a pair of brackets ([ ]) can be performed in one cycle and y can be computed in two cycles. Similarly, an N-point sum of products (see (10.10)) can be expressed as

$$
\begin{aligned}
y \quad &= \quad \sum_{k=1}^{B/2} F_N(a, x^{(2k-1)})2^{-(2k-1)} \\
&+ \quad \sum_{k=1}^{B/2-1} F_N(a, x^{(2k)})2^{-2k} - F_N(a, x^{(0)}).
\end{aligned}
$$
(10.16)

The above formulation allows odd-numbered and even-numbered bits to be processed in parallel.

Figure 10.5 shows the architecture for a modified RAC unit that processes data at double the speed. Data is still processed bit-serially, but two bits at a time. This design requires an additional adder and twice the ROM size of the original design. For 16-bit data, an array of eight modified RAC units can compute an eight-point DCT in eight cycles; that is, at the same speed as an array of four parallel multipliers.

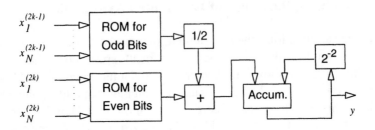

Figure 10.5   Block diagram of a modified RAC unit.

10.4   COMMERCIALLY AVAILABLE DCT PROCESSORS

SGS-Thomson makes three DCT (IDCT) ICs. The STV3200 operates at a maximum pixel rate of 15 MHz and supports block sizes from 4×4 to 16×16. The STV3208 is a dedicated 8×8 DCT IC and operates at pixel rates up to 20 MHz for double precision or up to 27 MHz for single precision. Both ICs use nine bits (two's complement) for input pixel data and 12 bits for coefficient and output data. The IMSA121 can perform 8×8 DCTs, 8×8 IDCTs, or low-pass filtering. It supports pixel rates up to 20 MHz. Its dynamic range is 12 bits for input and output data and 14 bits for the fixed coefficients. Internally, it uses distributed arithmetic with 16 bits of precision.

The Zoran ZR36020 is an 8×8 DCT-IDCT IC. It supports data rates of 15 and 21 Mbytes/s. The Zoran processor complies with the normative requirements of the JPEG standard regarding DCT or IDCT accuracy. It uses 16 bits of precision (two's complement) for the coefficients and either eight or nine bits of precision for the data (unsigned, with or without internal shift by 128, or in two's complement).

From LSI Logic, the L64735 is a DCT processor for 8×8 data blocks. It uses 12-bit signed DCT coefficients and supports a maximum data rate of 30 MHz. For higher data rates, the L64730 DCT processor can operate at 40 MHz. A separate DCT quantization processor, the L64740, is also available.

10.5 TO PROBE FURTHER

The efficient implementation of the 2-D DCT is still an area of active research. Wolter et al. [262] provided the basis of our classification of 2-D DCT algorithms. A detailed description of an architecture that uses two four-multiplier arrays is given in [119]. The authors also provide extensive details on the design and configuration of the transposition memory. A similar four-processor architecture is also used by Ruetz et al. [223] for the DCT processor in a multichip H.261 coder from LSI Logic. The SGS-Thomson chip-set for image compression is described in [13].

The first widely known description of distributed architecture was published in the early 1970s by A. Peled and B. Liu [202]. A tutorial by White [260] provides additional examples and optimization techniques. M. T. Sun et al. were among the first to present a DCT implementation based on distributed architecture [240]. A design that uses the modified RAC unit is presented in [251]. Distributed arithmetic can also be used in butterfly-based data flows. Such a design is described in [45]. The integration efficiency of distributed arithmetic is best demonstrated in the work by K. Aono et al. [10], where a DCT core is part of a single-chip video codec.

11

HARDWARE FOR MOTION ESTIMATION

11.1   INTRODUCTION

Typically, two consecutive frames of a video sequence are very similar. This observation is the basis of motion-compensated coding, where a frame is coded based on its difference from another frame. In practice, frames are divided into blocks. For a block in the current frame (called the *reference* or *source* block), motion estimation is the process of finding a block in another frame that best matches (according to a given criterion) its characteristics. A *motion vector* identifies the position of the best block relative to the reference block. The search for the best matching block is done in a rectangular area (called the *search window*) around the relative position of the reference block. Figure 11.1 shows a search window that extends p pixels to the left and top and $p - 1$ pixels to the right and bottom of an $M \times N$ reference block. This is usually referred to as a $[-p, p - 1]$ search window. In most video applications, the reference block is 16×16 pixels, and the search window is 31×31 pixels ($p = 8$).

There are many criteria and techniques for finding a motion vector. The mean absolute error (MAE) criterion offers a good tradeoff between accuracy and complexity. Under the MAE criterion, for a displacement vector (i, j), the distortion between two blocks is defined as

$$D(i, j) = \sum_{m=0}^{M-1} \sum_{n=0}^{N-1} |r_{m,n} - s_{m+i,n+j}|, \quad i, j \in [-p, p - 1], \qquad (11.1)$$

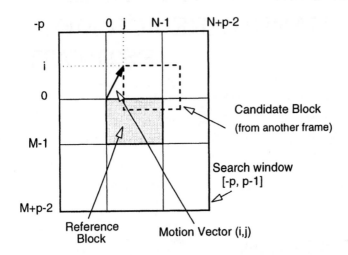

Figure 11.1 Search window in motion estimation with range [-p,p-1].

where r is the reference block and s is a candidate block. The motion vector is the displacement vector for which $D(i, j)$ is minimum. When the reference block is compared with all possible candidate blocks in the search area, the process is called *full-search block matching*.

Suppose the evaluation of $r = r + |a - b|$ requires five operations (two data loads, one addition, one subtraction, and one absolute value), using full-search, at least $5MN(2p)^2$ operations are required for one motion vector. For videoconferencing at 15 fps (using CIF frames, 16 × 16 blocks, and $p = 8$), close to two GOPS are required for motion estimation. Since no single processor can achieve such high throughput, a variety of parallel architectures have been used for the hardware implementation of motion estimators.

11.2 BLOCK MATCHING USING LINEAR ARRAYS

Without loss of generality, let us consider the problem of motion estimation for blocks of 16 × 16 pixels and a search range of [-8,7] pixels. Figure 11.2 shows the pixel coordinates for a 16 × 16 reference block and the corresponding

31×31 search window. Here, we have shifted the coordinate system so that there are only positive pixel indices. During the block-matching search, the

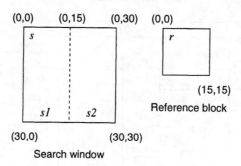

Figure 11.2 Modified pixel coordinates for a 16×16 reference block and its 31×31 search window.

four corner pixels of the search window are used only once, however, all other pixels are used in multiple search positions. For example, pixel $s_{0,15}$ is used in the evaluation of 16 different distortions: $D(0,0), D(0,1), \cdots, D(0,15)$. If the pixels from the search window memory are distributed to multiple processors, then one can compute multiple distortions in parallel.

11.2.1 Motion-vector-based Linear Arrays

Figure 11.3 shows an implementation of a motion estimator using a one-dimensional array processor with 16 elements. Each element in the array can compute a separate distortion value, and 16 distortion values can be computed in parallel. To compute all 256 distortions, the reference block has to be repeatedly scanned and processed 16 times. Since the width of the search window is double the width of the reference block, improved performance can be achieved by using two data ports for the search memory. Port $s1$ receives data from the left half of the search window, and port $s2$ receives data from the right half (see also Figure 11.2). Data from both ports are distributed to all processing elements that select the appropriate input through a multiplexer. We will refer to this design as a motion-vector-based linear array, since each processor computes a separate candidate motion vector (MV). In such arrays,

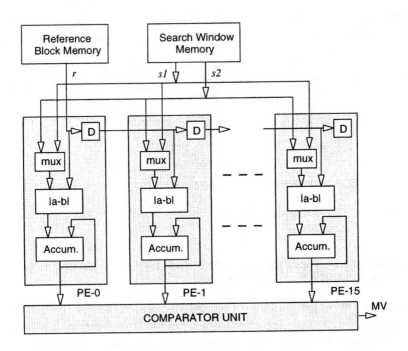

Figure 11.3 Architecture for motion estimation using a 16-processor array.

for best performance, the maximum number of processors depends on the size of the search window, and it is independent of the size of the source block.

Table 11.1 shows in more detail the pixel data flow for the parallel computation of the first 16 distortion values. Pixels from the reference block and the search window are processed sequentially, row by row. For a given vertical displacement i ($i = 0, 1, \cdots, 15$), PE-0 computes $D(i,0)$, PE-1 computes $D(i,1)$, and PE-15 computes $D(i,15)$. Pixels from the reference block move among processors through a series of delay latches. Hence, a pixel that is available to one processor at time t will be available to the next processor at time $t+1$.

Referring to Figure 11.3 and Table 11.1, the operation of this motion estimator can be described as follows. Let acc_j denote the accumulator register for

Cycle Time	Input Data			Processor Inputs		
	r	$s1$	$s2$	**PE-0**	**PE-1**	**PE-15**
0	$r_{0,0}$	$s_{0,0}$		$r_{0,0}, s_{0,0}$		
1	$r_{0,1}$	$s_{0,1}$		$r_{0,1}, s_{0,1}$	$r_{0,0}, s_{0,1}$	
2	$r_{0,2}$	$s_{0,2}$		$r_{0,2}, s_{0,2}$	$r_{0,1}, s_{0,2}$	
$\vdots$		$\vdots$		$\vdots$	$\vdots$	
14	$r_{0,14}$	$s_{0,14}$		$r_{0,14}, s_{0,14}$	$r_{0,13}, s_{0,14}$	
15	$r_{0,15}$	$s_{0,15}$		$r_{0,15}, s_{0,15}$	$r_{0,14}, s_{0,15}$	$r_{0,0}, s_{0,15}$
16	$r_{1,0}$	$s_{1,0}$	$s_{0,16}$	$r_{1,0}, s_{1,0}$	$r_{0,15}, s_{0,16}$	$r_{0,1}, s_{0,16}$
17	$r_{1,1}$	$s_{1,1}$	$s_{0,17}$	$r_{1,1}, s_{1,1}$	$r_{1,0}, s_{1,1}$	$r_{0,2}, s_{0,17}$
$\vdots$		$\vdots$		$\vdots$	$\vdots$	$\vdots$
30	$r_{1,14}$	$s_{1,14}$	$s_{0,30}$	$r_{1,14}, s_{1,14}$	$r_{1,13}, s_{1,14}$	$r_{0,15}, s_{0,30}$
31	$r_{1,15}$	$s_{1,15}$	$s_{0,31}$	$r_{1,15}, s_{1,15}$	$r_{1,14}, s_{1,15}$	$r_{1,0}, s_{1,15}$
$\vdots$		$\vdots$		$\vdots$	$\vdots$	$\vdots$
240	$r_{15,0}$	$s_{15,0}$	$s_{14,16}$	$r_{15,0}, s_{15,0}$	$r_{14,15}, s_{14,16}$	$r_{14,1}, s_{14,16}$
$\vdots$		$\vdots$		$\vdots$	$\vdots$	$\vdots$
255	$r_{15,15}$	$s_{15,15}$	$s_{14,31}$	$r_{15,15}, s_{15,15}$	$r_{15,14}, s_{15,15}$	$r_{15,0}, s_{15,15}$
256			$s_{15,16}$		$r_{15,15}, s_{15,16}$	$r_{15,1}, s_{15,16}$
257			$s_{15,17}$			$r_{15,2}, s_{15,17}$
$\vdots$		$\vdots$				$\vdots$
270			$s_{15,30}$			$r_{15,15}, s_{15,30}$

Table 11.1   Pixel flow for computing the first 16 distortions using a 16-processor array.

processor j. At time $t = 0$, PE-0 computes $acc_0 = |r_{0,0} - s_{0,0}|$. No data are available to the other processors.

At $t = 1$, PE-0 computes $acc_0 = acc_0 + |r_{0,1} - s_{0,1}|$. Pixel $r_{0,0}$ is now available to PE-1, which computes $acc_1 = |r_{0,0} - s_{0,1}|$.

At $t = 15$, pixel $r_{0,0}$ reaches the last processor, and from this time on all processors will be utilized 100 percent.

At $t = 16$, all three input ports are active. Port r receives the second row of the reference block, port $s1$ receives the second row of the left half of the search window, and port $s2$ receives the first row of the right half of the search window. PE-0 computes $acc_0 = acc_0 + |r_{1,0} - s_{1,0}|$ and PE-1 to PE-15 compute $acc_j = acc_j + |r_{0,16-j} - s_{0,16}|$, $j = 1, 2, ..., 15$.

At $t = 255$, PE-0 processes the last pixel from the reference block, and $D(0,0)$ will be available to the comparator unit at the beginning of the next clock cycle. Processing will continue in processors PE-1 to PE-15, and at each of the next 15 cycles a new distortion value will be generated. Since the distortion values are generated sequentially, the evaluation of their minimum can be computed with a single comparator.

From Table 11.1, the first 16 distortion values will be computed in 271 clock cycles. However, as shown in Table 11.2, processing for the next set of distortions $(D(1,0), D(1,1), \cdots, D(1,15))$ can begin at $t = 256$, as soon as $D(0,0)$ is computed. To compute the ith set of 16 distortion values, the

Cycle	Input Data			Processor Inputs			
Time	r	$s1$	$s2$	PE-0	PE-1		PE-15
256	$r_{0,0}$	$s_{1,0}$	$s_{15,16}$	$r_{0,0}, s_{1,0}$	$r_{15,15}, s_{15,16}$		$r_{15,1}, s_{15,16}$
257	$r_{0,1}$	$s_{1,1}$	$s_{15,17}$	$r_{0,1}, s_{1,1}$	$r_{0,0}, s_{1,1}$		$r_{15,2}, s_{15,17}$
$\vdots$		$\vdots$		$\vdots$	$\vdots$		$\vdots$
270	$r_{0,14}$	$s_{1,14}$	$s_{15,30}$	$r_{0,14}, s_{1,14}$	$r_{0,13}, s_{1,14}$		$r_{15,15}, s_{15,30}$
271	$r_{0,15}$	$s_{1,15}$	$s_{15,31}$	$r_{0,15}, s_{1,15}$	$r_{0,14}, s_{1,15}$		$r_{0,0}, s_{1,15}$

Table 11.2   Modified pixel flow for fully pipelined motion estimation.

reference block is scanned as before, but input from the search window starts at the ith row, with pixel $s_{i,0}$. Following this computational pipeline, all 256 distortion values will be computed in $16(16 \times 16) + 15 = 4,111$ cycles.

Consider now block sizes of 8×8 pixels. For the same search range, full-search motion estimation requires one-fourth of the operations required for block sizes of 16×16, but a frame has four times more 8×8 blocks than 16×16 blocks. Even though the computational requirements for each frame are independent of the block size, this may not be true for the I/O, especially in multiprocessor implementations. For example, for an 8×8 block and a [-8,7] search range, the width of the search window is three times larger than the width of the reference block. Hence, for 100 percent processor efficiency, the architecture of Figure 11.3 will require three (instead of two) parallel data ports for the search window.

11.2.2 Source-pixel-based Linear Arrays

A disadvantage of motion-vector-based arrays is that they require a wide data path between the array and the comparator unit. A different approach is shown in Figure 11.4, where each processor is assigned a pixel of the reference block. At each clock cycle, each processor computes the absolute value of the

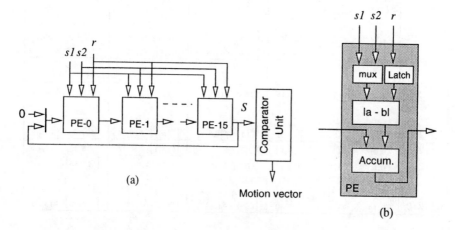

Figure 11.4 Source-pixel based array for motion estimation. (a) Linear array; (b) processing element.

difference between two pixels, adds this value to the sum received from its predecessor, and passes the new sum to the next processor. The last processor in the array outputs the total sum. In such architectures, the total number of processors depends on the number of pixels in the reference block, and it is independent of the size of the search area. Furthermore, there is a single data path between the array and the comparator unit.

Table 11.3 shows in more detail the flow of operations for the computation of the first 16 distortion values in PE-0, PE-1, and PE-15. Referring to Table 11.3

and Figure 11.4, the operations can be described as follows. Let acc_j denote

Cycle Time	Input Data			Processor Inputs		
	r	$s1$	$s2$	PE-0	PE-1	PE-15
0	$r_{0,0}$	$s_{0,0}$		$r_{0,0}, s_{0,0}$		
1	$r_{0,1}$	$s_{0,1}$		$[r_{0,0}], s_{0,1}$	$r_{0,1}, s_{0,1}$	
2	$r_{0,2}$	$s_{0,2}$		$[r_{0,0}], s_{0,2}$	$[r_{0,1}], s_{0,2}$	
$\vdots$	$\vdots$	$\vdots$		$\vdots$	$\vdots$	
14	$r_{0,14}$	$s_{0,14}$		$[r_{0,0}], s_{0,14}$	$[r_{0,1}], s_{0,14}$	
15	$r_{0,15}$	$s_{0,15}$		$[r_{0,0}], s_{0,15}$	$[r_{0,1}], s_{0,15}$	$r_{0,15}, s_{0,15}$
16	$r_{1,0}$	$s_{1,0}$	$s_{0,16}$	$r_{1,0}, s_{1,0}$	$[r_{0,1}], s_{0,16}$	$[r_{0,15}], s_{0,16}$
17	$r_{1,1}$	$s_{1,1}$	$s_{0,17}$	$[r_{1,0}], s_{1,1}$	$r_{1,1}, s_{1,1}$	$[r_{0,15}], s_{0,17}$
$\vdots$	$\vdots$	$\vdots$		$\vdots$	$\vdots$	$\vdots$
30	$r_{1,14}$	$s_{1,14}$	$s_{0,30}$	$[r_{1,0}], s_{1,14}$	$[r_{1,1}], s_{1,14}$	$[r_{0,15}], s_{0,30}$
31	$r_{1,15}$	$s_{1,15}$	$s_{0,31}$	$[r_{1,0}], s_{1,15}$	$[r_{1,1}], s_{1,15}$	$r_{1,15}, s_{1,15}$
$\vdots$	$\vdots$	$\vdots$		$\vdots$	$\vdots$	$\vdots$
240	$r_{15,0}$	$s_{15,0}$	$s_{14,16}$	$r_{15,0}, s_{15,0}$	$[r_{14,1}], s_{14,16}$	$[r_{14,15}], s_{14,16}$
$\vdots$	$\vdots$	$\vdots$		$\vdots$	$\vdots$	$\vdots$
255	$r_{15,15}$	$s_{15,15}$	$s_{14,31}$	$[r_{15,0}], s_{15,15}$	$[r_{15,1}], s_{15,15}$	$r_{15,15}, s_{15,15}$
256	$r_{0,0}$	$s_{1,0}$	$s_{15,16}$	$r_{0,0}, s_{1,0}$	$[r_{15,1}], s_{15,16}$	$[r_{15,15}], s_{15,16}$
257	$r_{0,1}$	$s_{1,1}$	$s_{15,17}$	$[r_{0,0}], s_{1,1}$	$r_{0,1}, s_{1,1}$	$[r_{15,15}], s_{15,17}$
$\vdots$	$\vdots$	$\vdots$	$\vdots$	$\vdots$	$\vdots$	$\vdots$
270	$r_{0,14}$	$s_{1,14}$	$s_{15,30}$	$[r_{0,0}], s_{1,14}$	$[r_{0,1}], s_{1,14}$	$[r_{15,15}], s_{15,30}$

Table 11.3 Pixel flow for computing the first 16 distortions using a source-pixel based 16-processor array.

the value of accumulator register for processor j. A sum that is computed by one processor at time t will be available to the next processor at time $t + 1$. At time $t = 0$, the input mux to the array is set to 0, PE-0 computes $acc_0 = |r_{0,0} - s_{0,0}|$, and the $r_{0,0}$ value is stored in PE-0 for future use (stored values inside PEs are denoted by square brackets).

At $t = 1$, PE-0 computes $acc_0 = |r_{0,0} - s_{0,1}|$, PE-1 computes $acc_1 = acc_0 + |r_{0,1} - s_{0,1}|$, and the value of $r_{0,1}$ is stored in PE-1 for future use.

At $t = 15$, the accumulated sum reaches the last processor, and from this time on all processors will be utilized 100 percent. Let

$$D_m(i,j) = \sum_{n=0}^{15} |r_{m,n} - s_{m+i,n+j}| \qquad (11.2)$$

be the m-th partial sum needed to compute $D(i,j)$. From (11.1) and (11.2),

$$D(i,j) = \sum_{m=0}^{15} D_m(i,j), \quad m = 0, 1, \cdots, 15. \qquad (11.3)$$

From Table 11.3, at the end of $t = 15$, the output of the linear array will be $S = \sum_{n=0}^{15} |r_{0,n} - s_{0,n}| = D_0(0,0)$.

At $t = 16$, the input mux to the linear array is set to S. PE-0 computes $acc_0 = D_0(0,0) + |r_{1,0} - s_{1,0}|$ and PE-1 to PE-15 compute $acc_j = acc_{j-1} + |r_{0,j} - s_{0,16}|, j = 1, 2, ..., 15$. At the end of $t = 16$, $S = D_0(0,1)$.

At $t = 255$, PE-15 starts processing the last pixel from the reference block $(r_{15,15})$ and the complete $D(0,0)$ will be available to the comparator unit at the beginning of the next clock cycle. From then on, until $t = 270$, a new distortion value will be generated at each cycle. Since the distortion values are generated sequentially, the evaluation of their minimum can be computed with a single comparator. In parallel, at $t = 256$, the input mux to the linear array is set to 0, and we begin processing for a new set of distortion values ($D(1,0)$ to $D(1,15)$). As in the motion-vector-based architecture, all 256 distortion values will be computed in 4,111 cycles.

11.3   HARDWARE FOR ONE-BIT MOTION ESTIMATION

In traditional block-based motion estimation schemes, motion estimation is performed at full pixel resolution, and the distance between two blocks is measured using either the minimum absolute error (l_1 metric) or the minimum mean square error (l_2 metric). Evaluating either of these metrics is computationally expensive, irrespective of the search strategy. Considerable savings can be realized if motion estimation is performed on a binary representation of the image blocks. Under this scheme, both the current and the reference frames are transformed first to frames of binary-valued pixels. Then, one can apply standard search strategies, and the l_1 metric reduces to computing the exclusive-or of a sequence of bits and adding up the number of ones in the result. One-bit motion estimation strategies are amenable to both single-processor- and multi-processor-based implementations. In this section we consider the design and performance of a linear-array architecture for one-bit block-based motion estimation.

Assuming again a [-8, 7] search range, Figure 11.5 shows the pixel coordinates for a 16×16 source block and a 31×31 search window. We assume that the

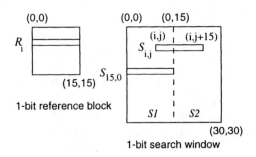

Figure 11.5 Pixel coordinates for a binary 16×16 source block and a [-8, 7] search range.

transform to convert pixels to binary data has been completed and all pixels are now only binary valued. For example, if $r_{i,j}$ denotes a pixel in the source block, then $r_{i,j}$ is either zero or one.

Let $R_i = [r_{i,0}, r_{i,1}, \cdots, r_{i,15}]$ denote the i-th row of the search block, and $S_{i,j} = [s_{i,j}, s_{i,j+1}, \cdots, s_{i,j+15}]$ denote a 16-bit vector from the search window, starting from location (i, j), where i and $j \in [0, 15]$. The problem of one-bit motion estimation can be expressed as finding the smallest distortion

$$D(i,j) = \sum_{m=0}^{15} f(R_m, S_{m+i,j}), i, j \in [0, 15], \qquad (11.4)$$

where

$$f(R_m, S_{m+i,j}) = \sum_{n=0}^{15} r_{m,n} \otimes s_{m+i,n+j} \qquad (11.5)$$

and $\otimes$ denotes the exclusive-or operation. In effect, the $f(\ )$ function computes the number of bits for which there is a match between the R and S binary vectors.

From (11.4), each S vector is used in the computation of multiple distortion values. For example, $S_{15,0}$ is being used in the computation of 16 distortions, namely, $D(0,0)$, $D(1,0)$, ..., $D(15,0)$. Hence, as in traditional motion-estimation schemes, multiple distortions can be computed in parallel.

For a hardware implementation, the designs of the previous section can easily be modified to support one-bit motion estimation schemes. For example, Figure 11.6 shows such an implementation for an array of 16 processors. Each processor now operates on 16-bit vectors (R and either $S1$ or $S2$) instead of on 8-bit pixels. The $f(\ )$ function, defined in (11.5), is computed using a 16-bit exclusive-or array, a dual-port lookup table (LUT) with 256 entries, and a 4-bit adder. Half of the xor-array operates on the eight most significant bits of the R and S vectors, and the other half on the eight least significant bits. The lookup table yields the total number of ones (or matches) at the output of the most- and least-significant halves of the exclusive-or array. These results are added to compute the total number of matches.

Table 11.4 shows in more detail the data flow of operations on the modified processors PE-0, PE-1, and PE-15 for the computation of the first 16 distortion values. At time $t = 0$, only PE-0 is active, with inputs the binary vectors R_0 and $S_{0,0}$. At $t = 1$, PE-0 processes R_1 and $S_{1,0}$, and PE-1 processes R_0 and $S_{1,0}$. Following this approach, $D(0,0)$, in PE-0, will be ready after $t = 15$, and all the first 16 distortion values will be computed in 16+15 cycles. However, as shown in Table 11.4, by using two ports for the search memory, processing of the next set of distortions can begin at $t = 16$. As shown in Figure 11.6, a multiplexor in each processor selects the appropriate input from the search memory. The complete set of 256 distortion values can then be computed in $16 \times 16 + 15 = 271$ cycles. From then on, a new set of distortion values can be computed every 256 cycles. In contrast, the traditional architecture of Figure 11.3 requires 4,096 cycles. Thus, in fully-pipelined mode, the one-bit transform allows for a 16:1 speed improvement. This is consistent with the

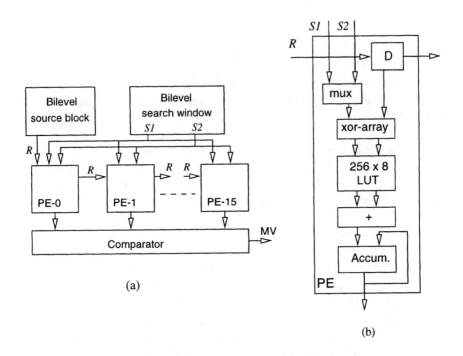

(a)

(b)

Figure 11.6 Architecture for one-bit motion estimation. (a) Linear array; (b) processing element.

fact that at each cycle we now process 16 binary pixels instead of one 8-bit pixel.

Consider now the case of motion estimation using a search range of [-16, 15] pixels. Then, the one-bit search window is 47 × 47 pixels and we need to compute 1,024 distortion values. Since this architecture can compute 16 distortion values in 16 cycles, all 1,024 distortion values can be computed in 16 × 64 + 15 = 1,039 cycles.

Cycle Time	Input Data			Processor Inputs		
	R	$S1$	$S2$	**PE-0**	**PE-1**	**PE-15**
0	R_0	$S_{0,0}$		$R_0, S_{0,0}$		
1	R_1	$S_{1,0}$		$R_1, S_{1,0}$	$R_0, S_{1,0}$	
2	R_2	$S_{2,0}$		$R_2, S_{2,0}$	$R_1, S_{2,0}$	
$\vdots$		$\vdots$		$\vdots$	$\vdots$	
14	R_{14}	$S_{14,0}$		$R_{14}, S_{14,0}$	$R_{13}, S_{14,0}$	
15	R_{15}	$S_{15,0}$		$R_{15}, S_{15,0}$	$R_{14}, S_{15,0}$	$R_0, S_{15,0}$
16	R_0	$S_{16,0}$	$S_{0,1}$	$R_0, S_{0,1}$	$R_{15}, S_{16,0}$	$R_1, S_{16,0}$
17	R_1	$S_{17,0}$	$S_{1,1}$	$R_1, S_{1,1}$	$R_0, S_{1,1}$	$R_2, S_{17,0}$
$\vdots$		$\vdots$	$\vdots$	$\vdots$	$\vdots$	$\vdots$
30	R_{14}	$S_{30,0}$	$S_{14,1}$	$R_{14}, S_{14,1}$	$R_{13}, S_{14,1}$	$R_{15}, S_{30,0}$
31	R_{15}		$S_{15,1}$	$R_{15}, S_{15,1}$	$R_{14}, S_{15,1}$	$R_0, S_{15,1}$

Table 11.4　Pixel flow for computing the first 16 distortions using a 16-processor array for one-bit motion estimation.

11.4　SUB-PEL MOTION ESTIMATION

Many video applications require motion estimation with half-pel or quarter-pel resolution. Sub-pel motion estimation can be performed in two stages. The first stage computes a motion vector with integer pixel resolution. The second stage refines the resolution to a sub-pel level. For example, for half-pel resolution, the second stage needs to perform 16 additional searches, corresponding to all possible horizontal and vertical displacements by -1, -0.5, 0, and 0.5 pixels. For quarter-pel resolution, 64 additional searches are required per motion vector. Since input data blocks have integer-only pixel resolution, sub-pel motion estimation requires additional circuitry for pixel interpolation.

Figure 11.7 shows a block diagram of a sub-pel motion estimator. Both the integer-resolution and the sub-pel-resolution motion estimation blocks can be designed using linear array processors, similar to the one we described in the previous sections. For pixel interpolation, Figure 11.8 shows a block diagram of a circuit that can be used for either half-pel or quarter-pel interpolation. In a motion estimator, two of these circuits should be used, one for interpolating between rows and one for interpolating between columns.

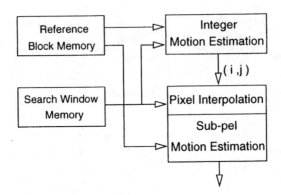

Figure 11.7 Architecture for sub-pel motion estimation.

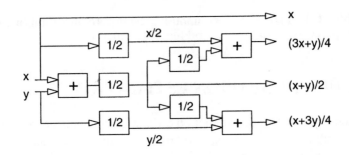

Figure 11.8 Circuit for quarter-pel interpolation.

11.5   IMPLEMENTATION EXAMPLES

The computational power of a multiprocessor motion estimator depends on the number of processors and the number of I/O ports. As the next four design examples demonstrate, the architectures we described in this chapter can easily be modified to support higher data rates and different coding requirements.

11.5.1   A 32-processor Array from LSI Logic

The L64720 from LSI Logic is a general-purpose motion estimation IC. It supports either 16 × 16 blocks and a [-8,7] search range or 8 × 8 blocks and a [-4,3] range. Multiple ICs can be used to increase either the block size or the search range.

The core of the L64720 is a linear processor array similar to the one shown in Figure 11.3, except that it includes 32 processors instead of 16. The L64720 has three input ports: one for the reference data and two for the search window data. On-chip memories double buffer the input and output data to minimize the main memory bandwidth requirements.

A motion vector is computed in 229 cycles for 8 × 8 blocks and in 2,237 cycles for 16 × 16 blocks. At 30 MHz, a single IC can process CIF frames and 16 × 16 blocks at 30 fps.

11.5.2   A 2-D Processor Array from SGS-Thomson

Like the IC from LSI Logic, the STI3220 from SGS-Thomson is a general-purpose motion estimator. The block height can be either eight or 16 pixels, and the block width can be any number that is a multiple of four. However, for eight-bit pixels, internal precision considerations restrict the maximum practical block size to 16 × 16. The maximum search range is [-8,7] pixels.

The STI3220 computes distortion values using a 2-D processing array with 256 elements. Each processing element computes a single distortion. The motion vector is computed using 16 comparators that operate in parallel. The processing array has four data ports: one for the reference block and three for the search window. As shown in Figure 11.9, the search window is divided into six blocks. The three left-most subwindows have a width of 15 pixels. The right-most subwindows are as wide as the reference block. Motion estimation is performed in two operational cycles. In the first cycle (initialization), the three left-most subwindows are loaded, but no computations are performed. In the second cycle (block sequence), the remaining three subwindows are loaded in parallel with the reference block. During this cycle, the processor computes the 256 distortions and their minimum.

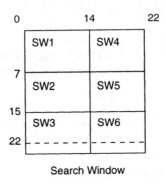

Figure 11.9   Subdivision of the search window for 8 × 8 blocks.

In video-frame processing, two consecutive (and on the same scan line) blocks share half of their search window pixels. For such blocks, only half of the search window memory needs to be updated for each motion vector. After a single initialization cycle in the beginning of each scan line, a motion vector can be computed every $M \times N$ cycles, the time required to load each reference block. The STI3220 supports pixel rates of up to 18 MHz. Both the LSI Logic and the SGS-Thomson ICs are best suited for H.261 implementations.

11.5.3   A Half-pel Motion Estimation IC for MPEG-1

Figure 11.10 shows a block diagram of an interframe prediction IC from NEC. This IC is part of a three-chip set for MPEG-1 video coding. The motion estimator part of the IC computes motion vector searches in a [-15,15] search window and supports predictive coding for both P-frames and B-frames. For half-pel motion estimation, this design uses a two-step search strategy. The first step is performed on a $\frac{1}{4}$ subsampled area and provides a motion vector with 2-pel accuracy. This stage uses a 16-processor array and a 24 × 24 local register array. Using the 18 × 18 window obtained from the first step, the second step generates a motion vector with half-pel accuracy. This stage uses another 16-processor array and a 18 × 18 local register array. Frames for I-pictures and P-pictures are stored on external SRAM, which is accessed via an on-chip frame controller and a local address generator.

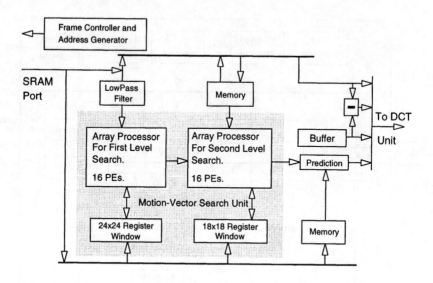

Figure 11.10   Block diagram of a sub-pel motion-estimation IC from NEC.

11.5.4   A Motion Estimation and Compensation Processor for MPEG-2

Motion estimation in MPEG-2 requires additional control and processing resources due to increased data rates, larger frame sizes, multiple prediction modes (field, frame, and dual prime), and the need to generate half-pel accurate motion vectors. Figure 11.11 shows ENC-M, a motion estimation and compensation processor from NTT for real-time MPEG-2 encoding.

From Figure 11.11, ENC-M includes a RISC-based controller, two memory interface modules, and several functional units for motion estimation and compensation. The internal RAM is divided into two buffers; one for the current macroblock data and one for the reference frame data. Incoming video data are converted from a 4:2:2 format to a 4:2:0 format and stored in an external VRAM.

Motion vectors are computed using a hierarchical block-matching algorithm. First, the subsampling filter module (SFM) subsamples the data by a factor of

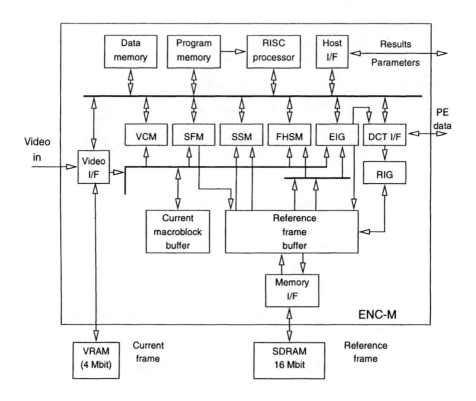

Figure 11.11   Block diagram of ENC-M from NTT.

four and stores them in local memory. Second, the subsampled-pel precision search module (SSM) performs a block-matching search for the first stage of the hierarchical algorithm. The SSM unit includes a search-pixel-based processing array in a 8×4 configuration. Using an 8×4 reference block and a $\pm 16 \times \pm 8$ search range, SSM yields motion vectors with a 2-pel precision. Finally, the full-pel and half-pel motion estimation unit (FHSM) performs block-matching operations for the second and third level of the hierarchical search algorithm. FHSM includes a motion-vector-based 3×3 array for full-pel precision motion estimation and an 8-processor array for half-pel motion estimation. The variance calculation module (VCM) calculates variance values for blocks in the current macroblock. These values are used for selecting the prediction and coding (intra or inter) modes.

ENC-M provides additional support for motion compensation via the error image generation module (EIG), the reconstructed image generation module (RIG), and the DCT interface module (DCT I/F). ENC-M operates at 81 MHz. For NTSC resolution video, it can compute either four field motion vectors in a $\pm 48.5 \times \pm 25.5$ search range or a single frame motion vector in a ± 32.5 search range.

11.6 TO PROBE FURTHER

Algorithms for full-search motion estimation are ideally suited for parallel implementations due to the regularity and the concurrency of operations. The architecture of Figure 11.3 was first described by Yang et al. [266]. Using a formal methodology for mapping the full-search motion-estimation algorithm into systolic arrays, Komarek and Pirsch [134] propose several 1-D and 2-D architectures. For example, for $N \times N$ blocks and displacement area $[-p, p - 1]$, an $N \times N$ processor array with N data ports can compute a motion vector in $(2p) \times (N + 2p - 1)$ cycles. Low-cost implementations require the minimum possible number of I/O ports. An N^2 systolic array by Hsieh and Lin [99] requires only two input data ports and can compute a motion vector in $(N + 2p - 1)^2 + 2 + log_2 N$ cycles. For HDTV rates, J.H. Lee et al. [144] present the design of a motion-estimation IC using a 2-D processing and register array. This IC has 16 input ports (eight ports for the reference frame and eight ports for the search window) and eight output ports. For 8×8 blocks, a horizontal search range of [-32,31], and a vertical search range of [-8,7], the authors use 16 chips. The maximum operating speed of each IC is 80 MHz. If the block size is $N \times N$, the search area is $[-p, p - 1]$, and $N = 2p$, then Yeo and Hu [267] prove that an $N \times N$ array of processor requires only three inputs and a motion vector can be computed in N^2 cycles.

The motion estimator from LSI Logic is described in [223]. A detailed description of the SGS-Thomson motion estimator is given by Artieri et al. [13], [44]. The NEC chip-set and the ENC-M processor are described in [241] and [238], respectively. Other examples of motion-estimation ICs for MPEG-2 can be found in [107] and [207].

The principles of designing full-search motion estimators can also be applied to the design of motion estimators that use different search algorithms. Jehng et al. [122] derive systolic architectures for the three-step hierarchical search algorithm. For 16×16 blocks and a [-7, +7] search range, each of the three stages of the hierarchical algorithm requires eight processors. Ogura et al. [187] describe an architecture for a matching criterion that is based on horizontal and vertical projections of the blocks to be compared. By modeling generic motion estimation algorithms as a tree search, Lin et al. [155] developed a programmable motion estimation IC. Here different search strategies are implemented through macrocommands that can be executed on-chip using a programmable 8×8 processor array. At 66 MHz, this IC can process CIF frames in realtime and at a fraction of the power (0.3 W) required by traditional hardwired architectures.

The discussion on architectures on one-bit motion estimation is based on recent work by Natarajan et al. [183]. A binary block-matching architecture is also described by Mizuki et al. [171]. In [148], Lee et al. describe an architecture based on a new two-bit matching criterion. Unlike the approach in [183], reference and search blocks are quantized to four levels, thereby improving motion-vector estimation.

12

HARDWARE FOR ENTROPY CODING

12.1 INTRODUCTION

The entropy coder is the last stage in the encoding pipeline of JPEG, H.261, H.263, and MPEG. Unlike transform-based coding, where compression is lossy, entropy coding is lossless. The entropy coder consists of two main data compression/decompression units: a run-length coder (RLC) and a variable-length coder (VLC).

The RLC encoder compresses an input stream by representing consecutive zeros by their run-length. When a zero run is present, the RLC counts the number of consecutive zeros until it reaches the last zero or the maximum zero run-length. The RLC decoder reverses the process by generating the appropriate number of zeros between two nonzero data. An RLC coder can easily be implemented using a counter, registers, and a few logic gates.

The VLC encoder maps the input source data into codewords of variable length, concatenates them together, and segments them into 16-bit words. Compression is achieved by assigning short codewords to input symbols of high probability and long codewords to input symbols of low probability. The expectation is that the average size of a codeword will be shorter than the size of the source input. For a given source-probability distribution, a Huffman entropy coder is optimal; that is, the average length of a codeword approaches the theoretical minimum, the entropy of the source.

Among the standards, H.261, MPEG and the baseline implementation of the JPEG specify only Huffman coding for the VLC coder. For improved compression, at a higher computational cost, JPEG also supports an arithmetic coder that is based on IBM's Q-Coder. Arithmetic coding may yield better compression ratios, but due to its complexity and the patent-related issues, most of the hardware JPEG variable-length coders are based on Huffman coding.

In Huffman coding, encoding can be done simply via lookup-tables. The coding tables are designed based on the statistics of the input source. The sizes and the contents of these tables are application dependent. For example, H.261 uses five VLC tables. In baseline JPEG, only the DCT coefficients are coded, and up to four tables can be used: two for the DC coefficients (one for luminance and one for chrominance data) and two for the AC coefficients.

Decoding is much harder, because codewords have variable length, and the receiver has no prior knowledge of the boundaries between two consecutive codewords. Huffman coding has the property that no codeword is a prefix of another code, thus each source symbol corresponds to a unique leaf in a binary decoding tree. For n symbols, a Huffman tree has $n - 1$ nodes. Decoding can be performed by tracing the decoding tree until a leaf of the tree is reached. Figure 12.1 shows an example of a VLC coding table with the corresponding decoding tree. The nodes of the tree are numbered from 0 to 6. From Figure 12.1, given the input sequence *bga*, the output of the VLC coder will be 011111000. In the decoder, given the encoded stream 110001110, the corresponding output will be the sequence *eaf*.

Both the encoding and the decoding processes can be implemented using finite-state machines that process one bit at a time. However, bit-serial architectures may not provide adequate throughput for real-time video applications. Both bit-serial and parallel implementations of VLC coders will be examined next.

12.2 ENCODER IMPLEMENTATION

As mentioned before, a variable-length encoder maps input data of fixed length into codewords of variable length. Consecutive codewords are concatenated together and the output is segmented again into words of fixed length (usually

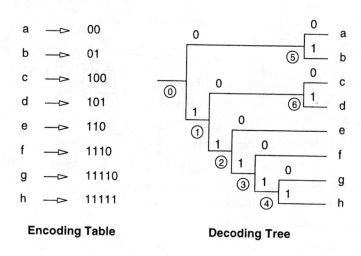

Figure 12.1 Example of a Huffman encoding table with the corresponding decoding tree.

16-bits). The mapping process can be performed either through table look-ups or bit-serially by tracing a Huffman encoding tree. Recently, Lei and Sun proposed a VLC encoder implementation that uses parallel operations to perform the encoding, the concatenation, and the codeword segmentation in one cycle, regardless of the length of a codeword.

Figure 12.2 shows a simplified block diagram of the Lei-Sun VLC encoder. A set of programmable logic arrays (PLAs) stores the Huffman tables and the length of each codeword. Instead of PLAs, one can also use ROM, RAM, or content-addressable (CAM) memories. The reason content-addressable memories work is related to the fact that no Huffman codeword is a prefix of another codeword. Two 16-bit registers (the upper and lower registers) are used to buffer the output data. The length of each codeword is accumulated in a four-bit adder whose output controls the operation of a barrel shifter. The barrel shifter places the output from the PLA next to the end of the current bit stream. An OR-logic operation then cascades the codeword with the existing bit stream. When the sum of the codelengths is larger than 15, the adder overflows and the carry-out bit is set to one. This indicates that a new 16-bit output is ready. The VLC encoder outputs the contents of the upper

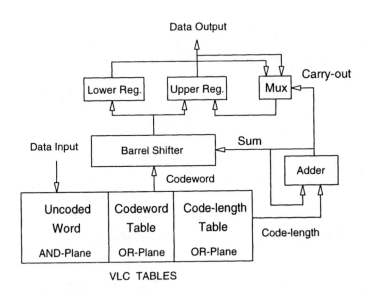

Figure 12.2   Block diagram of the Lei-Sun VLC encoder.

register, the content of the lower register is moved into the upper register, and operations continue as before.

Table 12.1 shows an example of the operation of the VLC coder for the codebook of Figure 12.1 and the input sequence *g, b, a, c, b, f, d*. The output of the barrel shifter is shown underlined. After the first input, *g*, the upper register is loaded with the corresponding codeword (11110), and the adder is incremented by five, the length of 11110. For *b*, the output of the VLC table, 01, is shifted by five bits and is appended to the prior codeword. The adder is incremented by two. After the input symbol *f*, the adder overflows. The carry-out bit is set to one and triggers an output of the data stored in the upper register (1111001001000111). Note that part of the output of the barrel shifter is stored in the lower register. After the output of the data in the upper register, the lower register is loaded into the upper register, and the operation continues as before.

Input	Upper Register	Lower Register	Sum	Carry Out
g	1111000000000000	0000000000000000	5	0
b	1111001000000000	0000000000000000	7	0
a	1111001000000000	0000000000000000	9	0
c	1111001001000000	0000000000000000	12	0
b	1111001001000100	0000000000000000	14	0
f	1111001001000111	1000000000000000	2	1
d	1010100000000000	0000000000000000	5	0

Table 12.1   Example of operation of the VLC encoder.

12.3   DECODER IMPLEMENTATION

The decoding of VLC codes is far more challenging than the encoding, because there are no predefined boundaries between codewords in the received stream. In this section we present two classes of VLC decoders: constant-input-rate decoders and constant-output-rate decoders. The constant-input-rate decoder processes input bits at a fixed rate, but codewords are decoded at a variable output rate. The constant-output-rate decoders have a variable input bit rate, but they decode a fixed number of codewords per cycle.

12.3.1   Constant-Input-Rate Decoders

The simplest constant-input-rate decoder processes the input bit-serially, one bit a time. Starting from the root, it traverses the branches of the decoding tree until it reaches a terminal node. At a terminal node, the codeword is fully decoded, and the corresponding symbol can be read at the output of the decoder. The process then restarts from the root of the tree.

This process can also be described by a model of a finite-state machine (FSM). Each node of the tree corresponds to a state in the FSM. The output of a state is either a decoded symbol or a pointer to another state. A single bit can indicate whether a state is a terminal node or not. The total number of states is one less than the number of the nodes. If the processing delay for decoding one bit is t_d seconds, then the maximum throughput of such an implementation is

$\frac{1}{t_d}$ bits/s. For a tree with n symbols, the average decoding time per codeword will be $log_2 n$ cycles.

Figure 12.3 shows a block diagram of a Huffman decoder using a ROM-based implementation of an FSM. The throughput of this architecture depends on

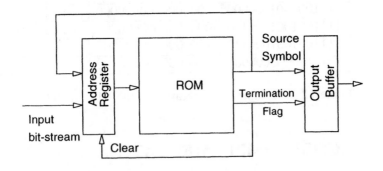

Figure 12.3   Block diagram of a bit-serial Huffman decoder.

the read cycle of the memory. Each ROM entry has two fields. The first field is either the next state or a decoded symbol. The second, one-bit, field (or termination flag) distinguishes between the two cases. The address of the ROM consists of two fields: the present state and the current input bit. When a codeword is decoded, the termination flag is 1, and it resets the input state address to zero. As an example of such a ROM, Table 12.2 shows the ROM entries for the decoding tree shown in Figure 12.1. In this example, a codeword will be decoded in $log_2 8 = 3$ cycles on average.

For improved performance, the above implementation can be modified so that the decoder traces multiple bits at a time. Figure 12.4 shows a modified version of the decoding tree of Figure 12.1 that allows the input to be processed two bits a time. On the average, a codeword will now be decoded in $\frac{log_2 8}{2} = 1.5$ cycles. Since the decoder processes two bits per cycle and not all codewords have codelengths that are a multiple of two, the model of our FSM requires an additional output field to indicate the number of bits decoded at each cycle. For example, Table 12.3 shows the entries for a ROM-based FSM implementation of the Huffman tree shown in Figure 12.4. In this example, the extra field (Next Shift) requires only one bit that indicates whether one (Next Shift = 0)

Address		Output	
Node	Input	Next-Node/ Codeword	Termination Flag
0	0	5	0
0	1	1	0
1	0	6	0
1	1	2	0
2	0	e	1
2	1	3	0
3	0	f	1
3	1	4	0
4	0	g	1
4	1	h	1
5	0	a	1
5	1	b	1
6	0	c	1
6	1	d	1

Table 12.2 Example of ROM entries for bit-serial Huffman decoding; input is one bit at a time.

or two bits were decoded. After a codeword is decoded, the input buffer is shifted by either one or two bits, depending on the value of this field.

12.3.2 Constant-Output-Rate Decoders

Constant-input-rate decoders are very simple; however, they don't provide a fixed symbol decoding rate. Furthermore, when multiple bits are processed at a time, the Huffman tree has to be reconfigured, and an efficient mapping of the decoding tree into memory is required. Lei and Sun proposed a PLA-based decoder implementation that decodes each codeword in a single cycle, regardless of its length.

Figure 12.5 shows a block diagram of the Lei-Sun VLC decoder. Like the VLC encoder of Figure 12.2, the decoder includes a set of the VLC tables, a barrel shifter, two data registers, and an adder. The upper register is 16-bits;

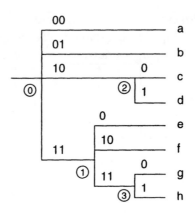

Figure 12.4 Example of a modified Huffman tree for decoding two bits at a time.

Address		Output		
Node	Input	Next-Node/ Codeword	Term. Flag	Next Shift
0	00	a	1	1
0	01	b	1	1
0	10	2	0	1
0	11	1	0	1
1	0X	e	1	0
1	10	f	1	1
1	11	3	0	1
2	0X	c	1	0
2	1X	d	1	0
3	0X	g	1	0
3	1X	h	1	0

Table 12.3 Example of ROM entries for bit-serial Huffman decoding; input is two bits at a time.

that is, as long as the assumed maximum codelength. This guarantees that one codeword will be decoded in each cycle. Since a codeword can be spread

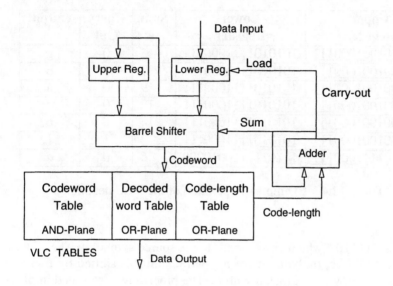

Figure 12.5 Block diagram of the Lei-Sun VLC decoder.

into two consecutive input segments, the decoder operates on two segments at a time, stored in the upper and lower registers. The barrel shifter operates like a sliding window on the contents of the two registers. The amount of shift is controlled by the adder, which accumulates the lengths of the decoded codewords.

At each cycle, the output of the barrel shifter is matched, in parallel, with all the entries in the PLA. This is analogous to the operation in a content-addressable memory. When a match is found, the PLA outputs the corresponding source symbol and the length of the decoded codeword, and the barrel shifter is shifted to the beginning of the next codeword. When the adder overflows, this indicates that the upper register has been fully decoded. The content of the lower register is transferred into the upper register, the VLC decoder loads a new 16-bit segment into the lower register, and operations continue.

Table 12.4 shows an example of the operation of the VLC decoder for the same codebook and encoded sequence described before. In Table 12.4, the output of the barrel shifter is shown underlined. Starting from the top, a search for a match of the upper register with a codeword in the codebook yields a

Upper Register	Lower Register	Sum	Carry Out	Output
1111001001000111	1010101111000111	5	0	g
1111001001000111	1010101111000111	7	0	b
1111001001000111	1010101111000111	9	0	a
1111001001000111	1010101111000111	12	0	c
1111001001000111	1010101111000111	14	0	b
1111001001000111	1010101111000111	2	1	f
1010101111000111	1110111110010000	6	0	d

Table 12.4 Example of operation of the VLC decoder.

match with "11110," which corresponds to the source symbol *g*. The adder is incremented by five, the window of the barrel shifter is shifted by five bits, and a new search yields a match with *b*. The procedure is repeated until the adder overflows (after *f*). Then the content of the lower register is moved into the upper register, and a new input is loaded into the lower register.

12.4 VARIABLE-LENGTH CODING IN JPEG

A case of special interest is the variable-length coding of the DCT coefficients in JPEG. In baseline JPEG, entropy coding is performed in two steps: (1) the quantized DCT coefficients are converted into pairs of symbols, and (2) these symbols are variable-length coded.

In the first step, after the run-length coder, each AC coefficient is represented by two symbols:

$$symbol1 \quad = \quad (\text{run-length, size})$$
$$symbol2 \quad = \quad (\text{amplitude})$$

where *size* is the number of bits required to encode the *amplitude* of the coefficient. Both *run-length* and *size* are four-bit numbers. The DC coefficients are also represented by two symbols, but *symbol1* has only *size* information.

In the second step, *symbol*1 is encoded using a variable-length code from a specified Huffman table, and *symbol*2 is encoded with a variable-length integer (VLI). Both codes have variable length, but the VLIs are not Huffman codes and cannot be changed. For example, an AC coefficient with a run-length of 1 and amplitude -2 is represented by the pair of symbols (1,2) (-2). Using the default Huffman tables, the corresponding VLC and VLI codes are (11011) and (01). An important difference between the two codes is that the length of the VLC is not known until it is decoded, but the length of a VLI is specified in the *size* field of the preceding VLC.

Like encoding, VLC decoding in JPEG is a two-step process. First, *symbol*1 is decoded; the four least significant bits of the decoded *symbol*1 specify the number of bits used to encode *symbol*2. If the most significant bit (MSB) of *symbol*2 is 1, then the *amplitude* is positive and the value of the extracted codeword represents the actual value of a DCT coefficient. If the MSB of *symbol*2 is 0, then the *amplitude* is negative, and its magnitude of the DCT coefficient is given by the one's complement of the extracted codeword. For example, if *symbol*2 = 101, then *amplitude* = 5. If *symbol*2 = 011, then *amplitude* = -4.

Figure 12.6 shows a modified version of the Lei-Sun decoder for VLC decoding in JPEG. The front end is the same as before, and the VLC tables store the codewords and their corresponding run-length, size, and codelength information. Additional circuitry uses the output *size* value to extract the VLI code from the barrel shifter. The MSB of the VLI code is used to select between the inverted and non-inverted VLI values. In parallel, the *size* value is added to the codelength so that the barrel-shifter is shifted to the beginning of the next VLC code.

12.4.1 Decoding of Programmable Huffman Codes

In many applications, the JPEG Huffman tables may be image-dependent to maximize compression. In such cases, the architecture of Figure 12.6 cannot be hardwired with a priori PLA or ROM based tables. Instead, one has to use more expensive CAM memory. However, one can take advantage of the numerical properties of Huffman codewords to perform fast VLC decoding using arithmetic instead of matching operations.

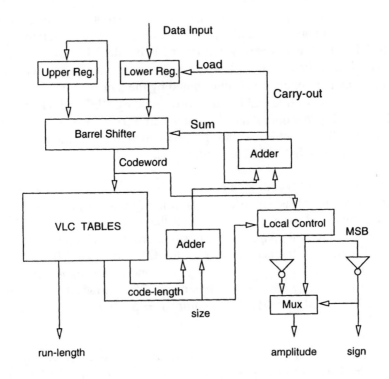

Figure 12.6   A modified Lei-Sun VLC decoder for JPEG.

Let $c_1, c_2, \cdots, c_N$ be codewords listed in order of increasing codelength, that is, $l_1 \leq l_2 \leq \cdots \leq l_N$, where l_i is the length of c_i. According to the JPEG interchange format, custom Huffman tables are defined as a series of two lists. The first list provides the values of $N_1, N_2, \cdots, N_{16}$, where N_k denotes the number of codewords with length k. The second one provides the *symbols* to be encoded, in order of increasing codelength. From these two lists, specific codewords are derived using the following procedure.

Begin
        c_1 has l_1 zeros.
        for i = 2 to N
                $c_i = (c_{i-1} + 1)2^{l_i - l_{i-1}}$
End

As an example, consider the codewords given in Annex K of the JPEG standard for the AC luminance coefficients. The two lists, in hexadecimal representation, are given by

00 02 01 03 03 02 04 03 05 05 04 04 00 00 01 7d

and

01 02 03 00 04 11 05 12 21 31 41 ... fa.

That is, there is no codeword with length one, there are two codewords with length two (for symbols 01 and 02), there is one codeword of length three (for symbol 03), and so on. Given these two lists and the code generation procedure defined above, Table 12.5 shows the first eleven symbols and their corresponding codewords.

Symbol	Codeword
01	00
02	01
03	100
00	1010
04	1011
11	1100
05	11010
12	11011
21	11100
31	111010
41	111011

Table 12.5   Example of codeword mapping in JPEG.

Obviously, these codes have monotonically increasing values. Before we describe an efficient decoding scheme for such codes, we need to present some of their properties. Let C_k^r denote the decimal value of the r-th codeword among all codewords of length k. Then

$$C_k^r = \sum_{j=1}^{k-1} 2^{k-j} N_j + r, \ r = 0, 1, \cdots, N_k - 1. \tag{12.1}$$

From (12.1), the minimum and maximum decimal values for the codewords in the k-th group are given by

$$C_k^{min} = C_k^0 = \sum_{j=1}^{k-1} 2^{k-j} N_j \; , \tag{12.2}$$

and

$$C_k^{max} = C_k^{N_k-1} = \sum_{j=1}^{k} 2^{k-j} N_j - 1 \; . \tag{12.3}$$

For example, for the codewords in Table 12.5, $C_4^{min} = 10$ and $C_4^{max} = 12$.

Thus, monotonic codes in the k-th group can be expressed in terms of C_k^{min} and an offset, their difference from C_k^{min}. If decoded symbols are stored in contiguous locations, then, given the length of the incoming codeword and the memory address of the symbol representing the minimum code in that length, one can easily compute the address of the decoded symbol.

The length of an incoming codeword can be computed as follows. Let $C_{k+i}^{min}(k)$ denote the decimal value of the k most significant bits of C_{k+i}^{min}, where $1 \le i \le 16 - k$. From (12.2) and (12.3),

$$
\begin{aligned}
C_{k+i}^{min}(k) &= \left[\sum_{j=1}^{k+i-1} 2^{k+i-j} N_j \right] 2^{-i} = \sum_{j=1}^{k+i-1} 2^{k-j} N_j \\
&\ge \sum_{j=1}^{k} 2^{k-j} N_j > C_k^{max}. \tag{12.4}
\end{aligned}
$$

In other words, for a k-bit codeword, C_k^{max} is always less than the k-prefix bits of all codewords with lengths greater than k. On the other hand, all codewords with length $l < k$ should be smaller than the l prefix bits of C_k^{min}. This property leads to the following decoding scheme.

- Feed bits 1 to 16 of an incoming codeword to 16 comparators. The k-th comparator detects if the word from the k prefix bits from the input is greater than or equal to its prestored value of C_k^{min}. If this is true for comparator L but false for comparator $L + 1$, then the length of the incoming codeword is L.

- Given the base address, say $M(C_L^{min})$, for the symbol corresponding to C_L^{min}, the address of the symbol corresponding to the input codeword c_i is given by

$$M(c_i) = M(C_L^{min}) + c_i - C_L^{min}$$
$$= c_i + \left[M(C_L^{min}) - C_L^{min} \right]. \qquad (12.5)$$

Note that the expression in the brackets can be precomputed and stored in memory for all 16 codelengths. In addition, all codewords larger than eight bits have at least eight leading ones. This property can be used to simplify the design of the high-order comparators.

Figure 12.7 shows a block diagram for a programmable JPEG VLC decoder using the decoding schemes discussed earlier. The core of this design is the same as the one in Figure 12.6, except that we replaced the VLC PLA with three smaller memories, the comparators, and an additional adder. For each of the Huffman tables, the minimum code memory stores the values of C_k^{min}, the base memory stores the offset values corresponding to $M(s_k^{min}) - C_k^{min}$, and the symbol memory stores the symbol alphabets in increasing codelength.

Example: Consider again the codewords given in Table 12.5. Assuming the symbol corresponding to codeword 00 is stored in memory location B, Table 12.6 shows the entries of the minimum code (C_k^{min}) and base memories for lengths 2 to 6. Given the input sequence 1011100110001001, we compare the

Code length	Minimum code memory	$M(C_k^{min})$	Base memory
2	0	B	B
3	4	B+2	B-2
4	10	B+3	B-7
5	26	B+6	B-20
6	58	B+9	B-48

Table 12.6 Examples of minimum code and base memory entries.

elements of {10, 101, 1011, 10111, 101110, ...} = { 2, 5, 11, 23, 46, ... } with their corresponding elements in C_k^{min} = { 0, 4, 10, 26, 58, ...}. Since

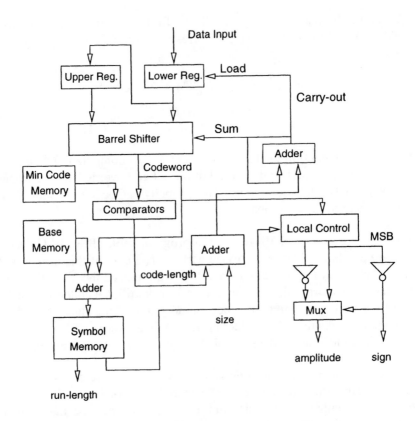

Figure 12.7 Block diagram of a programmable JPEG VLC decoder.

11 is greater than $C_4^{min} = 10$ but 25 is smaller than $C_5^{min} = 26$, the length of the incoming codeword is 4 bits and the input codeword is 1011 (11). From (12.5) and Table 12.6, $M(1011) = 11 + (B-7) = B+4$, which, from Table 12.5, corresponds to the memory address of symbol 04 (run = 0 and size = 4). Extracting the next four bits from the stream yields an amplitude value of 9 (1001).

The above decoding procedure is not restricted to JPEG codes only, but it can be applied to any application that uses monotonic Huffman codes.

12.5   TO PROBE FURTHER

A key problem in the design of any variable-length coder is how to efficiently map a coding tree into memory. In [176], Mukherjee et al. present such a mapping algorithm and describe its implementation in MARVLE, a single chip VLC coder. MARVLE has a 512 × 12 RAM, a comparator, an adder, and data and control registers. For eight-bit ascii data and at 83.3 MHz, it yields a 95.2 Mbits/s compression rate and a 60.6 Mbits/s decompression rate. This is under the assumption of a two-to-one data compression ratio.

Efficient mapping techniques and architectures are also described by Park and Prasanna [198], Hashemian [91], and Hsieh and Kim [100]. Park and Prasanna continue to use a single-bit decoder, but their implementation requires less memory. Hashemian applies his mapping algorithm to the design of a decoder that can process four bits at a time. Hsieh and Kim divide an MPEG-2 codebook into groups that have similar maximum likely bit patterns (MLBP). Basically, they try to extract the same bit pattern within codewords of the same length and use it as a reference to access the symbol memory. After extracting the MLBP pattern, the remainder bits of the codeword provide a second reference to the address of the corresponding symbol. Recently, Ooi et al. [191] presented a mapping scheme and an IC for the VLC decoding of MPEG-1 streams. Their architecture processes three bits at a time, and DCT coefficients can be decoded in four cycles or less. This IC is part of a three-chip set MPEG-1 codec from NEC [241]. At 27 MHz, it can process 162 Mbits/s.

The Lei and Sun architectures are described in [150]. The same designs are also used by Fujiwara et al. [62] in a chip-set for H.261, and by Yang et al. [265] in a multistandard VLC decoder IC with special hardware for JPEG VLC decoding. In [34], Chang and Messerschmitt describe parallel implementations of tree-based and PLA-based designs for VLC decoders. They also propose various PLA design-optimization techniques and show how the Lei-Sun decoder can be modified to decode multiple codewords per cycle. Parallel VLC decoders that take advantage of the numerical properties of monotonic Huffman codes are described in [162] and [256].

13

IMPLEMENTATION OF JPEG PROCESSORS

13.1 INTRODUCTION

Among the image compression standards, the baseline JPEG is the easiest to implement in hardware. The first single-chip JPEG processor was introduced by C-Cube Microsystems in 1990. JPEG processors can be found now in a variety of image and video processing systems, such as video editing equipment and digital cameras.

Figure 13.1 shows the block diagram of a typical JPEG implementation. The core of the design implements the baseline JPEG using a DCT unit, the quantizer, and the entropy coder. Up to four DCT quantization tables and four Huffman tables can be stored in local memory. In addition to the core processor, a JPEG IC includes memory and host interface units and local buffers for pipelined processing and I/O. Depending on the target applications, a JPEG processor may also include a color converter, a subsampler, and a level shifter. The processor supports a bidirectional data and processing pipeline for both compression and decompression.

The implementation details for DCT processors and entropy coders have already been covered in Chapter 10 and Chapter 12. In the next two sections we will discuss a few of the design issues for data I/O and color conversion. We will conclude this chapter with a short description of commercially available JPEG processors.

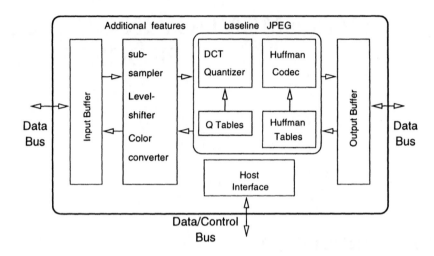

Figure 13.1 Block diagram of a typical single-chip JPEG processor.

13.2 DATA I/O AND MEMORY INTERFACE

From Figure 13.1, a typical JPEG processor supports at least two data buses, one for the uncompressed data stream and one for the compressed data stream. In many designs, an additional data/control bus may be used for interfacing and control with a local host.

Processing is usually performed on minimum coded units (MCUs). An MCU can be either an 8 × 8 block for grayscale images or a set of interleaved 8 × 8 blocks for color images. For example, for 4:2:0 YCbCr data, an MCU consists of two blocks of *Y*, one block of *Cb*, and one block of *Cr*. An MCU can contain up to ten 8 × 8 blocks made up from up to four components. For compression, the input buffer needs to store at least one MCU. For improved performance, most designs use double (or ping-pong) buffering. This allows one buffer to be loaded with new data while the data in the other buffer is processed by the JPEG unit. Since the output of the Huffman encoder has a variable rate, an output buffer is also needed to smooth the compressed data flow. The size of the output buffer depends on the worst-case compression ratio and the response time of the external memory.

In most applications pixels are stored in a raster-scan format; however, JPEG processes the data in 8×8 blocks. The raster-to-block conversion can be performed either by the host processor, an external dedicated unit, or by the on-chip memory interface unit, using a special address generator. During decompression, the data have to be reformatted again from 8×8 blocks into a raster-scan format.

13.3 COLOR CONVERSION

The JPEG standard makes no assumptions about the color representation of the input data. Color planes are treated as independent components, and each component is processed separately. Video sequences are usually stored in YCbCr format, but color still-images are usually scanned and stored in RGB format. Since the human visual system is more sensitive to the luminance rather than the chrominance component of an image, better compression is achieved if the image is first translated into a different color space, like YCbCr. Color translation requires the multiplication of the original data with a 3×3 color transformation matrix. For example, given three RGB values R, G, and B, the RGB-to-YCbCr conversion is given by

$$
\begin{bmatrix} Y \\ Cb \\ Cr \end{bmatrix} = \begin{bmatrix} 0.299 & 0.587 & 0.114 \\ -0.169 & -0.331 & 0.500 \\ 0.500 & -0.419 & -0.081 \end{bmatrix} \begin{bmatrix} R \\ G \\ B \end{bmatrix}. \tag{13.1}
$$

Using a parallel multiply-accumulate unit, color transformation can be completed in nine cycles per pixel. Figure 13.2 shows an alternative implementation using a barrel shifter and four adders. All operations in (13.1) are executed using shifts and additions. For example, 0.299 can be approximated by the sum $2^{-2} + 2^{-5} + 2^{-6} + 2^{-8}$. Hence, $0.299R$ can be approximated by the sum $R(2) + R(5) + R(6) + R(8)$, where $R(n)$ denotes a right shift of R by n bits. Following the same procedure, Y can be obtained from

$$
\begin{aligned}
Y = \ & R(2) + R(5) + R(6) + R(8) + \\
& G(1) + G(4) + G(6) + G(7) + \\
& B(4) + B(5) + B(6) + B(8).
\end{aligned} \tag{13.2}
$$

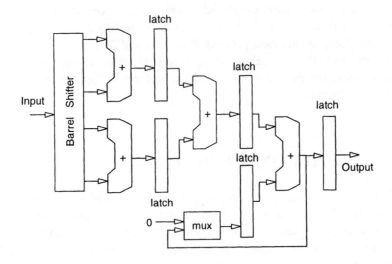

Figure 13.2   Block diagram of an adder-based color converter.

Data are loaded from the barrel shifter four at a time. $[R(2), R(5), R(6), R(8)]$ in the first cycle, $[G(1), G(4), G(6), G(7)]$ in the second cycle, etc. A Y value will be available after three cycles. Using pipelining, a color transformation can also be completed in nine cycles per pixel. The same circuitry can also be used for the reverse color transformation, that is, from YCbCr to RGB. For the same performance, the above implementation requires less hardware than a parallel multiplier-accumulator, but more complex control. Furthermore, a change in the color transformation matrix will require changes in the microcode that controls the barrel shifter.

13.4 COMMERCIALLY AVAILABLE JPEG PROCESSORS

13.4.1 ICs from C-Cube Microsystems

C-Cube offers two JPEG processors: the CL550 and the CL560. The CL550 is designed for PC multimedia and still-image-based systems. It runs at 30 or

35 MHz and can sustain a 2 Mbyte/s compressed data rate. CIF images can be compressed or decompressed in real time (30 fields/s). The chip has an on-chip raster-to-block converter and can also perform color conversion with a user-defined 3 × 3 transformation matrix. It supports eight-bit grayscale, RGB, YUV, or CMYK and 4:4:4:4 input and output.

The CL560 is a superset of the CL550 and is designed for video applications. It has an improved Huffman coder that allows codewords to be encoded and decoded in a single cycle regardless of their length. The CL560 can sustain a 60 Mbytes/s compressed data rate and can compress and decompress CCIR 601 frames in real time. It has a improved interrupt control and can support an external DMA processor.

13.4.2 ICs from LSI Logic

LSI Logic offers two JPEG processors: the L64745 and the L64702. The low-cost L64745 requires an external DCT processor, the L64735, and an external color and raster-to-block converter, the L64765. The L64745 can perform both lossy and lossless compression and decompression. At a clock rate of 20 to 30 MHz, it can process data at full-motion video rates (27 Mbytes/s). It contains two AC and two DC code tables and four quantization tables, and it can be used to generate event statistics for the creation of custom Huffman tables.

The L64702 is a single-chip JPEG processor and very similar to the CL560. Like the CL560, it can perform format conversion (that is, raster-to-block) and color conversion. Subsampling may also be user programmable. During compression, the processor buffers image data in one of two MCU buffers that operate in ping-pong fashion. Each buffer is 256 bytes deep. The output buffer is 32 × 16 bits. The processor provides direct support for VRAM and DRAM and an asynchronous system interface that gives it the capability to operate either as a slave device to an external host or as a peripheral device with an external DMA.

At 33 MHz, the L64702 can process up to 8.25 Mpixels/s for component sequential (CMYK) or 4.125 Mpixels/s for 2:1:1 RGB and 2.75 Mpixels/s for 1:1:1 RGB. For designs based on the L64702, LSI Logic provides the CW702 core in the CoreWare cell library. This is a reduced version of the L64702.

Designers can use the core with other libraries to design JPEG related ICs, customized for a particular application.

13.4.3 ICs from Zoran

Zoran offers two JPEG processors: the ZR36040 and the ZR36050. The ZR36040 is a DCT-based coder/decoder and requires an external DCT processor (the ZR36020) and an external controller (the ZR36045). It fully supports JPEG markers and can process CCIR 601 frames in real time. Data-processing rates can range from 15 to 21 Mbytes/s.

The ZR36050 is a superset of the ZR36040 and is a single-chip JPEG encoder and decoder. It is one of the few processors that supports lossless JPEG using one dimensional differential prediction followed by variable-length encoding. It supports data rates of 21 to 27 Mpixels/s, and, like the LSI IC, it has a slave/DMA bus interface. Unlike the ICs from C-Cube and LSI Logic, the Zoran processors require an external raster-to-block converter and an external color converter. However, they are the only JPEG processors that allow for bit-rate control and fast preview (thumbnail images).

Bit-rate control is very useful for systems that, due to constraints in memory sizes, require predictable file sizes (for example, a digital camera). Bit-rate control may also be required in networking applications where the time allocated to transmit a compressed image is fixed. Under bit-rate control, the Zoran ICs perform two computational passes. In the first pass, they collect image statistics and update the quantization tables. In the second pass, the image is compressed based on the size constraints and the statistics that were collected from the first pass.

The Zoran ICs also allow for fast image preview. By decoding only the DC coefficient in each block, they can generate a thumbnail image, $\frac{1}{64}$ the size of the original. Generation of thumbnail images can be up to 25 times faster than full-image decompression. This feature can be very useful when previewing files in an image database. The ZR36050 can be combined with the ZR36055, a motion-JPEG controller, for real-time JPEG compression of full-motion video.

13.4.4 ICs from Fujitsu and Atmel

The MB86357A JPEG processor from Fujitsu is the only processor that supports color components with either 8 or 12 bits per pixel. The MB86357A requires external memory for the Huffman tables. At 20 MHz, it can compress 6-8 YCbCr frames per second at 640 × 400 resolution (4:2:2).

The AT74001 is a low-cost JPEG processor from Atmel, targeted at low-cost, low-power, camera or video applications. Like the Zoran processors, the AT74001 does not support any color conversion or data subsampling, but it has on-chip video and microcontroller interfaces. At 25 MHz, it can compress 3-4 YCbCr frames per second at CCIR 720 × 480 resolution (4:2:2).

Table 13.1 summarizes the key features of the commercially available single-chip JPEG processors.

Processor	Clock (MHz)	Bit Rate (Mbytes/s)	Key Features
CL550	33-35	10.00	CIF at 30 fps, 2 Mbytes/s sustained output compressed rate, raster-to-block and color conversion.
CL560	33	16.50	CCIR 601 at 30 fps. 66 Mbytes/s sustained output compressed rate, raster to block and color conversion.
L64702	33	8.25	SIF at 30 fps, raster-to-block and color conversion.
ZR36050	21-27	21-27	CCIR 601 at 30 fps, bit-rate control, lossy and lossless JPEG, fast preview.
MB86357A	20	4	Support for 12-bit images.
AT74001	25	1.6	Small die-size.

Table 13.1 Key features of commercially available single-chip JPEG processors.

13.5 TO PROBE FURTHER

Ogawa et al. [186] provide an excellent report on the implementation of a single-chip JPEG processor from Sanyo Electric. The authors provide many details on JPEG marker code processing, the host interface, color conversion (their design was described in detail in section 3 of this chapter), and core JPEG processing. In the Sanyo processor, the DCT is computed using row-column transformations and two 14-bit parallel multipliers.

Another single-chip JPEG codec is described by Chen et al. [38]. In this implementation, the DCT is computed using distributed arithmetic, and quantization is performed using a parallel multiplier and lookuptables. A similar implementation for the DCT and the quantizer is also described by Bolton et al. [27] for a JPEG processor from SGS-Thomson.

For the commercially available processors, most of the information in this chapter was taken from their corresponding databooks. For the CL550, additional implementation details are described in [209], and a full description is given in [18]. The JPEG chipset from LSI Logic is also described in [15]. The Zoran bit-rate control algorithm is presented in [220]. Another JPEG chipset with bit-rate control capabilities was also developed by Nakagawa et al. [182] from Toshiba Corporation. The four-chip set (an encoder, a decoder, a memory controller, and a card controller) was developed for a digital camera. Like the Zoran processor, the authors use a two-pass approach to determine the scaling factors for the quantization tables.

14

INTEGRATED CIRCUITS FOR VIDEO CODERS

14.1 INTRODUCTION

Due to the computational requirements of the video standards, early hardware implementations of video coders were multichip designs, where each major component, such as the DCT processor or the variable-length coder, was a separate chip. For example, in 1992, NEC introduced a three-chip set for MPEG-1 video encoding and decoding. The three ICs were an interframe coder (required only for video encoding), a transform and quantization processor, and a variable-length coder. Operating at 27 MHz, the chip set could process SIF resolution frames at 30 frames/s.

In recent years, advances in circuit integration allow single-chip implementations of video coders. Recent hardware implementations of the video compression standards fall into three main design categories: video signal processors, multimedia coprocessors, and dedicated coders. Video signal processor designs are programmable processors with a DSP or RISC core and coprocessing units for compute-intensive operations, such as motion estimation. Multimedia coprocessors are also programmable, however, they also support multitasking for the simultaneous acceleration of multimedia tasks, including video, audio, and graphics. Dedicated coder implementations are video encoders or decoders that are hardwired for the data flow of a specific standard.

Programmable processors offer the highest flexibility in code design and implementation; however, for equivalent levels of performance, they require larger silicon area and dissipate more power than dedicated processors. Programmable processors also incur significant costs in software development tools and system integration. In the remainder of this chapter we examine in some detail various programmable and nonprogrammable video coders, and we present the key features of some of the commercially available processors.

14.2   VIDEO SIGNAL PROCESSORS

Programmable video signal processors are extensions of general-purpose digital signal processors (DSPs). They include either a RISC or DSP core for general purpose processing and control and dedicated coprocessing units for the efficient implementation of the DCT, motion estimation, and variable-length coding. The implementation of the DCT, motion estimation, and entropy coding has already been covered in Chapters 10 to 12. In this section, we present a system-level design of various video codecs, along with their key design and performance characteristics.

14.2.1   The VDSP2

VDSP2 is a second-generation video signal processor from Matsushita Electric. Figure 14.1 shows a block diagram of the VDSP2 processor. VDSP2 was designed for MPEG-1 and MPEG-2 (Main Profile at Main Level) encoding and decoding applications. From a programmer's point of view, processing of data is performed on pipelined macroblocks (16×16 blocks). Inside the processor, processing is done using a SIMD configuration at the block (8×8) level. From Figure 14.1, the SIMD array has four DSP processors (VPU0-3), controlled by a DSP-core controller. The DSP-core controller includes instruction and data RAM, a register file, an ALU, a barrel shifter, and a 16×16-bit multiplier. Each of the VPU processors has an enhanced ALU, a 24-bit adder, a 16×16-bit multiplier, and five memories (SBM, RBM, DBM, WBM, and CBM) comprising close to 23 Kbits of RAM per VPU processor. In addition to the SIMD array, VDSP2 has special circuitry for DCT/IDCT and variable-length coding/decoding. The DCT/IDCT unit is designed using distributed arithmetic.

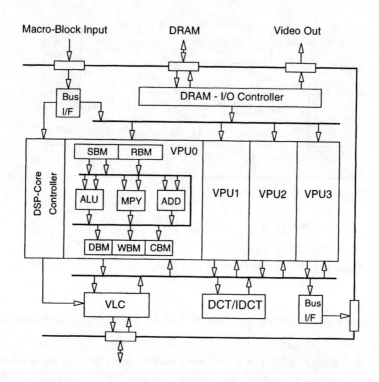

Figure 14.1 Block diagram of Matsushita's VDSP2 processor.

For MPEG-2 encoding, one needs two VDSP2 ICs and an external motion estimator. MPEG-2 decoding requires only one VDSP2. The VDSP2 was fabricated using 0.5 μm CMOS. It has approximately 2.5 million transistors and runs at 100 MHz. It supports a maximum bit rate of 497 Mbits/s.

14.2.2 The VideoRISC

The VideoRISC (VRP or CL4000) processor is the first generation architecture for video processing from C-Cube Microsystems. It is a 32-bit RISC-based processor and it is used in a variety of video encoding and decoding products. Figure 14.2 shows a block diagram of the VideoRISC processor. The heart of the processor is a 32-bit RISC processor with 4 Kbytes of instruction cache and 4 Kbytes of data cache. The videoRISC processor also has a variable-length

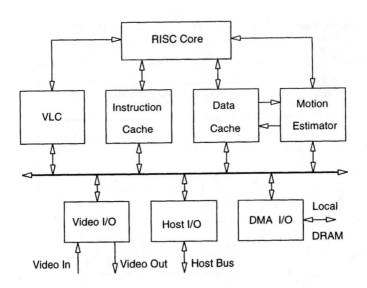

Figure 14.2   Block diagram of C-Cube's VideoRISC processor.

coder and decoder (VLC) and a motion estimator. There is no special DCT processor; however, special instructions in the RISC core allow efficient implementation of the DCT and other video-related operations.

The motion estimation unit has 16 subprocessors in a 4 × 4 array, and the specific motion estimation algorithm is programmable by the user. The chip also has video and host interfaces, 2 Mbytes of local DRAM, and a seven-channel DMA controller. For MPEG-1 SIF quality video, two processors are required. For MPEG-2 (in the Main Profile at the Main Level), eight (NTSC) to ten (PAL) processors are required. In multiprocessing applications, each video frame is divided into horizontal stripes, and each stripe is processed in a different processor. C-Cube provides a software development environment and multiprocessor boards to assist developers in testing and development of video processing systems.

The VideoRISC processor runs at 60 MHz and has 1.2 million transistors. The second generation of VRP, VRP2 or CL4100, has the same architecture but improved circuitry for variable-length encoding and decoding of MPEG-2 data streams. A single CL4100 can perform MPEG-1 encoding or decoding with downloadable code. For real-time MPEG-2 encoding, C-Cube offers the CLM47xx family of products. These are parallel designs with eight to 13 C-Cube VRP2 processors. C-Cube has already announced VRP3, with dual RISC engines, and is working on a new design based on a MicroSparc core from Sun Microsystems.

14.2.3 The VCP

Figure 14.3 shows a block diagram of the VCP processor from 8x8 Inc., formerly known as Integrated Information Technology (IIT). Like the C-Cube

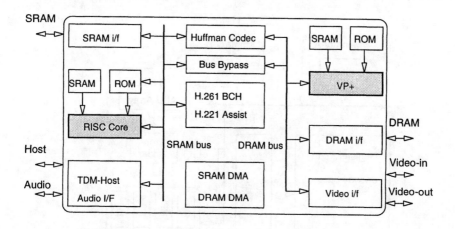

Figure 14.3 Block diagram of the VCP processor.

processor, the VCP processor has a RISC-like core (32 bits) and is fully user programmable. The processor has two main buses: (1) the SRAM bus, which handles instructions and data for the RISC processor, and (2) the DRAM bus, which handles uncompressed data or intermediate data. The two buses typically communicate via the Huffman coding-decoding unit; however, a special bus-interface unit allows direct communication too.

Data traffic is controlled by two DMA controllers under the supervision of the RISC core. The SRAM DMA is a ten-channel DMA controller, and the DRAM DMA is a five-channel DMA controller. The VCP has separate audio, host, DRAM, and video input and output buses with dedicated bus controllers.

Audio is processed externally (via a separate DSP); however, the audio bus allows the mixing (or separation) of video and audio streams.

The VCP is a superset of an earlier chip set: the Vision processor (VP) and the Vision controller (VC). The VCP has no instruction cache, but it has a full 32-bit interface to external SRAM and on-chip boot ROM. The H.261 and H.221 assist units allow for frame alignment and bit handling in video-conferencing applications. The core math processing of the VCP is performed in the VP+ processing unit.

The VP+ has functionality similar to that of the earlier VP processor. Figure 14.4 shows a block diagram of the VP processing core. Even though the VCP

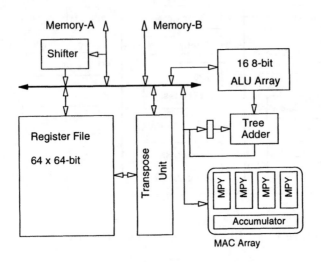

Figure 14.4   Block diagram of the core processor in the VP IC.

has no special circuitry for the DCT and motion estimation, the VP+ core (Figure 14.4) includes enough processing power to efficiently handle both of these operations. An array of 16 eight-bit ALUs (which can also be configured as eight 16-bit ALUs) together with the tree adder can be used for motion estimation. Direct access to two memories allows one of them to be used for the reference block and one for the search window.

The VP+ core also includes an array of four parallel 16×16-bit multipliers (MAC array), a register file, and a transposition unit. As we have seen in Chapter 10, the MAC array in combination with the transposition unit is ideally suited for computing the DCT. The VP+ can be programmed to do a variety of operations, including filtering, color conversion, and frame scaling.

The VCP runs at either 66 MHz or 80 MHz. A single 66-MHz VCP can perform MPEG-1 video encoding (SIF at 30 fps), or implement an H.261 codec (CIF/QCIF at 30 fps). An 80-MHz VCP can perform real-time MPEG-2 decoding (CCIR 601 at 60 fields/s). Based on the architecture of the VCP, 8x8 Inc. offers a whole family of integrated circuits that includes: the MPPex (MPEG playback processor), the MEP (MPEG encode processor), and LVP (low bit rate videophone processor). LVP is capable of executing all components of the H.324 standard. It can also be programmed as an MPEG-1 encoder or decoder.

14.2.4 The Multimedia Video Processor

The Multimedia Video Processor (MVP) or TMS320C80 is the latest entry from Texas Instruments in the DSP market. Unlike other video processors, the MVP was not designed specifically for the implementation of the video standards. For example, it does not have any special function units for the DCT, motion estimation, or Huffman coding. It is a single-chip parallel processor that can be used in a variety of applications in graphics, signal, and video processing.

Figure 14.5 shows a block diagram of the MVP. It includes a master processor (MP) with a 100-MFLOP IEEE floating-point unit, four parallel digital signal processors (DSP1-DSP4), a transfer controller for data transfers and memory control, local memory, and two video controllers for data capture and display. MP is a general-purpose RISC processor with an internal floating-point unit. It uses a 32-bit instruction word and can load or store 8-bit to 64-bit data sizes. It has a 4 Kbyte instruction cache and a 4 Kbyte data cache. The MP can be used for control, floating-point operations, audio processing, or 3D graphics transformations.

Each DSP unit performs all the typical operations of a general-purpose DSP and can also perform bit-field and multiple-pixel manipulations. Internally,

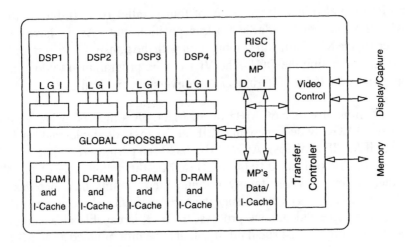

Figure 14.5   Block diagram of the MVP from Texas Instruments.

it has a splittable (one 16×16 or two 8×8) multiplier and ALU, 44 local registers, a barrel shifter, two address generators, and a program-flow control unit, all controlled by very long 64-bit instruction words. Each DSP has three parallel data ports for local data, global data, and instructions (L, G, and I). A global crossbar switch allows data transfers among the DSP processors and the RISC core. All processors share 50 Kbytes of SRAM that is divided into data RAM and instruction caches. The transfer controller manages the hardware interface between the on-chip and the off-chip memories. The two frame-buffer controllers support programmable video timing to control both display and capture.

At 50 MHz, the MVP executes more than 2 GOPS. It is fabricated in 0.5 μm CMOS and includes more than 4 million transistors. A single MVP can perform real-time MPEG-1 encoding or decoding (SIF resolution). Since the MVP has an architecture very similar to that of the VDSP2 and no special hardware for motion estimation, we estimate that multiple MVPs are needed for real-time MPEG-2 encoding.

The TMS320C82 is the first derivative of the MVP architecture. It maintains the core architecture of the TMS320C80 but uses only two DSP units instead of four. Each DSP uses a 4 Kbyte I-Cache and a 12 KByte data RAM.

The TMS320C82 is targeted for low-cost videoconferencing applications. For example, the complete H.324 standard can be implemented on a single chip, where the master processor can perform all high-level processing, DSP1 can process video (H.263), and DSP2 can process audio (G.723) and also be used as a V.34 modem data pump.

14.2.5 The AVP

The AV4400A is the latest version of the AVP audio/video processor from Lucent Technologies (formerly AT&T Microelectronics). Figure 14.6 shows a simplified block diagram of the AVP processor. It features a programmable

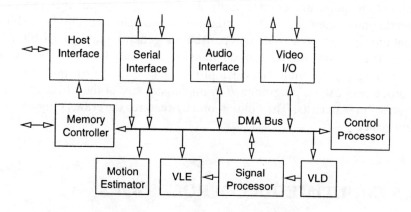

Figure 14.6 Block diagram of the AVP.

32-bit RISC control processor, a programmable signal processor, a motion estimator, a variable-length encoder (VLE) and decoder (VLD), a memory controller, and host, video, and audio interface circuitry.

The RISC processor includes a 16×32-bit hardware multiplier, a $3K \times 32$-bit instruction SRAM, a 256×32-bit data cache, and a 142×32-bit register file. The signal processor has six execution units, each one with two local 128×16-bit caches, a 96×16-bit three-port register file, an ALU, and a 16×16-bit multiply-accumulate unit. The signal processor performs video pre- and post-processing, DCT and IDCT, and other filtering operations. The motion estimator uses 72 processing elements and can be programmed to perform a variety of motion estimation algorithms. The video I/O processor can also perform filtering, cropping, and resizing operations.

The AV4400A runs at 44 MHz, but speeds as high as 67 MHz have been reported in the literature. The AVP has approximately 2.27 million devices, and, at 67 MHz, it provides 16 GOPS of processing power. At 44 MHz, it supports H.261 simultaneous video encoding and decoding (15 CIF frames/s), or H.263 simultaneous video and audio (G.723) encoding and decoding (15 QCIF frames/s), or MPEG-1 audio and video decoding (30 SIF frames/s), or MPEG-1 video-only encoding at 30 SIF frames/s.

In summary, programmable video processors now allow real-time H.261, H.263, or MPEG-1 encoding. However, for MPEG-2 encoding no single-chip solutions are presently commercially available. Programmable processors integrate more than 1.2 million transistors and can execute 1-2.5 billion operations per second. (In comparison, a general purpose 64-bit RISC processor has more than 3 million transistors.) Table 14.1 shows the key features of the programmable video processors we described in this section. All processors can be programmed to implement any of the image and video compression standards. The column on key features describes only real-time performance.

14.3   MULTIMEDIA COPROCESSORS

Multimedia implies an amalgamation of various data types such as text, audio, 2D and 3D graphics, animation, images, and video within a computing system or within a user application. Recently, there has been a major effort to provide multimedia "on a chip," and this class of processors is widely referred to as *media processors*.

The notion of a single chip performing multimedia functions is not new. We have already seen that programmable video processors can execute most of the video standards. However, video processors are not designed to perform more than one multimedia task at a time. Furthermore, they are not typically designed for graphics functions. The I/O and memory interfaces on video and DSP chips are also restricted and do not permit easy integration with a general purpose processor in a desktop computing system. Media processors possess

Processor	Clock (MHz)	GOPS	Key Features
VDSP2	100		MPEG-2 coder. Encoder requires two ICs and external motion estimator.
VideoRISC	60	2.5	Two ICs for MPEG-1 and 8-10 ICs for MPEG-2 encoding.
VCP	80		MPEG-1 coder or MPEG-2 decoder.
LVP	80		H.324 coder.
MVP	50	2.0	MPEG-1 coder, MPEG-2 decoder, H.324 coder.
AVP	67	16	H.320 and H.324 coder, MPEG-1 coder.

Table 14.1 Key features of programmable video signal processors.

features that differentiate them from DSPs and video processors. Standard DSPs typically do not include frame buffers with GUI acceleration. The media processor has features that facilitate its integration onto the PC motherboard. Furthermore, DSPs typically operate in stand-alone fashion, whereas media processors exhibit tight integration with the host operating system and often implement task load sharing with the host. A media processor motivated from DSP architectures was recently announced by Samsung (MSP). This processor combines a vector processor and a RISC core and provides 3.2 GOPS for 16-bit integer data.

The media processor, like the general purpose processor, is software-programmable and thus provides some insurance against the changing requirements and standards of multimedia processing. Its high bandwidth and fast integer performance can simultaneously accelerate the different media types, and thus provide for the seamless integration of the different media types as will be required in the next generation multimedia computing systems. However, the media processor is not a general purpose microprocessor in that it depends on the host processor for memory management and general purpose computing. Media processors also readily adopt architectural techniques usually found in supercomputers, such as very long instruction word (VLIW), SIMD, and vector processing. Also, critical to a media processor's performance are off-chip memory bandwidth, a real-time kernel, and a well-integrated software environment. In some media-processor designs, dedicated logic blocks that perform highly performance-sensitive operations, such as those found in graphics, are selectively used to complement the media processor's programmable core.

Recently, several media-processor designs have emerged including architectures from IBM (M.f.a.s.t), Hitachi, and others. In this section we review two of these media processors: the Mpact architecture from Chromatic Research and the Trimedia processor, TM-1, from Philips. Both of these processors have VLIW architectures and special function units to accelerate multimedia operations. Both TM-1 and Mpact have built-in PCI interface and dedicated video and audio ports.

14.3.1   The Mpact Architecture

Figure 14.7 shows a block diagram of the Mpact architecture. It has a single, but multi-port 512 × 72-bit SRAM for both instructions and data, four ALU groups, and a special motion-estimation unit. The ALU groups can process eight 9-bit bytes in parallel. ALU group 1 is a shift and align unit. ALU groups 2 and 3 are general purpose ALUs. ALU group 4 in combination with ALU group 3 can be used for hardware multiplication. The multipliers can produce either eight 9 × 9-bit multiplications with 18-bit results, or four 18 × 18-bit multiplications, or two 24 × 24 multiplications with 44-bit results Mpact issues two VLIW-like instructions per cycle and supports multiple data types, such as 9-bit (× 8), 18-bit (× 4), or 24- and 36-bit (× 2). A single instruction may control multiple ALU groups.

In addition to the PCI interface, Mpact has support for direct communication to the video monitor. Mpact is not user-programmable. It has a proprietary instruction set, and system integrators have to depend on Chromatic Research to provide code for the different multimedia functions they want to support in their systems.

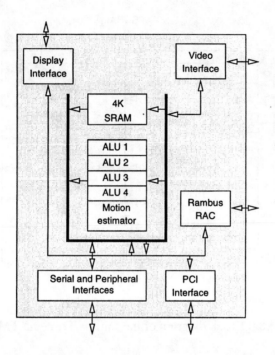

Figure 14.7 Block diagram of the Mpact architecture.

Without taking into account the motion estimation unit, at 62.5 MHz, Mpact provides a sustained performance of 2-3 GOPS. Chromatic has already announced a second-generation design of Mpact. The new processor includes a new floating-point ALU group for 3-D graphics and adds support for floating-point arithmetic in the other ALU groups. It also has separate instruction and data caches and an internal RAMDAC. The new Mpact should provide double the processing power of the original design.

14.3.2 The Trimedia Architecture

Figure 14.8 shows a block diagram of the Philips TM-1 Trimedia processor. At the core of TM-1, there is a VLIW core processor with up to 27 functional units. However, only five of those can receive new instructions at any given time. Each of these units can provide special support for multimedia

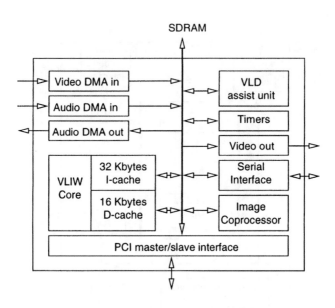

Figure 14.8   Block diagram of the Philips Trimedia TM-1 architecture.

applications. For example, there are arithmetic and logic units, multiply-accumulate units, floating-point arithmetic units, and a divider. In addition to this core, TM-1 includes coprocessors for variable-length decoding (VLD) and imaging operations. The VLD is optimized for decoding MPEG-1 and MPEG-2 video streams with minimal assistance from either the host or core CPUs. The image coprocessor can be used for either block moves or filtering and scaling operations. As a block mover, it allows up to four different streams of data to move from the external SDRAM to the video-audio ports or PCI peripherals. As a processing engine, it can perform horizontal and vertical filtering, resizing, and color transformations.

The input and output video DMA engines can interface directly to CCIR 601/656-compliant devices. They can also perform horizontal subsampling or upsampling. Similarly, the audio input and output units can interface directly to ADC and DAC chips and support a variety of sampling rates.

In contrast to Chromatic Research, Philips plans to provide a C compiler for the TM-1, for user-friendly programmability. A 100 MHz TM-1, can perform up to 4 GOPs. In a desktop computing system employing a general purpose processor and a media processor such as the TM-1 or Mpact, one can provide simultaneous support on the coprocessor for video teleconferencing, audio (speaker phone), and a 28.8 kb/s (software-based) modem along with the general purpose processing on the host.

14.4 DEDICATED CODERS

Dedicated coders provide limited, if any, programmability and have a dedicated architecture or control logic for a specific video encoding or decoding standard. There is an expectation that applications using video compression (such as set-top boxes for interactive TV, high-definition TV, digital cameras and VCRs, and desktop video teleconferencing) will see a huge growth in the next few years. As a consequence, there is a proliferation of announcements of products and partnerships related to video compression and decompression.

Figure 14.9 shows a block diagram of a typical video decoder. In addition to the main video decompression core, it also includes a memory controller for glueless interface to external data RAM, input and output buffers (FIFOs), input and output buses, and a display controller. The display controller performs operations like color conversion (YCbCr to RGB), vertical and horizontal filtering, and frame rate conversion (NTSC, PAL, and film).

Table 14.2 shows a representative list of commercially available video processing ICs and their key features. In addition to these processors, related product announcements have been made by Zoran, and others. Japanese companies like Mitsubishi and NEC are also developing new generations of video coders.

From Table 14.2, most of the commercially available dedicated processors are audio/video (a/v) decoders. This is not surprising, since real-time video decoders are easier to implement (they don't require motion estimation), and they will be in higher demand than real-time video encoders.

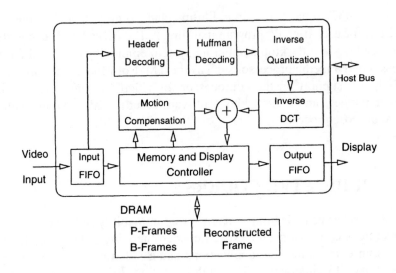

Figure 14.9   Block diagram of a typical video decoder IC.

From C-Cube, the CL480VCD and CL480PC are similar, but the CL480VCD has builtin a serial CD interface and supports CD-ROM decoding for low-cost consumer electronics products, such as video CD players.

IBM is one of the few companies that offer chip sets for MPEG-2 video encoding. IBM's chip set (MPEGSE10, MPEGSE20, and MPEGSE30) provides real-time encoding for CCIR-601 resolution in 4:2:2 chroma format (Main Profile at Main Level). The MPEGSE10 is an I-frame-only video encoder and is the anchor chip for the complete chip set. The MPEGSE20 is called the Refine chip. Together with the MPEGSE10, it enables encoding of I- and P-frames. The MPEGSE30 is called the Search chip. It supports a higher search range and allows encoding of I-, P-, and B-frames.

The MPEGCD20 MPEG-2 audio/video decoder supports 4:2:0 resolution for all frame types and and 4:2:0 resolution for I-frames only. The MPEGCD21 is an enhanced version and supports 4:2:2 resolution for all frame types.

TI's AV7000 integrates a complete television set-top decoder in a chip. It includes a 32-bit RISC controller, MPEG-2 audio and video decoder, traffic interface manager, and an advanced graphics accelerator. It is fully compliant

Company	Part No.	Key Features
C-Cube	CL450	MPEG-1 video decoder.
	CL480VCD	MPEG-1 a/v decoder.
	CL480PC	MPEG-1 a/v decoder.
	AViA-500	MPEG-2 a/v decoder.
	AViA-502	MPEG-2 a/v decoder, AC-3.
Hyundai Electronics	HDM8211M	MPEG-2 a/v decoder.
IBM Microelectronics	MPEGSE1	MPEG-2 I-frame encoder.
	MPEGSE10/20/30	MPEG-2 encoder (3 chips).
	MPEGCD10	MPEG-2 video decoder.
	MPEGCD20,21	MPEG-2 a/v decoders.
LSI Logic	L64002	MPEG-2 a/v decoder.
	L64005	MPEG-2 a/v decoder.
	L64020	MPEG-2 a/v decoder, AC-3.
SGS-Thomson	STi3400	H.261, MPEG-1 decoder.
	STi3430	MPEG-1 a/v decoder.
	STi3520	MPEG-2 a/v decoder.
TI	AV7000	Set-top decoder.
Toshiba	TC81201F	MPEG-2 video decoder.
	TC81211F	MPEG-2 a/v decoder.

Table 14.2 Key features of commercially available dedicated audio/video processors.

with worldwide digital satellite system (DSS) and digital video broadcast (DVB) standards.

14.5 TO PROBE FURTHER

Digital video processors have evolved from simple extensions of digital signal processors to powerful multiprocessing architectures. For example, Enomoto and Yamashina [55] describe the evolution of the NEC video signal processors: from the 14.5-MHz P-VSP, introduced in 1987, to the 300-MHz VSP3, introduced in 1993 [104]. A separate MPEG chip set from NEC is described by Tamitani et al. [241]. The VDSP processor from Matsushita [10] was among the first video processors with an on-chip DCT unit and an extended instruction set for video compression and decompression. The VDSP2, a second-generation processor from Matsushita, is described in [245] and [5].

For commercially available processors, most of the information in this chapter is based on data sheets, company Web pages, and product catalogs. Trade magazines, like *EDN, Computer Design,* and *Electronic Design* [31] also provide timely overviews and performance characteristics. Descriptions of the Vision processor and Vision controller from 8x8, Inc. can also be found in [17]. In [83], Guttag et al. describe the design philosophy and implementation details for TI's MVP processor. Performance characteristics of the MVP in a real imaging and graphics system (the UWGSP from the University of Washington) are presented by Kim et al. [129]. A description of an H.324 coder using a TMS320C82 is given in [80].

A short overview of the MSP processor from Samsung is given in [185]. Descriptions of the Mpact architectures are given in [60] and [208]. The TM-1 is described in [236] and [217]. Another VLIW processor for multimedia applications, from Mitsubishi, is described in [95].

Additional implementation details for the AVP processor can be found in [30]. IBM's MPEG-2 (I-frame-only) video decoder is described in [237]. An MPEG-1 audio and video decoder IC from C-Cube and JVC is described in [67]. The chip integrates 305,000 logic transistors and 485,000 memory transistors. At 40 MHz, it can decode two channels of 48 kHz audio and one SIF video stream at 30 fps. Descriptions of MPEG-2 video decoders from Toshiba and Texas Instruments (the AV7100) are given in [48] and [72], respectively. A single-chip H.261 codec from SGS-Thomson is described in [90].

Single-chip MPEG-2 encoders were also announced recently by Philips [252] and NEC [172]. The Philips processor integrates 4.5 million transistors and operates in Simple Profile, Main Level mode. Its target application is video encoding for storage. The NEC processor can encode MPEG-2 streams at Main Profile, Main Level. It integrates 3.1 million transistors. A survey paper by Pirsch et al. [206] presents an overview of VLSI implementations for video decoders and compares them in terms of speed and silicon area.

15

MULTIMEDIA ENHANCEMENTS FOR GENERAL PURPOSE PROCESSORS

15.1 INTRODUCTION

The JPEG and MPEG standards make it easier now to encode and exchange images and video streams across different computing platforms and networks. Video in particular has the potential to become just another *data type*. Real-time video processing on PCs and workstations is now handled with special video boards and custom circuits. This is similar to the way most of the advanced graphics processing was performed in the 1980s. Special hardware may yield the highest performance but at a considerable cost to the end user.

General-purpose RISC and CISC processors have significant computing power but are not well suited for a software-only solution. For example, some of their main features, such as 32-bit or 64-bit data paths and on-chip floating-point processing, are not needed in video processing. Therefore, there is a strong desire among computer manufacturers to enhance existing architectures so that video processing is integrated into the next generation general-purpose processors just as graphics processing has been integrated into today's architectures.

Manufacturers that have introduced multimedia-enhanced processors include Hewlett-Packard, Sun Microsystems, and Intel. The specific features of these processors are examined next.

15.2   GENERIC OPERATIONS IN VIDEO PROCESSING

It is always tempting to enhance a general-purpose processor by adding special function units. However, this approach violates the fundamental design principle of a RISC architecture, which states that only the most commonly used operations are implemented in hardware. Hence, integration of multimedia enhancements requires first an analysis and understanding of the generic operations in the video-processing pipeline. Table 15.1 shows the arithmetic operations required for the main functional blocks of the image and video compression standards. From Table 15.1 and an analysis of the

Function	Operations
Color transformation, preprocessing, and postprocessing	$\sum c_i x_i$, $clip()$ $(x_i + x_j)/2$, $(1/4) \sum_{i=1}^{4} x_i$
DCT, IDCT	$ax + b$, $\sum c_i x_i$
Quantization	x_i/c_i
Dequantization	$x_i c_i$
Huffman coding	data shifts, comparisons
Motion estimation/ compensation	$\sum |x_i - y_i|$ or $\sum (x_i - y_i)^2$, $min(a,b)$, $x_i + c x_j$

Table 15.1   Generic operations in video compression.

processing pipeline in video compression one can conclude the following:

- Input data and coefficients have usually eight to 16 bits of precision.

- There is no need for floating-point operations.

- The multiply-accumulate operation is very common; however, most of the multiplications are with constants.

- Saturation arithmetic, where a result is clipped (*clip()*) to the maximum or minimum value of a predefined range, is common in many operations (such as, color transformation).

In addition to these arithmetic operations, video processing requires efficient data addressing and I/O. The above analysis indicates that RISC processors need to handle fixed-point arithmetic more efficiently (specially multiplication) and to support a more efficient utilization of their longer than necessary data paths. The latest processors from Hewlett-Packard, Sun Microsystems, and Intel attempt to do exactly that.

15.3 MULTIMEDIA ENHANCEMENTS IN THE HP-PA

In early 1994, Hewlett-Packard introduced the first general-purpose RISC processor with built-in support for multimedia instructions. The new processor (the PA7100LC) and algorithmic enhancements in the decoding algorithms allowed for the first time real-time MPEG-1 video and audio decoding on a desktop computer without using any special hardware.

The 7100LC is a 32-bit superscalar RISC processor with two fixed-point ALUs, a floating point unit, and a memory and I/O controller integrated into a single chip. This is a low-cost processor with off-chip instruction and data caches and only a small on-chip instruction buffer. Since video operations never require more than 16 bits of precision, the key approach for multimedia enhancements in the HP design was to allow for parallel operations at the half-word (16-bit) level. For instance, each 32-bit integer ALU was partitioned so that it could execute a pair of 16-bit arithmetic operations in a single cycle with a single instruction. Arithmetic operations that were accelerated using this strategy include *add, subtract, average, shift-left-and-add,* and *shift-right-and-add.*

Consider the standard ADD instruction:

$$\textbf{ADD } Ra, Rb, Rt \, ,$$

where the ALU performs a 32-bit addition on the contents of registers Ra and Rb and places the result in register Rt. Assume now that two 16-bit operands ($a1$ and $a2$) are stored into the left and right halves of register Ra and another two 16-bit operands ($b1$ and $b2$) are stored in register Rb. Figure 15.1 shows an example of how a 32-bit adder can be used to perform either a 32-bit addition or two 16-bit additions. The 32-bit adder is divided into two 16-bit halves. When the output carry (c_out) of the lower half is blocked (c_in = 0), then two 16-bit additions can be performed in parallel. Two 16-bit additions can be

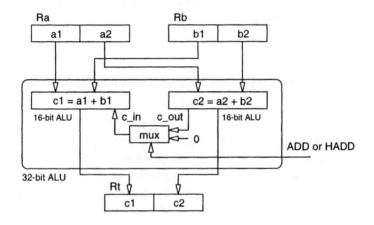

Figure 15.1   Half-word addition using a 32-bit ALU.

invoked with the new half-word-add instruction:

$$\textbf{HADD } Ra, Rb, Rt,$$

where if $Ra = [a1, a2]$ and $Rb = [b1, b2]$, then $Rt = [c1, c2] = [(a1 + b1), (a2 + b2)]$.

This instruction can also be invoked with a signed or unsigned saturation option. This option clips the 16-bit results so that they do not exceed a preset maximum and minimum value. This can be expressed as

$$
\begin{aligned}
if & \quad c1 > max \quad then \quad c1 = max, \\
if & \quad c1 < min \quad then \quad c1 = min, \\
if & \quad c2 > max \quad then \quad c2 = max, \\
if & \quad c2 < min \quad then \quad c2 = min.
\end{aligned}
\tag{15.1}
$$

For signed saturation, $max = 2^{15} - 1$ and $min = -2^{15}$. For unsigned saturation, $max = 2^{16} - 1$ and $min = 0$. This operation is particularly useful in pixel-related operations and replaces 10 conventional add and shift operations. Similar to the HADD instruction is a half-word subtract instruction (HSUB) that allows two pairs of 16-bit values to be subtracted in parallel.

Another new instruction is the half-word average instruction:

$$\textbf{HAVE } Ra, Rb, Rt,$$

where $Rt = [c1, c2] = [(a1 + b1)/2, (a2 + b2)/2]$. This instruction is particularly useful for half-pixel interpolation, since two new pixel values can be computed in a single cycle. Without this multimedia instruction, this operation would require four cycles.

The PA7100LC CPU has no integer multiplier, hence it is very hard to support parallel multiplication on half-words. However, in video coding most of the multiplications involve small constants, and thus they can be performed by simple combinations of shifts and adds. The original PA-RISC architecture already had shift-left-and-add operations. These operations perform a left shift of one operand by 1, 2, or 3 bits before adding the other operand. To implement this operation on half-words required a simple blocking of the shifts and the carries between two halves of the integer ALU. The new half-word left-shift-and-add operation is defined as

$$\textbf{HSHLADD } Ra, Rb, Rt, k,$$

where $Rt = [c1, c2] = [(a1 << k) + b1, (a2 << k) + b2]$. The symbol "$<< k$" denotes a left shift by k bits, where k is either 1, 2, or 3. A similar half-word shift-right-and-add (HSHRADD) instruction was also added. Table 15.2 summarizes the key operations in the first set of PA-RISC multimedia instructions (MAX-1).

In 1995, Hewlett-Packard introduced its first 64-bit processor, the PA-8000, and a new set of multimedia extensions (MAX-2). MAX-2 includes all instructions of MAX-1 (Table 15.2) and new instructions for parallel shifts and the rearrangement of subwords. These new instructions are shown in Table 15.3.

Instruction (Ra = [a1,a2], Rb = [b1,b2])	Operation (Rt = [c1,c2])
HADD Ra, Rb, Rt	Rt = [(a1+b1),(a2+b2)]
HSUB Ra, Rb, Rt	Rt = [(a1-b1),(a2-b2)]
HAVE Ra, Rb, Rt	Rt = [(a1+b1)/2,(a2+b2)/2]
HSHLADD Ra, Rb, Rt, k	Rt = [(a1 << k)+b1,(a2 << k)+b2]
HSHRADD Ra, Rb, Rt, k	Rt = [(a1 >> k)+b1,(a2 >> k)+b2]

Table 15.2 The MAX-1 set of PA-RISC multimedia instructions.

Instruction	Operation
HSHR HSHR,u	Parallel shift right; extend sign. Parallel shift right; zero fill on left.
HSHL	Parallel shift left.
MIXH, MIXW	Rearrange subwords from two source registers.
PERMH	Permute subwords within a source register.

Table 15.3 New PA-RISC multimedia instructions in MAX-2.

The HSHR and HSHL instructions can be used for data alignment and simple multiplications and divisions with powers of two. The MIX instructions can interleave subwords from two different registers. They are particularly useful for simple matrix transpositions. The PERMH instruction can rearrange the half-words in a given register in all possible permutations. Compared to other architectures, the HP-PA has the smallest number of multimedia instructions. However, it is a powerful mix of simple operations that can be combined to develop efficient macros for image and video coding applications.

15.3.1 I/O and Postprocessing Enhancements

In video decoding, a large percentage of the computations is spent in post-processing, which includes color transformations and display I/O. A new graphics IC allowed HP to move many of these display-related operations into the graphics controller and thus reduce the CPU load. The main processor connects directly to the graphics controller via a high-speed system bus. With a peak available bandwidth of 100 Mbytes/s, this direct connection allows for fast data transfers between the processor and the display.

The new graphics IC can handle YCbCr data, perform the up-sampling of the chrominance components and convert YCbCr data to RGB. A special color-compression unit translates and stores 24-bit RGB pixels into eight-bit values. Before the final display, the eight-bit values are translated back into true-color 24-bit values. This allows real-time decompressed video to be displayed in true-color even on entry level systems that have only eight-bit frame buffers.

15.4 MULTIMEDIA ENHANCEMENTS IN THE ULTRASPARC

In late 1994, Sun Microsystems provided the first details on the architecture of its new 64-bit processor, the UltraSPARC. Like the HP processor, the UltraSPARC incorporates architectural enhancements for multimedia processing and supports Sun's visual instruction set (VIS), a comprehensive set of instructions for the acceleration of media processing.

The UltraSPARC is a 64-bit superscalar RISC processor with two integer ALUs, an address-generation unit, and a special floating-point and graphics unit. The processor can issue up to four instructions per cycle, but there is a limit of only two integer operations and two floating-point or graphics operations per cycle. Multimedia instructions are executed by two specialized graphics execution units in the floating-point data path (Figure 15.2).

From Figure 15.2, the floating-point execution unit integrates a floating-point adder, a floating-point multiplier, and a floating-point divider (or square root). The graphics execution unit includes a graphics adder and a graphics multiplier. All processing units share a set of 32×64-bit floating-point registers. The graphics adder consists of four 16-bit adders and a custom shifter that can be used to perform single cycle additions, subtractions, merge, expand, and logical operations. By controlling the carry propagation among the adders one

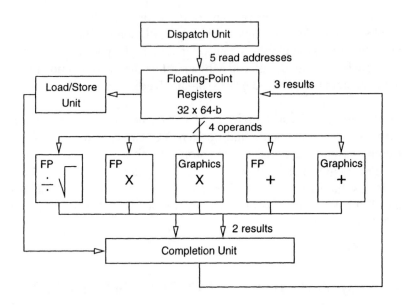

Figure 15.2 Block diagram of the floating-point and graphics unit in the UltraSPARC.

can compute either four 16-bit additions or two 32-bit additions in a single cycle. This technique is identical to the one used in the HP-PA, except that in the HP-PA the partitioning of the data path is performed in the integer ALU. The core of the graphics multiplier is an array of four 8×16-bit multipliers. They perform three-cycle partitioned multiplications, comparisons, and pixel distance operations. For graphics applications, the four parallel multiplications allow the Sun processor to manipulate in parallel up to four color components of a color pixel.

VIS includes close to 30 multimedia instructions that can be classified into five categories: data-conversion instructions, arithmetic and logical instructions, address-manipulation instructions, memory-access instructions, and the motion estimation instruction. The data-conversion instructions convert between the various data types used in the graphics unit. The arithmetic and logical instructions allow partitioned add, subtract, multiply, and logical operations using either single or double precision. Table 15.4 shows a synopsis of the VIS instructions. For example, a 16×16-bit

Instruction	Operation
Arithmetic and logical	
fpadd 16/32 fpsub 16/32	Four 16-bit or two 32-bit partition additions or subtractions.
fmul8x16	Four 8×16-bit multiplications.
flogical	One of ten logical boolean functions.
fone, fzero	Fill destination with ones or zeros.
fnot	Boolean inverse.
fsrc	Copy source to destination.
Data conversion	
fexpand, fpack 16/32, fpackfix, fpmerge	Pack, expand, and merge instructions.
Address-handling	
fcmpcc 16/32	Packed comparisons.
edge8/16/32, array8/16/32, alignaddr, faligndata	Address-handling instructions. Data alignment.
Memory access	
pst, fld, stf bld, bst	8 or 16-bit loads and stores. 64-byte block loads and stores.
Pixel distance	
pdist	Pixel distance computation.

Table 15.4 Synopsis of instructions in VIS.

multiplication requires two of the *fmul8x16* instructions and an extra addition. The address-manipulation instructions operate on the register files and can be used to concatenate data or extract data fields. Memory-access instructions allow for partial or block load and store operations. For example, a block load causes the transfer of 64 bytes from memory to a group of eight consecutive double-precision floating-point registers. The new block instructions allow data to be transferred from the processor directly to the screen at peak rates of up to 600 Mbytes/s.

Of special interest is the new *pdist* instruction that can be used for block-based motion estimation in MPEG, H.261, and H.263 encoding. For eight-bit data, this instruction computes eight subtractions, eight absolute values, and eight additions in a single cycle. If register Ra has 8-bit data a_0 to a_7 and register Rb has 8-bit data b_0 to b_7, the output Rc is given by

$$Rc = Rc + \sum_{i=0}^{7} |a_i - b_i|. \tag{15.2}$$

Thus, computing the minimum absolute error between two 16×16 blocks requires only 32 such instructions, compared to close to 1,500 conventional instructions.

15.5 MULTIMEDIA ENHANCEMENTS IN THE INTEL ARCHITECTURE

In 1996, Intel introduced MMX, a new technology that extends the Intel x86 architecture to accelerate multimedia operations. Like MAX and VIS, MMX applies a single instruction to multiple data elements (8×8-bits, 4×16-bits, or 2×32-bits) that are packed into 64 bits. Like the UltraSPARC, Intel's MMX-enabled processors share the floating-point data path between floating-point instructions and MMX instructions. Thus, applications cannot simultaneously use the floating-point registers for both MMX and floating-point code.

MMX includes 57 new instructions that share many of the features in MAX and VIS and can be classified into seven categories: arithmetic, logical, comparison, shift, conversion, data transfer, and state management. Table 15.5 shows a summary of the MMX instructions. In addition to the usual packed add, subtract, and shift operations, MMX is the only set that provides support for 16-bit fixed-point multiplication. PMULxx performs four 16×16-bit multiplications, and the user can select either the low-order (PMULLW) or the high-order (PMULHW) parts of the 32-bit products. Another unique arithmetic instruction is PMADDWD, which performs four 16×16-bit multiplications and two 32-bit additions. If register Ra has 16-bit data a_0 to a_3 and register Rb has 16-bit data b_0 to b_3, the output Rc is given by

$$Rc = (a_3 \times b_3 + a_2 \times b_2, a_1 \times b_1 + a_0 \times b_0). \tag{15.3}$$

Conversion instructions facilitate conversion among the different packed-data types. Logical operations support packed AND, ANDNOT, OR, and XOR.

Instruction	Operation
Arithmetic and logical	
PADD, PSUB	Packed addition and subtraction.
PMULLW, PMULHW	Four 16 × 16-bit multiplications.
PMADDWD	Four 16 × 16-bit multiplications and two additions.
PAND, PANDN POR, PXOR	Logical operations.
Comparison	
PCMPEQ, PCMPGT	Packed comparison.
Data shifts	
PSRA, PSRL, PSLL	Parallel shifts, right or left.
Data conversion	
PUNPCKL, PUNPCKH	Unpack (interleave) data.
PACKSS	Pack data.
Data transfer	
MOVD, MOVQ	Move 32 or 64 bits between MMX registers and memory.
State management	
EMMS	Empty MMX state.

Table 15.5   Summary of instructions in MMX.

Memory transfer instructions can be used for the efficient transfer of data to and from memory.

MMX maintains full compatibility with all existing PC-based operating systems, which view MMX instructions like floating-point instructions. During run-time, the software can query the microprocessor to determine MMX support and execute either regular or MMX code depending on the answer.

Hewlett-Packard, Sun, and Intel are not the only ones with multimedia-enhanced general purpose processors. Digital Equipment Corporation (DEC) has extended the Alpha instruction set with 13 special instructions for motion estimation, min/max, and pack/unpack operations (the MVI extensions). The new pixel-error instruction (PERR) computes the sum of the absolute value of the differences of eight pairs of bytes in two cycles. The MIN and MAX vector instructions can be used to clamp the values of eight 8-bit words or four 16-bit words with a single instruction. All MVI instructions are implemented in the integer unit. DEC hopes to be the first one to offer MPEG-2 video encoding on a general purpose processor. MIPS has announced its own set of multimedia enhancements (the MDMX instructions) in the MIPS V architecture. By packing two 32-bit floating-point values into a single 64-bit register, MIPS plans to provide support for parallel single-precision floating-point operations for improved graphics performance. Parallel computations on integer data will also be supported. Cyrix is also planning its own x86-compatible processor with full MMX compatibility.

15.6 TO PROBE FURTHER

Designers of programmable video processors were the first to analyze the video compression pipeline for generic operations [178]. Of particular interest are the architectural enhancements and the instruction set of the vector pipeline digital signal processor (VDSP) from Matsushita [10]. Based on a DSP core, the VDSP is a programmable video processor suitable for JPEG and H.261 coding applications. A special vector-pipeline controller allows the VDSP to execute arithmetic operations on vector data with a single instruction.

RISC cores are used extensively in video codecs. In [95], Holmann er al. present a dual-issue RISC processor for real-time MPEG-2 decoding. Similarly, Nadehara et al. [181] describe a low-power 32-bit RISC processor with DSP capabilities. This processor can achieve real-time MPEG-1 video decoding while dissipating only 160 mW.

The HP PA7100LC chip-set is described in [130] and [250]. Additional details regarding the software-based MPEG-1 decoder from Hewlett-Packard can be found in [146] and [24]. A detailed description of MAX is given in [147]. Sun's UltraSPARC architecture and the VIS instruction set are described in [246], [132], and [247]. A detailed description of MMX is given in [203]. An Alpha architecture with the first implementation of the MVI extensions is described in [117].

16

STANDARDS FOR AUDIO COMPRESSION

16.1 INTRODUCTION

Digital audio is an integral part of any video or multimedia application and may consume a considerable portion of the overall bandwidth. Therefore, efficient compression of the audio data is as important as the efficient compression of the video data. There are two main parameters that control the quality and the bit rate of digital audio: the sampling resolution (that is, bits per audio sample) and the sampling frequency.

The sampling resolution determines the dynamic range of a signal. For example, audio compact discs (CDs) use 16-bit samples and provide 96 dB of dynamic range ($96 = 20 \, log_{10} 2^{16}$). The sampling frequency determines how much of the frequency spectrum of the original analog signal can be faithfully reproduced. According to the Nyquist theorem, the sampling frequency has to be twice as large as the maximum frequency component. For example, CD recordings use a 44.1 kHz sampling frequency, approximately twice the maximum frequency response of the human auditory system (22 kHz). In digital telephony, where high audio fidelity is not as important, signals are bandlimited to approximately 3.5 kHz, and the sampling frequency is 8 kHz. Table 16.1 shows some typical audio applications and the corresponding sampling rates.

According to the CCIR recommendations, digital audio in broadcast studio format requires a sampling resolution of 16 bits and a sampling frequency of 48 kHz. For stereo signals, the corresponding bit rate is $2 \times 768 = 1.54$

Sampling Frequency (kHz)	Application
8.00	Digital Telephony
22.05	Personal Computers
32.00	Digital audio and TV
44.10	Compact Discs
48.00	Digital Audio Tapes, HDTV

Table 16.1   Sampling frequencies for typical audio applications.

Mbits/s. Similarly, the bit rate from a stereo CD is close to 1.4 Mbits/s, and a ten-minute stereo audio clip requires close to 105 Mbytes of storage. The above data indicate that efficient storage and transmission of digital audio require some form of data compression.

Audio compression can be achieved by taking into consideration the characteristics of the audio source (for example, the human vocal tract), the human perception system, or both. The telecommunications industry has applied compression to speech signals for many years. Of special interest are the coding schemes described in the G.72x series of Recommendations from the International Telecommunications Union (ITU) for coding speech and audio over band-limited channels. For toll-quality speech, up to 20:1 lossy compression is now possible. However, the quality requirements for generic audio are far greater than speech over toll-quality telephone channels. There are two main compression techniques for high-fidelity audio: the MPEG audio standard and the AC audio compression algorithms developed by Dolby Laboratories. Both techniques use psychoacoustic models and can achieve CD-like quality at 128 kbits/s per audio channel (5.5:1 compression).

Both MPEG and AC-3 make no assumptions about the source of the input signal, and they can be applied to both speech and high-fidelity audio. Compression is achieved by transforming the input signal into the frequency domain and by using psychoacoustic models to remove information that is perceptually irrelevant. On the other hand, coders in the G.72x family of standards achieve compression by taking into consideration the special characteristics of speech and the human vocal tract.

In the next sections we provide a short overview of the G.72x recommendations and we describe in more detail the MPEG-1 and the AC-3 audio standards. Some fast algorithms for the implementation of MPEG audio and future directions in MPEG are also discussed. We conclude with a short description of commercially available audio decoders.

16.2 THE G.72X STANDARDS

These standards have been developed by the ITU specifically for the efficient coding of speech over band-limited telephone channels. They are applicable to a variety of telecommunication applications, including teleconferencing (for example, together with the H.261 or H.263 video compression standards) and voice-data modems. These standards achieve compression by using predictive coding.

G.721 was the first standard for the coding of speech at 32 kbits/s on channels with a 300 to 3400 Hz bandwidth. In 1990, G.721 was replaced by the G.726 standard, which allows for compressed data rates at 16, 24, 32, and 40 kbits/s. Both G.721 and G.726 encode the input data using adaptive differential pulse code modulation (ADPCM).

For broadband telephone channels (0 to 8 kHz), ITU has developed the G.722 standard on "7 kHz audio coding within 64 kbit/s." This is an enhanced version of the G.726 standard and uses a subband filter as a front end to two separate ADPCM coders. The input is sampled at 16 kHz, and the subband filter separates the input stream into the two subbands of 0 to 4 kHz and 4 to 8 kHz. The higher subband is coded at 16 kbits/s, and the lower subband is coded at 48 kbits/s. The two streams are multiplexed for a total bit rate of 64 kbits/s.

For lower bit rates, the G.728 standard achieves speech quality at least as good as the G.721 but at 16 kbits/s. The improved performance is achieved by replacing the ADPCM coder in G.721 with a low-delay code excited linear predictor (LD-CELP). The main principle of a linear predictive coder (LPC) is to model the human vocal tract as a linear filter of the form

$$y_n = \sum_i a_i y_{n-i} + G e_n, \tag{16.1}$$

where a_i are the filter coefficients, G is the gain of the filter, e_n is the input to the filter, and y_n is the output of the filter. In a CELP coder, first, the encoder determines the best filter coefficients by analyzing short segments of the input data. Next, it excites the vocal tract filter using excitation patterns taken from a codebook, and it computes the difference between the original and the synthesized speech signals. After a perceptual weighting of the error signals, the encoder determines the excitation pattern in the codebook that provides the best match between the original and the synthesized speech. Compression is achieved by transmitting only the index of the optimum excitation pattern. Note that, if the encoder determines the LPC filter coefficients from past data, then their values need not to be transmitted to the decoder. Instead, the decoder can compute the same values by mimicking the operations in the encoder.

G.723.1 is an audio coding standard that has been developed as part of the H.324 standard for videoconferencing. This standard specifies a dual-rate coder that can be used for compressing either speech or other audio signals. G.723.1 specifies two bit rates: 5.3 and 6.3 kbits/s. Both rates are mandatory for the encoder and the decoder, and it is possible to switch between the two rates at any frame interval. For clear speech, the 6.3 kbits/s mode provides speech quality equivalent to that provided by the 32 kbits/s G.726 standard (toll-quality). The 5.3 kbits/s mode performs better than the IS54 digital cellular standard. The coder encodes speech and other audio using linear predictive analysis-by-synthesis coding. For the high rate, it uses a multipulse maximum likelihood quantizer (MP-MLQ) coder. For the low rate, it uses an algebraic codebook excitation linear predictor (ACELP) coder.

16.3   FUNDAMENTALS OF PERCEPTUAL CODERS

Coding techniques for high-fidelity audio achieve compression by using knowledge of both the audio signal and the characteristics of human perception. The basic idea of perceptual coding is to take advantage of the perceptual weaknesses in the human ear, so that any distortion introduced by the coder is perceptually irrelevant.

One of the best understood psychoacoustic effects is the phenomenon of auditory masking. For example, in time-domain masking, the ear is less sensitive immediately before and after a loud signal such as a percussion strike. However, most coding techniques take advantage of masking effects in the frequency domain. In frequency domain masking, a faint signal may be completely masked if it is in the vicinity of louder signals with similar frequency content. A simple example is shown in Figure 16.1, where, due to the two narrow-band tones, all signals below the masking threshold will be inaudible. In Figure 16.1, signal power is denoted as sound pressure level

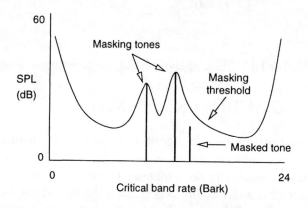

Figure 16.1 Example of frequency-domain masking.

(SPL) and is measured in dB. Frequency is shown in a logarithmic-like scale, defined as critical band rate, and is measured in Bark. Critical bands reflect the resolving power of the ear as a function of frequency.

Figure 16.2 shows the block diagram of a basic perceptual coder. The input signal is processed first by an analysis filterbank, also known as a subband filter, which translates the input into the frequency domain. Using rules from psychoacoustics, a perceptual model estimates the masking thresholds. This information is passed to the quantizer so that all quantization noise is below the masking thresholds. The output coded bit stream consists of the encoded quantized data plus some side information, such as the bit allocation scheme. Both the MPEG and the AC-3 coding standards follow this model and are examined next.

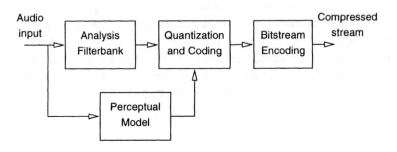

Figure 16.2   Example of frequency-domain masking.

16.4   MPEG AUDIO ENCODING

The MPEG audio compression algorithm is part of the MPEG-1 standard that addresses not only the compression of video and audio but also their synchronization. Audio can be compressed down to 32 kbits/s for a single channel. The MPEG audio coding algorithm is the first international standard for the compression of digital audio signals. It can be applied to streams that combine both audio and video information or to audio-only streams. Like the MPEG video standard, the MPEG audio standard specifies the syntax of the coded bit stream, defines the decoding process, and provides compliance tests. This allows for different encoding algorithms, provided that the encoded data can be decoded by an MPEG-compliant decoder. The MPEG audio compression standard achieves compression by exploiting the spectral and temporal masking effects of the ear. It uses subband coding and a psychoacoustic model to eliminate information that is irrelevant from a human sound-perception viewpoint.

Figure 16.3 shows a block diagram of the MPEG encoder for a single audio channel. In multichannel systems, the process is repeated independently for each channel. Input pulse code modulated (PCM) samples have 16 to 20 bits of precision and are sampled at 32, 44.1, or 48 kHz. The first stage of the encoder is a subband analysis filter that can be viewed as filter bank with 32 band-pass filters. The frequency response of each 512-tap band-pass filter is derived by frequency shifting the response of a prototype low-pass filter. The output of each filter is decimated by a factor of 32 so that for every 32 input samples the subband filter generates 32 output samples.

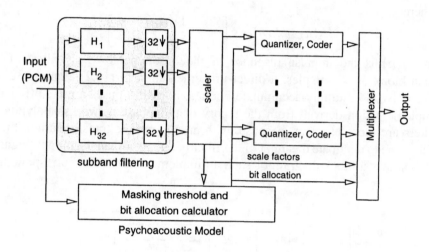

Figure 16.3 Block diagram of the MPEG audio encoder.

In the next stage (the scaler), the output of each subband filter is normalized by scale factors that will be transmitted together with the compressed bit stream as side information. Scale factors correspond to the maximum absolute value of every 12 consecutive output values in each subband. Based on the frequency characteristics of the input signal and the desired bit rate of the compressed signal, a psychoacoustic model computes masking thresholds and bit allocations. This information is used by the quantizer and coder units. Finally, the quantized subband samples, the scale factors, and the bit-allocation information are multiplexed to form the compressed bit stream.

16.4.1 Subband Analysis

As mentioned before, the subband analysis filter can be viewed as a filter bank with 32 filters. Let $x(t)$ denote an audio sample at time t and $H_i(t)$ denote the impulse response of the i-th filter. Then the output of each filter can be expressed as

$$s(i) = \sum_{n=0}^{511} x(t-n) H_i(n), \qquad (16.2)$$

where

$$H_i(n) = h(n)\cos\left[\frac{(2i+1)(n-16)\pi}{64}\right], \qquad (16.3)$$

and $h(n)$ is the impulse response of the prototype low-pass filter. For each block of 32 samples, a direct implementation of (16.2) requires $32 \times 512 = 16{,}384$ multiply-accumulate operations. MPEG uses a more efficient implementation derived from the theory of polyphase networks. Polyphase filters are more efficient, because, with an appropriate transformation, they allow one to decimate the input of the filter bank rather than the output. Figure 16.4 shows a block diagram of such an implementation. The corresponding

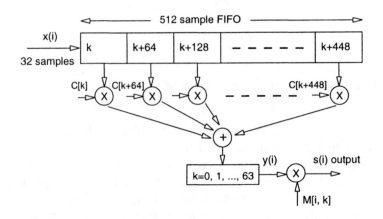

Figure 16.4 MPEG implementation of the subband analysis filter.

flow of operations as defined by the MPEG standard is as follows:

- *Input:* 32 new samples $x(i)$ are shifted into the FIFO buffer.

- *Windowing:*
$$z(i) = C[i]x(i), i = 0, 1, \cdots, 511, \qquad (16.4)$$
where $C[i]$ is one of 512 coefficients defined in the standard.

- *Partial calculation:*
$$y(i) = \sum_{j=0}^{7} z(i + 64j), i = 0, 1, \cdots, 63. \qquad (16.5)$$

- *Matrixing operation:*

$$s(i) = \sum_{k=0}^{63} y(k)M[i,k], i = 0, 1, \cdots, 31,$$ (16.6)

where $M[i,k] = \cos\left[\frac{(2i+1)(k-16)\pi}{64}\right]$.

- *Output:* Output 32 subband samples $s(i)$.

From Figure 16.4 and the above flow of operations, one can derive the following equation for the output of the i-th filter.

$$s(i) = \sum_{k=0}^{63}\sum_{j=0}^{7} M[i,k]\left(C[k+64j]x(k+64j)\right), i = 0, 1, \cdots, 31,$$ (16.7)

The above implementation requires $512 + 32 \times 64 = 2,560$ multiplications (512 for the windowing operation and 2,048 for the matrixing operation) and $64 \times 7 + 32 \times 63 = 2,464$ additions. The matrixing operation (16.6) is the most time-consuming with 2,048 multiply-accumulate operations.

16.4.2 Fast Matrixing Computation

The computational complexity of the matrixing operation can be reduced further using a fast inverse DCT (IDCT) transform. Let $y(k)$ be the output of the partial calculation step (16.5). If

$$y'(k) = \begin{cases} y(16) & k = 0 \\ y(k+16) + y(16-k) & k = 1, 2, ..., 16 \\ y(k+16) - y(80-k) & k = 17, 18, ..., 31 \end{cases}$$ (16.8)

then, it can be shown that the output of the matrixing operation can be computed as

$$s(i) = \sum_{k=0}^{31} y'(k)\cos\left[\frac{\pi}{64}(2i+1)k\right], i = 0, 1, \cdots, 31,$$ (16.9)

where $s(i)$ is (within a scale factor) the 32-point IDCT of y'. Using a fast IDCT algorithm, such as Lee's, the IDCT can be computed with only 80 multiplications and 209 additions.

16.4.3   The Psychoacoustic Model

A key, and computationally intensive, component of the MPEG encoding standard is the psychoacoustic analysis of the input signal. This analysis is performed in parallel with the subband filtering operation. The first step in this process is the computation of the masking curve according to the frequency-masking properties of the human ear. From the masking curve, a set of masking thresholds is derived for each subband. Each of them determines the maximum energy acceptable for the quantization noise in each subband (that is, below this level the noise will not be perceived). The computation of the masking curve requires a very accurate spectral analysis of the input signal. In MPEG, the spectrum of the input signal is usually computed using a 1,024-point fast Fourier transform (FFT) of the input samples. From the spectral lines, the psychoacoustic model determines tonal and nontonal components and determines the final masking curve.

For a given bit rate and perceptually lossless compression, all of the quantization noise should be below the masking thresholds. However, for low bit rates this is usually not the case. In such cases, the psychoacoustic model uses an iterative algorithm that allocates more bits to the subbands where increased resolution will provide the greatest benefit.

Note that MPEG is not enforcing the use of a psychoacoustic model. For example, applications that can afford high bit rates can use simpler bit allocation schemes, such as assigning more bits to the subbands with the lower signal to noise ratios.

16.4.4   Layer Coding Options in MPEG

The MPEG audio compression standard defines three layers of compression. Each successive layer offers better compression performance but at a higher complexity and computational cost. All three layers support bit rates as low as 32 kbits/s for a single channel.

Layer I is the simplest of all and provides good quality audio at 192 kbits/s. The highest supported bit rate for Layer I is 448 kbits/s. Layer II codes data in larger groups, uses a more complex psychoacoustic model, and provides

CD-audio quality at 128 kbits/s per channel. The highest supported bit rate for Layer II is 384 kbits/s.

Layer III is the most complex of the three and can be used in low-bandwidth or noisy channels. Like Layers I and II, it uses a subband filter, but it provides additional frequency resolution by processing the output of the filter bank by a modified DCT. Other distinct features of Layer III include nonuniform quantization, adaptive segmentation, and entropy coding. Layer III is also performing a better compression of stereo signals by exploiting the cross-correlation between the left and right channels. The highest supported bit rate for Layer III is 320 kbits/s. Layer III is the only Layer that can operate in a variable-rate mode. Due to the additional complexity of Layer III, most of the existing applications use either Layer I or Layer II.

16.4.5 MPEG-2 Audio Coding

The MPEG-2 audio compression standard is an extension of MPEG-1 with the following added features.

- *Multichannel input*: MPEG-2 supports up to five high-fidelity audio channels plus a low-frequency enhancement channel (these six channels are also known as 5.1 channels). Thus, it is suitable for the compression of audio in HDTV or digital movies.

- *Multilingual audio support*: Up to eight commentary channels are supported.

- *Lower bit rates*: Compressed bit rates down to 8 kbits/s are now supported.

- *Additional sampling rates*: In addition to the original sampling rates of 32, 44.1, and 48 kHz, MPEG-2 accommodates sampling rates at 16, 22.05 and 24 kHz.

In many ways, the MPEG-2 audio standard is compatible with the MPEG-1 audio standard. For instance, MPEG-2 audio decoders can decode MPEG-1 audio bit streams, and MPEG-1 audio decoders can decode the two main channels of the MPEG-2 audio bit streams.

16.5   MPEG AUDIO DECODING

Figure 16.5 shows a block diagram of an MPEG audio decoder. The MPEG

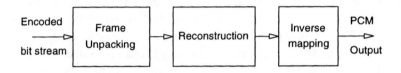

Figure 16.5   Block diagram of the MPEG audio decoder.

audio decoder reverses the encoding operation by performing the following steps.

1. *Frame unpacking and parsing:* The bit stream is parsed, and the various pieces of coding information are demultiplexed.

2. *Reconstruction:* The bit allocation information is decoded, and the scale factors are extracted.

3. *Inverse mapping:* Data samples are dequantized and denormalized and then processed by a subband synthesis filter.

4. *Output:* The output of the subband filter is translated into PCM audio samples.

Among these steps, subband synthesis is the most compute intensive. Figure 16.6 shows a block diagram of the subband synthesis filter as defined by the MPEG standard. From Figure 16.6, there are three main operations in subband synthesis: (1) the matrixing operation, (2) data shifting, and (3) the windowing operation. In most implementations, the shifting and the windowing operations are combined into a single step.

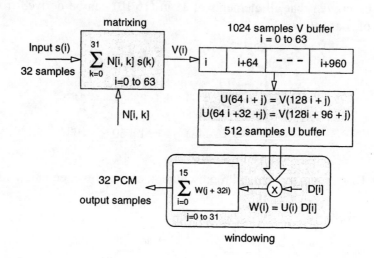

Figure 16.6 MPEG implementation of the subband synthesis filter.

16.5.1 Fast Matrixing Operation

From Figure 16.6, the matrixing operation in subband synthesis is defined as

$$
V(i) \;=\; \sum_{k=0}^{31} N[i,k]s(k) \;= \tag{16.10}
$$

$$
\;=\; \sum_{k=0}^{31} \cos\left[\frac{\pi}{64}(2k+1)(i+16)\right] s(k), i = 0, 1, \cdots, 63,
$$

where $s(k)$, $k = 0$ to 31 denote input encoded audio samples. Brute force evaluation of $V(i)$ requires $32 \times 64 = 2,048$ multiply-accumulate operations. Let S denote the 32-point DCT of s, defined as

$$
S(i) = \sum_{k=0}^{31} \cos\left[\frac{\pi}{64}(2k+1)i\right] s(k), i = 0, 1, \cdots, 31. \tag{16.11}
$$

It can be proven that all elements of V in (16.10) can be derived from the values of S as follows:

$$V(i) = \begin{cases} S(i+16) & i = 0, 1, \cdots, 15 \\ 0 & i = 16 \\ -V(32-i) & i = 17, 18, \cdots, 32 \\ -S(48-i) & i = 33, 34, \cdots, 47 \\ -S(0) & i = 48 \\ -S(i-48) & i = 49, 50, \cdots, 63 \end{cases} \tag{16.12}$$

Figure 16.7 shows a graphical representation of the above equation. If $S(i)$,

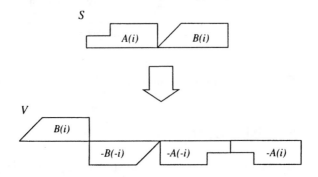

Figure 16.7   Graphical representation of data dependencies in the output of the matrixing operation in subband synthesis.

the output of the 32-point DCT, is divided in two parts $A(i)$ and $B(i)$, then V can be composed by the appropriate shifting and manipulation of these parts. Shapes of data in Figure 16.7 were chosen arbitrarily to better show the symmetries within V.

16.6   THE AC-3 CODING STANDARD

The AC-3 audio coding scheme from Dolby Laboratories was developed in response to a need to provide high-quality, multichannel, digital audio for HDTV and film. For example, a digital film sound system has 5.1 channels (left, center, right, left and right surround, and a subwoofer.) This audio information has to be placed in the film without interfering with either the picture or the analog sound area. It was determined that one could use the film area between the spocket hole perforations to place a 320 kbits/s error corrected audio stream. In 1991, *Star Trek VI* was the first film coded with the AC-3 system.

In 1993, after successful field tests, an FCC advisory committee recommended the use of Dolby AC-3 for the Grand Alliance HDTV system. Interestingly enough, the Grand Alliance system uses the MPEG-2 standard for video compression. In HDTV, the sampling resolution is 18 bits, and the sampling frequency is 48 kHz. For six channels, AC-3 compresses the original 5 Mbits/s bit stream down to 384 kbits/s (13:1 compression). The AC-3 standard has also been adopted for video/audio delivery using the digital versatile disk (DVD) system. An AC-3 standard is currently under development by the FCC advisory committee on advanced television services (ATSC).

16.6.1   The AC-3 Encoder

Like MPEG, the AC-3 system uses transform-based coding and a psychoacoustic model to eliminate information that is irrelevant to the human hearing system. A block diagram of an AC-3 encoder is shown in Figure 16.8. The audio input, sampled at 48 kHz, is first transformed from the time domain into the frequency domain using a filter bank. The block size is 512 points, and overlapping blocks of 512 windowed samples are transformed into 256 frequency domain points. A 512-point modified DCT with 50% overlap can be used for these operations.

Each output point of the filter bank is represented by an exponent and a mantissa. Coding of the exponent allows for a wide dynamic range, while quantization of the mantissa results in quantization noise. The exponents from each channel represent the overall signal spectrum and are referred to as the *spectral envelope*. Exponents from each block are encoded using one of three differential coding schemes. The coding efficiency of these schemes ranges from 2.33 bits per exponent to 0.58 bits per exponent. The AC-3 encoder selects the appropriate scheme based on the signal characteristics. However, it is common for multiple blocks to use the same exponent set. In a typical case, the spectral envelope is sent every six audio blocks (32 milliseconds) for an average bit rate of less than 0.39 bits per exponent. After the transformation

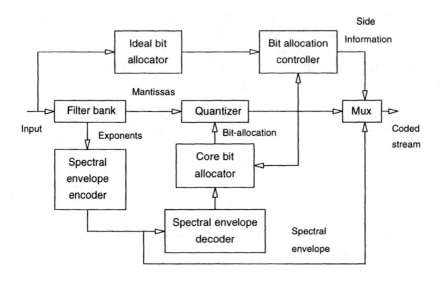

Figure 16.8   Block diagram of the AC-3 audio encoder.

of the input samples, the encoder determines the optimum bit allocation and quantizes the mantissas. There are two main bit-allocation strategies for psychoacoustic based coders: forward-adaptive or backward-adaptive bit allocation.

In forward-adaptive bit allocation the encoder uses only the input signal to determine the bit allocation, which is later explicitly coded in the output bit stream. For example, this is the strategy used by the MPEG audio coder. This is the most accurate method; however, it consumes a significant portion of the available bandwidth. For example, MPEG (Layer II) may allocate close to 4 kbits/s for this side information. Another advantage of forward bit allocation is that the decoder requires no prior knowledge of the psychoacoustic model. This allows the encoder to be updated as needed without affecting the design of the decoder.

In backward-adaptive bit allocation, the decoder can determine the bit allocation directly from the encoded data without any side information from the encoder. This requires that both the encoder and the decoder have identical copies of the bit-allocation scheme. This technique is not as accurate as forward-adaptive bit allocation, and the psychoacoustic model cannot be updated after the decoders are built.

AC-3 uses a hybrid forward/backward model. There is a core backward-adaptive bit allocator that resides on both the encoder and the decoder. However, the encoder can use any other psychoacoustic model to determine the optimum bit allocation. If a comparison between the outputs of the optimum and the core allocator shows that the default bit allocation needs to be improved, then the encoder tries first to modify some of the parameters used in the core model. If no further improvement is possible with the core model, then the encoder can explicitly transmit some additional bit-allocation information. Since the core algorithm performs well in most cases, only small changes in bit allocation need to be transmitted.

The coding efficiency of AC-3 improves as the number of source channels increases, due to two features of the AC-3 coder: a global bit pool and high-frequency coupling. The audio bit pool allows the bit allocator to allocate the available bits among the audio channels on an as-needed basis. Coupling is used to compress the data rate for the high-frequency range of the audio spectrum. This is achieved by reducing the high-frequency components of correlated channels to a single coupling channel and then generating additional data that describe the spectral envelope of each channel.

16.6.2   The AC-3 Decoder

A block diagram of the AC-3 decoder is shown in Figure 16.9. The coded stream is first demultiplexed to generate the spectral envelope, the quantized mantissas, and any bit-allocation side information. The spectral envelope is decoded and the mantissas are dequantized using the core bit allocator and the side bit-allocation information. The decoded mantissas and exponents are combined to form the frequency coefficients and an inverse filter bank translates the frequency data into PCM audio signals.

The AC-3 bit stream syntax supports the coding of one main audio service with one to 5.1 channels. Other services, such as information for visually and hearing impaired, may also be embedded in the AC-3 bit stream. Single channels may be coded at bit rates as low as 32 kbits/s. The maximum supported bit rate

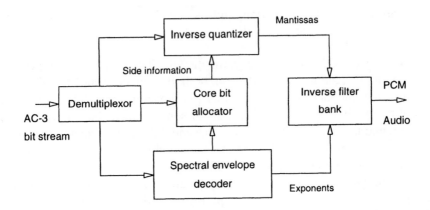

Figure 16.9   Block diagram of the AC-3 audio decoder.

is 640 kbits/s. The higher bit rate allows for the incorporation of additional services without compromising the audio quality of the main service.

Table 16.2 summarizes the main characteristics of the G.72x, MPEG, and Dolby AC-3 audio coding standards.

16.7   FUTURE DIRECTIONS IN MPEG

16.7.1   MPEG-2 AAC

MPEG has already completed work on a non-backward compatible extension of MPEG-2, referred to as MPEG-2 NBC or MPEG-2 advanced audio coding (MPEG-2 AAC). This work reflects the need for aggressive bit-rate reduction in multichannel audio at bit rates close to 320 kbits/s for five channels. In addition to the MPEG-2 requirements, MPEG-2 AAC also supports (but without backward compatibility):

- Sampling rates of 48, 44.1, and 32 kHz.

- Mono, stereo, and other multichannel configurations, up to 5.1 channels.

Coder	Bit Rate	Quality
G.726	16-40 kbits/s	toll-quality at 32 kbits/s.
G.722	64 kbits/s	high-quality at 48 kbits/s.
G.728	16 kbits/s	toll-quality.
G.723.1	5.3 or 6.3 kbits/s	toll-quality at 6.3 kbits/s.
MPEG-1 Layer I	32-448 kbits/s	transparent at 192 kbits/s per channel.
MPEG-1 Layer II	32-384 kbits/s	transparent at 128 kbits/s per channel.
MPEG-1 Layer III	32-320 kbits/s	transparent at 96 kbits/s per channel.
MPEG-2	8-448 kbits/s	same as in MPEG-1.
AC-3	32-640 kbits/s	transparent at 384 kbits/s for 5.1 channels.

Table 16.2   Main characteristics of the G.72x, MPEG, and AC-3 audio coding standards.

- Broadcast quality audio at 384 kbits/s in a 3/2 channel configuration (left, right, center, and left and right surround).

- A predefined audio access unit for a minimum granularity needed for editing purposes.

- Improved error resilience and some error concealment.

Work on MPEG-2 AAC is expected to be completed by April of 1997 as the international standard ISO/IEC IS 13818-7. MPEG-2 AAC will also constitute the kernel of the forthcoming MPEG-4 audio standard. A short description of MPEG-2 AAC is given next.

Unlike MPEG-1 and MPEG-2, MPEG-2 AAC is not a monolithic standard, but a collection of tools. Users can make tradeoffs between quality and implementation requirements by using these tools in any of the three MPEG-2 AAC profiles: (1) the main profile, (2) the low complexity profile, and (3) the sample rate scalable profile. The MPEG-2 AAC tools are divided into required and optional tools. Required tools are mandatory in any of the three profiles.

Optional tools may be skipped in some profiles. Figure 16.10 shows a block diagram of an MPEG-2 AAC encoder, where optional blocks shown in gray.

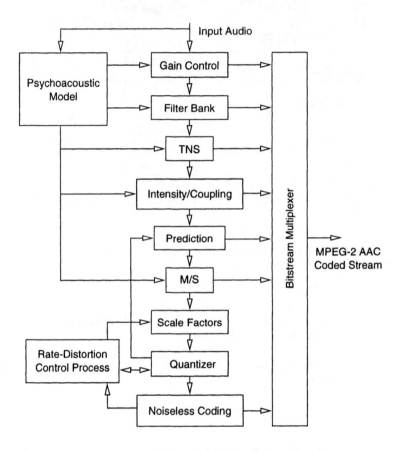

Figure 16.10   Block diagram of the MPEG-2 AAC audio encoder.

The main profile yields the highest quality and utilizes the following tools.

- **Filter bank tool:** This block is used to decompose the input signal into subsampled spectral components. The MPEG-2 AAC filter bank is based on the modified DCT (MDCT) and time domain aliasing cancelation (TDAC). Thus, it is similar to the one used in Dolby AC-3. Using data from either the time-domain signal or the output of the filter bank, the encoder estimates masked thresholds using a psychoacoustic model. MPEG-2 AAC uses a model similar to the second psychoacoustic model used in MPEG-1 and MPEG-2.

- **Temporal noise shaping (TNS) tool:** This tool is used to control the temporal shape of the quantization noise. This is done by filtering parts of the spectral data in each channel. TNS is particularly useful when coding transient or pitched signals. Certain parameters of this tool are profile dependent.

- **M/S, intensity stereo, and coupling tools:** In traditional coding of multiple channels, a pair of channels, say L and R, can be coded either separately or jointly using Mid/Side (M/S) coding, where $M = \frac{L+R}{2}$ and $S = \frac{L-R}{2}$.

  In addition to M/S coding, MPEG-2 AAC allows for joint coding using the intensity stereo and coupling tools. The idea behind intensity coding is that a pair of channels may share a single set of spectral values. In that case, the original energy-time envelopes of the coded channels are preserved approximately by simple scaling operations so that each channel can be reconstructed with its original levels. The coupling tool is far more general and allows channel spectra to be shared across channel boundaries or among different channel pairs. It also allows other sound objects, such as commentary, to be mixed with the coded signal.

- **Prediction tool:** This tool applies a second-order backward-adaptive predictor to the spectral components, so that each predictor is working on the spectral components of the two preceding frames. By predicting the values of spectral components the overall coding efficiency improves.

- **Scale factors, quantization, and noiseless coding tools:** While all of the preceding blocks preprocess the audio data, the actual data reduction is performed in these blocks. Scale factors are used to shape the quantization noise in the spectral domain. The spectrum is divided into scalefactor bands and each scalefactor band has a scale factor, which represents a gain value that has to be applied to all spectral coefficients in that band. After scaling, MPEG-2 AAC uses a non-uniform quantizer for quantization and Huffman coding for noiseless (lossless) coding. Note that, the standard specifies only the bitstream syntax and does not specify the strategy to optimize the quantization process. A typical encoder will have to use some sort of an iterative scheme, where the scale factors and the quantization steps are iteratively modified until the requirements for error distortion and bit rate are jointly satisfied.

- **Bitstream formatting tool:** This tool assembles the bit stream, which includes control parameters and the quantized and coded spectral data.

The low complexity profile does not use the gain control and prediction tools, and TNS order is limited. The sampling rate scalable profile is the only profile that uses the gain control tool, but it does not use the prediction and coupling tools, and TNS order and bandwidth are limited. The gain control tool includes several gain compensators, overlapp/add processing and an inverse polyphase quadrature filter (IPQF) stage.

A block diagram of an MPEG-2 AAC decoder is shown in Figure 16.11. The decoder reverses the encoding process and has the option to bypass any of the optional tools that are specified in the bitstream syntax.

16.7.2   MPEG-4

MPEG-4 intends to become the next major standard for multimedia applications. Like MPEG-4 video, it is focused on an object-oriented approach, where speech, high-fidelity audio, and synthetic computer music are all coded using the same scalable approach. This will be accomplished by using the MPEG syntax description language (MSDL). MPEG-4 does not plan to introduce any new tools for coding audio at rates above 64 kbits/s per channel. Instead, the new functionalities of MPEG-4 will include

- Scalability. This is one of the main features of MPEG-4 and means that some parts of the bit stream are sufficient for decoding and generating a meaningful signal with low fidelity, bandwidth, or content.

- Limited time audio streams. A user can control the reproduction of multiple audio channels by selectively mixing them into a single or multiple data streams.

- Pitch change and time-scale change. This functionality allows for easier synchronization of audio events.

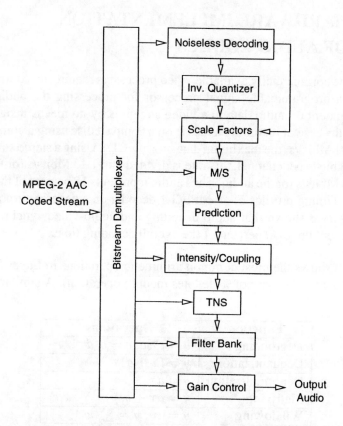

Figure 16.11   Block diagram of the MPEG-2 AAC audio decoder.

- Editability. That is, flexible access to the bit stream and if possible direct manipulation of the data in the compressed domain.

- Flexible, application-dependent, delay.   For example, noninteractive applications can tolerate large delays, but videoconferencing applications cannot.

Since no single proposal is expected to be able to accomplish all these requirements, MPEG-4 will consist of a family of tools and algorithms with a high degree of configurability. Work on MPEG-4 is expected to be completed by the end of 1998.

16.8   HARDWARE IMPLEMENTATION
OF AUDIO CODECS

Most of the commercially available video processors require an external audio coder or a programmable signal processor for processing the audio. Note that the amount of audio data in a video stream is by no means insignificant. Consider the case of recording a movie on a compact disc using a compression rate of 1.41 Mbits/s (the maximum data rate of a CD). Using a stereo soundtrack with 128 kbits/s per channel, this rate is divided into 1.15 Mbits/s for the video and 0.256 Mbits/s for the audio. Thus audio represents 18 percent of the overall data rate. Timing profiles of an MPEG-1 decoder on a general purpose RISC processor have shown that audio decoding and audio-video synchronization may represent up to 30 percent of the overall decoding time.

Table 16.3 shows the most common arithmetic operations in Layer II MPEG audio decoding (the symbol % denotes modulo operation). A similar analysis

Function	Operations
Degrouping	$y = c \% a, c = c/d$
Dequantization	$y = (x + a)b$
Denormalization	$y = ax$
Matrixing	$y = ax + b, y = \sum_i x_i c_i$
Windowing	$y = xa, y = \sum_i w_i$

Table 16.3   Arithmetic operations in MPEG audio decoding.

of the operations in other audio coding algorithms shows that multiplication and addition are the most common operations. Hence, general-purpose DSPs are ideally suited for audio processing. Because audio samples have 16 bits of precision, encoders should use either floating-point or 24-bit precision.

16.8.1   Commercially Available Audio Decoders

Commercially available audio processing ICs provide a low-cost solution to the decoding of compressed audio. They are either dedicated or programmable

DSPs and provide a glueless interface to memory and a host controller (if needed).

The CS4920 from Crystal Semiconductors is a programmable DSP for either MPEG-1 or AC-2 (precursor to AC-3) audio decoding. It has a 24-bit architecture and is the only audio decoder with an on-chip programmable PLL clock generator and dual 16-bit digital to analog converters (DACs) for the direct generation of stereo audio from a compressed digital stream. Its audio transmitter is compatible with the Sony/Philips digital interchange format (SPDIF). The CS4922 will be pin-compatible with the CS4920 and will support both MPEG-1 and MPEG-2 (Layers 1 and 2) audio decoding. In addition, it will offer PCM synthesis for auxiliary audio and ancillary data support.

The L64111 from LSI Logic is a two-channel MPEG audio decoder. It can decompress both MPEG-1 and MPEG-2 (two channels) streams (Layers I and II). It supports bit rates from 32 (mono) to 384 kbits/s (stereo). Data can be loaded either bit-serially or in parallel. The L64111 has a 24-bit architecture and runs at 25 to 30 MHz. The processor uses a serial divider for the degrouping operation and a two-cycle, 24-bit multiplier-accumulator for all other decoding operations.

The STI4510 from SGS-Thomson is another MPEG-1 audio decoder. At 24 MHz, it supports input rates of up to 20 Mbits/s. It supports all MPEG-1 sampling rates and has a microcontroller interface.

Most of the DSPs from Texas Instruments (TI) could be used for audio coding or decoding. For example, results from TI indicate that an MPEG-1 (Layers I and II, mono or stereo) encoder can run on a 40 MHz TMS320C30 using a 32K × 32-bit, zero wait state, external SRAM. The decoder requires at least a 27 MHz TMS320C30 with 8K × 32-bit, one wait state, external SRAM. (The TMS320C30 is a floating-point processor). TI also provides a dedicated MPEG-1 audio decoder (Layers I and II), the TMS320AV110.

Another audio decoder is the ZR38000 programmable DSP from Zoran. This is a 20-bit processor with a 20 × 20-bit multiplier, a barrel shifter, and a 48-bit accumulator. The processor has a 16-word instruction cache and supports block floating-point operations. The ZR38000 can execute a radix-2 FFT butterfly in four cycles. This allows it to compute a 1,024-point radix-2 FFT in 0.877 ms (using a 25-MHz clock). Using the ZR38000 core, Zoran plans to introduce dedicated processors for six channel AC-3 decoding (the ZR38500), two-channel AC-3 decoding (the ZR38501), and two-channel MPEG audio decoding (the ZR38511). According to Zoran, the ZR38500 is the only IC that can decode in real-time a six-channel AC-3 stream.

16.9   TO PROBE FURTHER

The textbook on data compression by K. Sayood [225] provides additional details and examples on predictive coders for speech signals. The G.72x recommendations for speech coding are described in the ITU publications [65], [63], [66], and [64]. In addition, for the G.722 standard, an extended summary is given in [165], and a custom hardware implementation is described in [259]. A hardware implementation of the G.728 on a general-purpose DSP is reported in [7]. The G.723.1 coding standard is actually the TrueSpeech coding method developed jointly by the DSP Group, France Telecom, AudioCodes Ltd., and the University of Sherbrooke.

The MPEG audio standard is based on the MUSICAM subband coder with some elements (in Layer III) from the ASPEC transform coder. Musmann [179] provides details on the early efforts by the International Standards Organization (ISO) to define an audio compression standard and discusses similarities and differences between the MUSICAM and ASPEC coding algorithms. Another description of the MUSICAM algorithm can be found in [47].

The MPEG-1 and MPEG-2 audio standards are described in the third part of the ISO documents [109] and [111]. In [193] and [231], Pan and Shlien provide additional references and details on the polyphase filter bank, the psychoacoustics models, and the bit-allocation algorithms. An excellent tutorial on MPEG audio compression is presented in [195]. Proofs for the DCT-based matrixing operations in subband coding are given in [194] and [135]. Lee's fast DCT algorithm is described in [140]. Other fast implementations for MPEG audio are reported in [199] and [89]. A good overview of the AC-3 system is given in [243], and the final AC-3 standard is available from ATSC [14]. An article by Brandenburg and Bosi [29] presents an excellent overview of the differences and similarities among several audio coding schemes, including MPEG and AC-3, and provided us with most of the material in this chapter regarding the future work in MPEG audio. The MPEG-2 AAC standard is described in [180] and [28].

For additional details on the hardware implementation of an MPEG audio decoder, Maturi [167] provides a description of the architecture of the MPEG decoder from LSI Logic. MPEG implementations on a programmable TI processor are presented in [197]. An implementation of an AC-3 coder on a Zoran processor is presented in [253]. For those interested in how AC-3 is integrated into the Grand Alliance HDTV system, a general overview of that system is given by Hopkins in [96].

17

NONSTANDARD COMPRESSION TECHNIQUES

17.1 INTRODUCTION

Standard compression schemes, such as JPEG and MPEG, have been widely supported by the industry; however, this has not stopped the development and marketing of several proprietary image and video compression schemes. Examples of such schemes include the Kodak Photo-CD compression and storage format, the QuickTime environment from Apple, and Indeo technology from Intel. There are various reasons for the introduction of proprietary compression schemes:

- Certain applications are self-contained, and there is no need to exchange their data with other platforms (such as a stand-alone image database system).

- Computational requirements of the standard algorithms are beyond the capabilities of a particular platform (such as the software-based video playback on low-end personal computers).

- New market areas open (such as digital photography).

- There is dissatisfaction with the capabilities of the standard algorithms (such as video teleconferencing).

- There is concern about royalties (such as arithmetic coding in JPEG).

Even though details on proprietary schemes are seldom published, most of the nonstandard techniques seem to be based either on vector quantization

(VQ) or wavelet-based subband coding. After a short overview of VQ and subband coding, we briefly present a few of the commercially available and nonstandard compression techniques.

17.2   VECTOR QUANTIZATION

In transform-based coders, such as JPEG and MPEG, blocks of spatial data are first transformed into the frequency domain. The image energy is now packed into a few of the frequency components that are carefully quantized to remove information that is irrelevant to the human visual system. Decompression is performed by reversing the encoding process. After an inverse quantization, the dequantized frequency components are translated back to spatial data by an inverse transform.

A main characteristic of such transform-based algorithms is that both the encoder and the decoder have similar levels of computational complexity. However, most multimedia applications are CD-ROM based and require only real-time decoding of the embedded video clips. Unfortunately, even with fast transform algorithms, real-time decompression may be beyond the capabilities of low-end machines. An alternative compression scheme is vector quantization (VQ) coding. VQ processes image blocks directly in the spatial domain. It has a far more complex encoder than transform-based coders, but it allows for very fast decoding.

Figure 17.1 shows a block diagram of a VQ coder for image compression. An input image is first segmented into nonoverlapping blocks. Each block is regarded as an M-dimensional *vector*, where M is the number of pixels in the block. For example, a 4×4 block corresponds to a vector of dimension 16. The principle of VQ is to match each input vector to a codeword (or reproduction vector) from a codebook so that the distortion between them is minimized. The codebook is generated a priori from a collection of representative images and should be a good representation of all possible image vectors. Compression is achieved by transmitting the index of the best codeword. For example, if the input vector has 16×8 bits and the codebook has $2^8 = 256$ entries, then the corresponding compression ratio is 16:1 since each 4×4 block can be represented by an 8-bit codeword index.

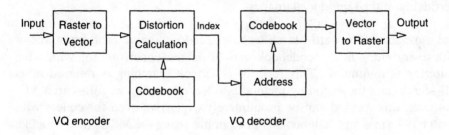

Figure 17.1 Block diagram of a VQ image coder.

Decoding in VQ is much simpler than encoding. The decoder, which has an identical codebook, uses the codeword index and a simple table look-up operation to reconstruct the original image. The simplicity of the VQ decoder is the principle reason that VQ is the method of choice for image and video coding on computing platforms with limited resources, such as PCs and portable electronic note pads. The complexity of the VQ encoder depends on the distortion measure and the search algorithm used for determining the best codeword.

17.2.1 Codebook Search Algorithms

Given an input vector $x = x_1, x_2, \cdots, x_M$ and a set of N M-dimensional codewords $\{c^1, c^2, \cdots, c^N\}$, a commonly used distortion measure in VQ is the mean square error

$$MSE(x, c^i) = \sum_{k=1}^{M}(x_k - c_k^i)^2, i = 1, 2, \cdots, N. \qquad (17.1)$$

From (17.1), each evaluation of the MSE requires M data loads, M subtractions, M multiplications, and $M - 1$ additions, for a total of $4M - 1$ operations.

An alternative measure is the mean absolute error

$$MAE(x, c^i) = \sum_{k=1}^{M}|x_k - c_k^i|, i = 1, 2, \cdots, N. \qquad (17.2)$$

The MAE criterion requires fewer computations; however, it may compromise the fidelity of the reproduced images.

The most straightforward search technique is to compare each input vector with every entry in the codebook and select the codeword for which the distortion is minimum. This form of codebook searching is referred to as full-search, and the encoding method is often referred to as full-search VQ. However, this method can be prohibitively expensive even for coders with small block sizes and codebooks. For example, using the MSE criterion and a codebook with 256 entries, this approach requires at least $256 \times 63 = 16{,}128$ operations per 4×4 block.

The amount of processing can be significantly reduced if one uses a binary tree-search vector quantizer (TSVQ). In binary TSVQ, the codebook is organized in a tree structure. Figure 17.2 shows such a tree with two levels. Starting

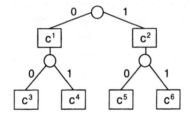

Figure 17.2   Block diagram of a VQ image coder.

from the root, at each level of the tree the input vector is compared with only two codewords. Based on the comparison, one of the branches is chosen and an index bit of either 0 or 1 is transmitted, The process continues until a leaf of the tree is reached. For N codewords, this approach requires the evaluation of only $2 \times log_2 N$ distortions. For a codebook with 256 entries, a TSVQ requires only $2 \times 8 \times 63 = 1{,}008$ operations per 4×4 block.

TSVQ encoding is much faster than full-search; however, it may also result in lower image quality. TSVQ is suboptimal because it may lead into a local minima in the search space, whereas full-search VQ finds the global minima. These techniques can be further refined or improved by using adaptive codebooks or pruned tree-search algorithms.

17.2.2   Hardware Considerations

Consider now the hardware implementation of a TSVQ coder. Given an input vector x, at each level of the tree we need to compute the difference

$$d_{ij} = MSE(x, c^i) - MSE(x, c^j). \tag{17.3}$$

From (17.1) and (17.3), after expanding the quadratics and some simplifications,

$$d_{ij} = \sum_{k=1}^{M} \left[(c_k^i)^2 - (c_k^j)^2 \right] + \sum_{k=1}^{M} 2x_k(c_k^i - c_k^j). \tag{17.4}$$

From (17.4), the first summation is a constant that is independent of x; it can be precomputed and stored in memory. The differences in the second summation can also be precomputed. Thus, if a_{ij} denotes the first summation in (17.4), and $e^{ij} = 2(c^i - c^j)$, then d_{ij} can be expressed as

$$d_{ij} = a_{ij} + \sum_{k=1}^{M} x_k e_k^{ij}. \tag{17.5}$$

Evaluation of (17.5) on a general-purpose RISC-like processor requires $M+1$ data loads, M multiplications, and M additions, for a total of $3M + 1$ operations per level of the tree. For M=16 (such as 4×4 blocks), 49 operations are required per level of the code tree. Table 17.1 shows the MOPS requirements for real-time TSVQ (30 fps), 4×4 blocks, various frame sizes (4:2:2 color images), and two codebooks: one with 256 entries and one with 1,024 entries. For comparison, a DSP processor with a multiply-accumulate unit can perform a data load and a multiply-accumulate operation in one cycle; thus (17.5) can be computed in $M + 1$ clock cycles. Assuming again 4×4 blocks and a codebook of only 256 entries, real-time TSVQ encoding of CIF images on a DSP requires a clock frequency of at least 51 MHz. This assumes no delays in accessing the memory (approximately 4 Kbytes) and does not take into consideration any other computations, such as color transformations or lossless compression of the VQ output using an entropy coder.

One approach to reducing the clock speed is to use multiple processors and pipelined processing. In TSVQ, only one node is active at each level of the coding tree. Furthermore, computations at one level do not require codewords from another level. This localization of operations allows the codeword memory to be partitioned into separate memories for each level of the tree.

Codebook size = 256	
Frame Size	MOPS
QCIF	37.25
CIF	149.02
SIF	124.19
CCIR	508.03
Codebook size = 1024	
Frame Size	MOPS
QCIF	46.57
CIF	186.28
SIF	155.23
CCIR	635.04

Table 17.1   MOPS requirements for TSVQ encoding at 30 fps for 4 × 4 blocks and different codebook and frame sizes.

Figure 17.3 shows the implementation of a TSVQ encoder using a linear array of $L = log_2 N$ processors, where each processor is mapped to a level of the coding tree. An output "Ready" signal from processor P_i resets the

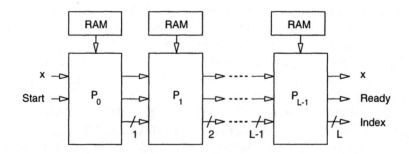

Figure 17.3   Block diagram of a pipelined TSVQ image coder.

accumulator of P_{i+1} and initiates processing at the next level. Each processor adds the result of its computation to the index data path. The last processor returns the index of the best codeword. Note that each processor requires different sizes of memory. Since each level i ($i = 0, 1, ..., L - 1$) of the tree has 2^i nodes, and each level is mapped into a single processor, from (17.5), processor P_i requires $(M + 1)2^i$ words of memory. This pipelined scheme allows the clock frequency of each processor to be reduced by L.

17.3   SUBBAND CODING

The main principle behind transform-based coding is that data in the transform domain have a more compact representation than data in the spatial domain. The transformation between the two domains can be performed either on a block by block basis (as in JPEG) or as in the case of a subband coder using a filter bank.

Figure 17.4 shows a block diagram of a subband coder. In the encoder, the

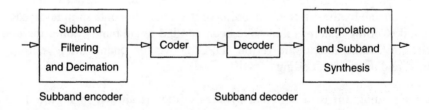

Figure 17.4   Block diagram of a subband coder.

image is first divided into multiple frequency bands using a combination of low-pass and high-pass filters. For example, Figure 17.5 shows a four-band filter bank and the corresponding partition of the image spectrum in octave bands. Filtering is followed by a 2×2 decimation operation where each subband output is subsampled by a factor of two in each dimension. In effect, decimation translates each subband back into a baseband image but with only one-fourth of the original pixels. Thus, even though there are four output channels, the output bit rate is the same as the input bit rate. The process of frequency subdivision can be iterated on each of these four subbands.

Subband filtering by itself is a lossless operation. The input data is simply decomposed into multiple data streams, without any loss of information.

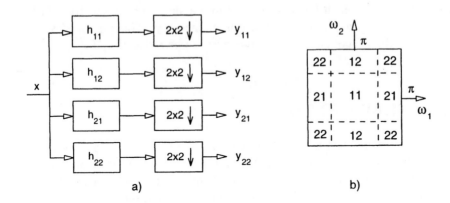

Figure 17.5   (a) Block diagram of a four-band filter bank. (b) Partition of the image spectrum.

Data compression is achieved by discarding subbands that are irrelevant for an application and by the efficient encoding of the remaining data samples in each subband. Coding schemes that have been applied to subband coding include predictive coding and quantization (DPCM), vector quantization, entropy coding, and arithmetic coding.

The block diagram of a subband synthesis filter is shown in Figure 17.6. After the decoding of the subband data, each subband is upsampled by a $2 \times$

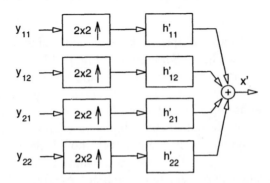

Figure 17.6   Block diagram of a four-band synthesis filter bank.

2 interpolator and then filtered by a band-pass filter that eliminates aliasing produced by the upsampling. The subbands are finally summed to reconstruct the original image.

As an example, Figure 17.7 shows the spectrum of a signal at the various stages of a two-band subband coder. The input signal is first decomposed into two bands using a low-pass and a high-pass filter. The bandwidth of these signals is half of the original bandwidth; hence, according to the Nyquist theorem their sampling rate can also be reduced by half. This is accomplished by downsampling each signal by a factor of two (in other words, by throwing away every other sample). With downsampling, the aliased output of the high-pass filter is shifted back into the baseband.

In the receiver, upsampling (a zero is inserted between every two samples) causes alias spectral signals (shown in gray) that can be suitably filtered with a low-pass and a band-pass filter. The filtered signals are then added to form the reconstructed signal.

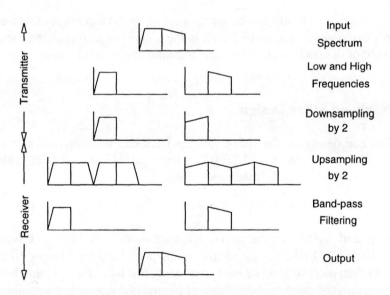

Figure 17.7 Signal spectrum in subband coding.

Subband coding is more computationally expensive than block-based coding; however, it has many other advantages, including the following:

- Since the subband filters operate on the whole image, subband coded images display fewer artifacts (such as blockiness) than block-coded images.

- Subband coding is more robust under transmission or decoding errors because errors on a particular subband may be masked by the information on the other subbands.

- Subband coders allow for better data rate control during encoding or decoding by the selective transmission or decompression of subbands. For example, to reduce the data rate in a video teleconferencing application, the image of the active speaker may be encoded at full resolution, while the images of the other participants may be encoded and decoded at lower resolutions.

Since the coding part of a subband coder uses compression techniques that we have already discussed in earlier chapters, in the remaining of this chapter we examine some of the subband filter design and implementation issues.

17.3.1   Subband Filter Design

The analysis and design of the subband filter bank can be simplified if we assume that all two-dimensional (2-D) filters are separable. A 2-D filter with impulse response $h(m, n)$ is defined as separable if

$$h(m, n) = h_1(m)h_2(n), \qquad (17.6)$$

where $h_1(m)$ and $h_2(n)$ are the impulse responses of 1-D filters. Using separable filters, a 2-D filtering operation can be performed by filtering first each row and then each column of the image using the 1-D filters. Following this approach, the four-band subband filter of Figure 17.5 can be implemented using a combination of low-pass (h) and high-pass (g) 1-D filters, as shown in Figure 17.8. Using a similar approach, the corresponding synthesis filter bank is shown in Figure 17.9.

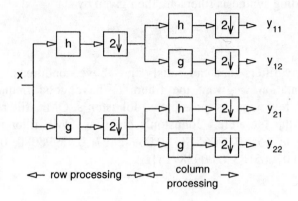

Figure 17.8   Analysis filter bank using row-column filtering.

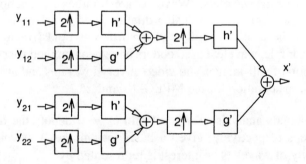

Figure 17.9   Synthesis filter bank using row-column filtering.

Ignoring the effects of coding and decoding on the input signal, we want the synthesis filters h' and g' to cancel the aliasing effects of the interpolation. Let h be an L-tap (L is even) finite impulse response (FIR) filter; then the aliasing can be canceled if h and g form a quadratic mirror filter (QMF) pair. By definition, h and g form a QMF pair if in the time domain

$$g(n) = (-1)^{n+1}h(L-1-n), \tag{17.7}$$

and in the frequency domain

$$|H(\omega)|^2 + |G(\omega)|^2 = 1. \tag{17.8}$$

The corresponding synthesis filters are then given by

$$h'(n) = h(L - 1 - n) \tag{17.9a}$$
$$g'(n) = g(L - 1 - n) \tag{17.9b}$$

There are many filters that can satisfy the above conditions. In a practical implementation we want the filters to have good frequency separation and as few taps as possible. Johnston's QMF filters are used quite often in the literature. Johnston's eight-tap QMF filter is given by $h(n) = \{0.00938715, -0.07065183, 0.06942827, 0.48998080, 0.48998080, 0.06942827, -0.07065183, 0.00938715\}$.

17.3.2   Subband Coding and Wavelets

A case of particular interest in subband coding is the implementation of the subband filters using wavelets. Wavelets are families of functions that are generated from a single function using dilations and translations. They have been proven to be a useful tool in many different fields, including image coding. For example, wavelet based coders have been applied successfully for the compression of still-images and video at very low rates and are now being proposed for consideration for the MPEG-4 standard.

For discrete signals analysis, if $a_o > 1$ and $b_o \neq 0$ denote the dilation and translation steps respectively, given a basis (or "mother") wavelet function $\psi(t)$, the family of wavelets of interest is represented by

$$\psi_{mn}(t) = a_o^{-m/2}\psi(a_o^{-m}t - nb_o). \tag{17.10}$$

For $a_o = 2$ and $b_o = 1$ one can find mother wavelets so that $\psi_{mn}(t)$ constitute an orthonormal basis, which allows a given signal $f(t)$ to have a wavelet decomposition given by

$$f(t) = \sum_m \sum_n c_{mn}\psi_{mn}(t), \tag{17.11}$$

where

$$c_{mn} = \int f(t)\psi_{mn}(t)dt. \tag{17.12}$$

In intuitive terms, the decomposition of a signal using orthonormal wavelets yields a representation that lies in between the spatial and the Fourier representations.

Wavelet analysis is also related to multiresolution representation. Under that context, one can consider that the a_o^{-m} term of a wavelet function corresponds to magnification, and that the nb_o term corresponds to location. If $a_o = 2$, then the analysis of a signal is performed octave by octave. In that case, given a fixed magnification (or scale) m, one can find a "mother" *scaling* function $\phi(t)$ such that the family of functions

$$\phi_{mn}(t) = 2^{-m/2}\phi(2^{-m}t - n) \tag{17.13}$$

forms an orthonormal basis. If $f_m(t)$ denotes the value of $f(t)$ at resolution level m, then $f_m(t)$ can be expressed as

$$f_m(t) = \sum_n d_{(m+1)n}\phi_{(m+1)n} + \sum_n c_{(m+1)n}\psi_{(m+1)n}. \tag{17.14}$$

In other words, if $\phi(t)$ is considered a low-pass function, and the wavelet $\psi(t)$ is considered a band-pass function, then at a given resolution level m the function can be expressed as the sum of low and high frequency components.

In 1989, Mallat showed that this multiresolution representation of a signal can be computed using a pyramidal filter structure of QMF filter pairs, as shown in Figure 17.10. The filters h and g form again a QMF pair and are related to

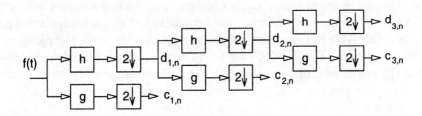

Figure 17.10 Filter bank for wavelet-based signal decomposition.

the scaling and wavelet functions as follows:

$$\Phi(2\omega) = H(\omega)\Phi(\omega), \tag{17.15}$$

$$\Psi(\omega) = G(\omega)\Phi\left(\frac{\omega}{2}\right), \tag{17.16}$$

where $\Phi(\omega)$ and $\Psi(\omega)$ are the Fourier transforms of $\phi(t)$ and $\psi(t)$. As an example, using the Daubechies four-tap wavelet, $h(n) = \{\frac{11}{32}, \frac{19}{32}, \frac{5}{32}, -\frac{3}{32}\}$

and $g(n) = \{\frac{3}{32}, \frac{5}{32}, -\frac{19}{32}, \frac{11}{32}\}$. In the synthesis band, $h'(n) = h(3 - n)$ and $g'(n) = g(3 - n)$.

The above wavelet decomposition of one-dimensional signals can be generalized for 2-D signals. Using row-column filtering, a 2-D wavelet decomposition of an image can be computed using the filter structure of Figure 17.8. The 2-D wavelet decomposition can be interpreted as an image decomposition in a set of independent *spatially oriented* frequency representations: horizontal, vertical, and diagonal. For a four-band wavelet decomposition, the four output bands in Figure 17.5(b) can be interpreted as follows. Band 1,1 corresponds to the lower horizontal and vertical frequencies of the image, band 2,1 gives the vertical high frequencies and horizontal low frequencies (horizontal edges), band 1,2 gives the vertical edges, and band 2,2 gives the high frequencies in both directions (corners).

Figure 17.11 shows the wavelet decomposition of an image into seven subbands. After the original image was divided into four bands, the lower subband was further subdivided into four bands. The h and g filters are the seven-tap QMF filters with $h(n) = \{0.00525, -0.05178, -0.25525, 0.60355, -0.25525, -0.05178, 0.00525\}$ and $g(n) = \{-0.00525, -0.05178, 0.25525, 0.60355, 0.25525, -0.05178, -0.00525\}$.

For real-time wavelet-based video compression, Analog Devices offers the ADV601, an application specific video compression and decompression processor. The ADV601 supports real-time compression or decompression of CCIR 601 frames. The ADV601 can be set to provide either fixed compression ratios (from lossless compression to 256:1), or "constant quality" with variable bit rate. The processor includes a wavelet transform block, a quantizer, a run-length coder and modeler, and a Huffman coder. The ADV601 can be combined with the AD1843 audio codec for fully synchronized audio/video coding.

In summary, from an implementation viewpoint, wavelet decomposition can be considered as a special case of subband filtering. In a subband coder, compression efficiency and overall performance depend not only on the choice of the analysis/synthesis filters but also on the compression method that will be used on the subband filtered data. Similar considerations apply to image and video compression methods based on wavelet decompositions.

Figure 17.11 Example of wavelet decomposition into seven subbands.

17.3.3 Computational Requirements for Subband Filtering

The computational requirements of a subband coder may vary widely, depending on the complexity of the filter bank (number of bands, filter types, and so on) and the complexity of the coder (such as VQ, DPCM, and Huffman). As an example, we estimate the number of multiply-accumulate (MAC) operations in a filter bank where all filters are L-tap FIR filters.

Consider an $M \times N$ image and the subband filter shown in Figure 17.8. In the first stage, each of the two filters requires $\frac{ML}{2}$ MAC operations per row of data for a total of NML MAC operations for the whole image. The second stage has twice as many filters; however, the columns are also half as many. Hence, the number of operations remains the same, and the total number of

MAC operations for a four-band 2-D filter is $2MNL$, or $2L$ operations per pixel. Assuming we iterate this scheme on the lower band, the number of MAC operations is still $2L$ per pixel; however, the input image at this stage has one-fourth of the pixels in the original image. Hence, for multiple iterations, the total number of MAC operations is bounded by

$$MAC_{total} = 2NML(1 + \frac{1}{2^2} + \frac{1}{2^4} + \cdots) < \frac{8MNL}{3}. \qquad (17.17)$$

For example, if $L = 6$, then the total number of MAC operations is bounded by 16 operations per pixel. In comparison, in the 8×8 fast DCT, the number of operations can be less than seven operations/pixel. The number of multiplications can be reduced if the filter coefficients are powers of two or if the filters are symmetric. When h is symmetric, $g(n) = (-1)^n h(n)$. Therefore, a multiplication with a filter coefficient needs to be performed only once, and the product can be added or subtracted to the appropriate partial sum of products.

17.4   VIDEO CODING SCHEMES FOR THE DESKTOP

One of the first commercially available techniques for the compression and storage of digital video was Intel's DVI (Digital Video Interleaved). DVI was based on image color subsampling and vector quantization, and processing could be accelerated using Intel's i750 processor.

DVI has been replaced now by Indeo video technology, which allows software-only playback of video on desktop personal computers. Using frame sizes of 320×240 pixels, Indeo allows software playback at 30 frames/s. Early Indeo coders were based on color subsampling, pixel differentiation (interframe coding), run-length coding, VQ, and lossless coding. The latest version uses wavelets and both forward and backward motion-compensated prediction. On slower processors, instead of dropping frames, Indeo video interactive dynamically varies visual quality according to the available processing power.

In 1991, Apple Computer introduced QuickTime. This is a multimedia extension to the Macintosh operating system and allows for the manipulation of dynamic data types, such as sound, video, and animation. Within a year, Apple introduced QuickTime for Windows and Microsoft introduced Video for Windows. QuickTime supports a variety of compression schemes, including JPEG, MPEG playback, and Apple-proprietary video coders that incorporate both spatial and temporal compression. With Windows 95, Video for Windows is now part of the operating system and supports Intel's Indeo, Apple's QuickTime, and MPEG playback.

Another popular video compression scheme for multimedia applications is motion-JPEG (M-JPEG). Under M-JPEG, each frame of the video is compressed independently using the standard JPEG algorithm. Real-time compression and decompression is possible using low-cost video boards with any of the commercially available JPEG processing ICs. However, there is no standard syntax for M-JPEG coded streams, and encoded data may not be able to be decoded across different platforms.

17.4.1 Videoconferencing on the Desktop

Early efforts on software-only videoconferencing systems were either focused on high-end workstations or were based on simple coding schemes, best suited for low-end computing platforms. Examples include the INRIA videoconferencing system (IVS) and CU-SeeMe from Cornell University. IVS was one of the first standard-based (H.261) videoconferencing systems over the Internet.

CU-SeeMe is another free desktop videoconferencing system for use over the Internet. Video output is 4-bit grayscale (16 levels) at resolutions of either 160 × 120 or 640 × 480 pixels. CU-SeeMe uses a non standard coding scheme based on image subsampling, change detection, and spatial compression.

The captured 640 × 480 video frame is first subsampled down to 160 × 120 4-bit pixels and then it is divided into 8 × 8 blocks. A block is transmitted only if it differs substantially from its corresponding block that was transmitted most recently. Blocks are compared using a weighted sum of absolute differences between pixels. If a block is selected for transmission, then it is transmitted using simple differential lossless coding. Let $r[i]$ denote the 32-bit word of the i-th row in a block to be transmitted. Assuming similarity among rows within a block, $r[i]$ is expressed as $r[i] = r[i-1] + d[i]$. This scheme allows for a high degree of parallelism on 32-bit architectures and yields close to

1.67:1 compression. An enhanced version of CU-SeeMe that supports color and audio is available from White Pine Software.

The adoption of the H.324 videoconferencing standard and the availability of more powerful Pentium processors helped remove some of the limitations of early software-based videoconferencing systems. The Intel VideoPhone is one of the first products from Intel based on its ProShare technology. It is H.324 compliant (H.263 QCIF video, G.723 audio) and allows for up to 12 frames/s and full duplex speakerphone on a 166 MHz Pentium processor.

Apple's VideoPhone enables audioconferencing and collaboration with limited video over the Internet with a 28.8 kbits/s modem and full videoconferencing with connections faster than 128 kbits/s. It works with Netscape's CoolTalk software and supports a variety of coding formats, including H.261, H.263, G.723, and Apple Video.

Microsoft offers its own video-phone software with NetMeeting. NetMeeting supports the H.326 standard and can interoperate with Intel's VideoPhone. It also supports multi-user audio and video conferencing using a multipoint conferencing bridge service. There are a number of other videoconferencing systems, including FreeVue from FreeVue Communications and VDOPhone from VDOnet.

17.4.2   Streamed Audio and Video

With the growth of the worldwide Web (WWW) and increased connectivity among computing systems, multimedia information is not constrained any more to local media (such as a local CD-ROM), but may reside in a remote information server. Unfortunately, the Internet was never designed to handle real-time video and audio data. Internet's transmission control protocol (TCP/IP) breaks data into packets. Packets are routed to their destination along whatever channel is available and there is no guarantee that successive packets will arrive in the order transmitted. If there is network congestion, packets are dropped, and TCP/IP requests the information to be retransmitted. This works quite well for basic file transfers. Out-of-order transmissions are corrected by simple buffering of the incoming data, and retransmissions merely delay the transfer time. However, even compressed audio and video files are orders of magnitude larger than typical ascii files, and users cannot tolerate long waiting times for short audio or video clips. In the past, one would have to wait for the whole video or audio file to be transferred before one could begin playback; however, a variety of vendors now provide solutions for streaming the data. Streaming allows a user to begin playback of audio and video data before all the information has arrived and times for file transfer and playback overlap.

Streaming transfers usually use the user datagram protocol (UDP) on top of TCP/IP. UDP is not as reliable as TCP/IP, but guarantees faster data transfers. For lost packets, vendors use a variety of proprietary schemes for error correction or error mitigation, depending on the data to be transferred. For example, it is not uncommon to replace lost audio packets with white noise, or to freeze video until new data arrives. We expect that as the WWW matures it will adopt a standard protocol for the transfer of real-time information. In the mean time, we will continue to see a variety of propriety transmission and coding schemes. Table 17.2 provides a summary of a few of the software tools that are available for multimedia creation and playback on the desktop.

17.5   TO PROBE FURTHER

Vector quantization for signal compression was first popularized in the 1980s by Gersho and Gray. Their textbook [73] describes the fundamentals of VQ theory and various codebook design and search techniques.

A multiprocessor implementation of a VQ coder is described in [49]. It consists of a linear array of multiplier-accumulator ICs terminated by a next-address-selector processor. A single-chip VQ encoder for pruned tree search was presented in [164]. This 80-MOPS chip can compute one eight-dimensional inner product in four clock cycles. The pipelined architecture of the TSVQ encoder was first introduced by Kolagotla et al. in [133]. A low-power implementation of the same design is also described in [154]. Another hardware implementation of a very low-power VQ decoder for real-time decompression on a portable terminal is presented in [33].

Subband coding was first introduced in 1976 for the efficient coding of speech signals. Woods and O'Neil [263] were among the first to combine subband filtering and a DPCM coder for compressing images. A good overview of wavelet theory and its application to signal processing is given by Rioul and

Product	Company	Coding method
Multimedia Authoring		
Indeo	Intel	Proprietary (wavelets).
QuickTime	Apple	Proprietary.
Video for Windows	Microsoft	
M-JPEG	Various	JPEG.
Videoconferencing		
IVS	INRIA	H.261.
CU-SeeMe	Cornell Univ.	Proprietary.
VideoPhone	Intel	H.324.
VideoPhone	Apple	H.261, H.263, and others.
NetMeeting	Microsoft	H.323.
CoolTalk	Netscape	Audioconferencing.
FreeVue	FreeVue	Proprietary.
VDOPhone	VDOnet	Proprietary.
Communique!	InSoft	Proprietary.
Streamed Audio and Video		
Vosaic	Vosaic Corp.	MPEG, H.263.
StreamWorks	Xing Technology	MPEG.
RealVideo/Audio,	Progressive Networks	Proprietary.
VivoActive	Vivo Software, Inc.	H.263, G.723.
VDOLive	VDOnet	Proprietary.
INTV	InSoft	Proprietary.

Table 17.2 Software for multimedia development and playback on the desktop.

Vetterli [222]. Mallat [166] was the first to show the relationship between subband filtering and wavelet decomposition. The book on multiresolution and signal decomposition by Akansu and Haddad [4] provides a unified coverage of subband filtering and wavelet transforms and many examples on filter design for image coding.

There is a variety of wavelet-based coders. In [153], Lewis and Knowles use the four-tap Daubechies wavelets and a quantization model based on the human visual system. The subband-filtering operation for this system can easily be implemented using a multiplier-free architecture [152]. A coding scheme that combines wavelet decomposition and VQ is described in [9]. In [152], Lewis et al. describe a low-bit video coder using wavelet transforms and differential coding. Another architecture for a low-power subband video decoder using 4-tap wavelet-based filters is described in [82]. At 20 MHz, the chip can process 10 million color pixels per second while dissipating only 16 mW. A single IC can process images up to 352 pixels wide and performs both subband decoding and YUV to RGB color transformations.

Trade magazines and the WWW are a good source of information for commercially available video coders. A general overview of an early version of QuickTime is given in [94]. In addition, most of the QuickTime related documentation is available from Apple's Web site. Descriptions of IVS and CU-SeeMe are given in [248] and [51], respectively.

COLOR TRANSFORMATIONS

Images and video are usually displayed in the RGB color space; however, the original data are typically processed and compressed in some type of a luminance-chrominance color space, such as YCbCr or CIELAB. In this Appendix we present color transformations between RGB and three commonly used color spaces: YCbCr, YUV, and CIELAB.

YCbCr

This is the most commonly used color coordinate system for the compression of image and video signals. Y is the luminance component and Cb and Cr are the chrominance components. Given the primary RGB inputs (R, G, and B in [0,1]),

$$
\begin{aligned}
Y &= 0.299(R - G) + G + 0.114(B - G), \\
Cb &= 0.564(B - Y), \\
Cr &= 0.713(R - Y).
\end{aligned}
$$

Alternatively,

$$
\begin{bmatrix} Y \\ Cb \\ Cr \end{bmatrix} = \begin{bmatrix} 0.299 & 0.587 & 0.114 \\ -0.169 & -0.331 & 0.500 \\ 0.500 & -0.419 & -0.081 \end{bmatrix} \begin{bmatrix} R \\ G \\ B \end{bmatrix}.
$$

Given a YCbCr input (Y in [0,1] and Cb, Cr in [-0.5, 0.5]),

$$\begin{bmatrix} R \\ G \\ B \end{bmatrix} = \begin{bmatrix} 1 & 0.0 & 1.4021 \\ 1 & -0.3441 & -0.7142 \\ 1 & 1.7718 & 0.0 \end{bmatrix} \begin{bmatrix} Y \\ Cb \\ Cr \end{bmatrix}.$$

YUV

This color coordinate system is used in the PAL TV system. Given the primary RGB inputs (R, G, and B in [0,1]),

$$\begin{aligned} Y &= 0.299(R - G) + G + 0.114(B - G), \\ U &= 0.493(B - Y), \\ V &= 0.877(R - Y). \end{aligned}$$

Alternatively,

$$\begin{bmatrix} Y \\ U \\ V \end{bmatrix} = \begin{bmatrix} 0.299 & 0.587 & 0.114 \\ -0.148 & -0.289 & 0.437 \\ 0.615 & -0.515 & -0.100 \end{bmatrix} \begin{bmatrix} R \\ G \\ B \end{bmatrix}.$$

Given a YUV signal,

$$\begin{bmatrix} R \\ G \\ B \end{bmatrix} = \begin{bmatrix} 1 & 0.0 & 1.140 \\ 1 & -0.395 & -0.581 \\ 1 & 2.032 & 0.0 \end{bmatrix} \begin{bmatrix} Y \\ U \\ V \end{bmatrix}.$$

CIELAB

CIELAB or CIE $L^*a^*b^*$ is a color space defined by CIE (Commission Internationale de l'Eclairage). In CIELAB, L^* is the luminance component, and a^* and b^* are the chroma components. Translation from RGB to CIELAB requires two steps: (1) a linear transformation from RGB to CIE XYZ, and (2) a non-linear transformation from CIE XYZ to CIELAB.

The RGB to XYZ transformation is device dependent. For RGB inputs defined by the ITU-R Recommendation BT.709, given R, G, and B in [0, 100], the

RGB to CIE XYZ transformation is given by

$$
\begin{bmatrix} X \\ Y \\ Z \end{bmatrix} = \begin{bmatrix} 0.412453 & 0.35758 & 0.180423 \\ 0.212671 & 0.71516 & 0.072169 \\ 0.019334 & 0.11919 & 0.950227 \end{bmatrix} \begin{bmatrix} R_{709} \\ G_{709} \\ B_{709} \end{bmatrix}.
$$

Let X_n, Y_n, and Z_n denote the XYZ values for reference white. For example, for the CIE standard illuminant D_{50} (used in color facsimile), $X_n = 96.422$, $Y_n = 100$, and $Z_n = 82.521$. Then

$$
L^* = 116 f\left(\frac{Y}{Y_n}\right) - 16,
$$

$$
a^* = 500\left[f\left(\frac{X}{X_n}\right) - f\left(\frac{Y}{Y_n}\right)\right],
$$

$$
b^* = 200\left[f\left(\frac{Y}{Y_n}\right) - f\left(\frac{Z}{Z_n}\right)\right],
$$

where

$$
f(r) = \begin{cases} r^{\frac{1}{3}} & \text{if } r > 0.008856 \\ 7.7867r + \frac{16}{116} & \text{otherwise} \end{cases}.
$$

The above equations will yield L^* in [0, 100] and a^* and b^* in [-100, 100]. For CIELAB input data in the same gamut range, the RGB output can be obtained as follows: (1) translate the CIELAB data to CIE XYZ, and (2) translate the CIE XYZ data to RGB.

If $L^* < 8.0$, then

$$
\frac{Y}{Y_n} = \frac{L^*}{903.3},
$$

else

$$
\frac{Y}{Y_n} = \frac{1}{100}\left[\frac{L^* + 16}{25}\right]^3.
$$

Given Y,

$$
\frac{X}{X_n} = \left[\frac{a^*}{500} + \left(\frac{Y}{Y_n}\right)^{\frac{1}{3}}\right]^3.
$$

$$
\frac{Z}{Z_n} = \left[\left(\frac{Y}{Y_n}\right)^{\frac{1}{3}} - \frac{b^*}{200}\right]^3.
$$

Finally,

$$\begin{bmatrix} R_{709} \\ G_{709} \\ B_{709} \end{bmatrix} = \begin{bmatrix} 3.240479 & -1.53715 & -0.498535 \\ -0.969256 & 1.875991 & 0.041556 \\ 0.055648 & -0.20404 & 1.057311 \end{bmatrix} \begin{bmatrix} X \\ Y \\ Z \end{bmatrix}.$$

B

ABBREVIATIONS AND SYMBOLS

$\lceil x \rceil$ The smallest integer larger or equal to x.

$\lfloor x \rfloor$ The largest integer smaller or equal to x.

1-D One-dimensional.

2-D Two-dimensional.

4:2:0 A color subsampling format in which the chrominance components are subsampled by a factor of two in both dimensions. Example: If Y is 720 $\times$ 480, then Cb and Cr are 360 $\times$ 240.

4:2:0 macroblock A 4:2:0 macroblock has four 8 $\times$ 8 blocks of luminance and two 8 $\times$ 8 blocks of chrominance. For example, a 4:2:0 YCbCr macroblock has four blocks of Y, one block of Cb, and one block of Cr.

4:2:2 A color subsampling format in which the chrominance components are subsampled by a factor of two in the horizontal only dimension. Example: If Y is 720 $\times$ 480, then Cb and Cr are 360 $\times$ 480.

AC component of the DCT Any component of the DCT output except the (0,0) (or DC) component.

ADPCM Adaptive differential pulse code modulation.

AGU Address generation unit.

ALU Arithmetic and logic unit.

ANSI American National Standards Institute.

bps Bits per second.

CAM Content-addressable memory.

CCIR The International Radio Consultative Committee of the International Telecommunications Union (now ITU-R).

CCIR 601 CCIR recommendation 601 is a specification for digital TV. A CCIR 601 image is defined as 720 pixels × 480 lines at 60-Hz interlaced (NTSC) or 720 pixels × 576 lines at 50-Hz interlaced (PAL). This recommendation is now called ITU-R Recommendation BT.601-4.

CCITT The telecommunication standardization sector of the ITU. It is now called ITU-T.

CD Compact disc.

CELP Code excited linear prediction.

CISC Complex instruction set computer.

CIF Common intermediate format. In a YCbCr CIF frame, Y is 352 pixels × 288 lines, and Cb and Cr are 176 pixels × 144 lines.

CPU Central processing unit.

DC component of the DCT The (0,0) component of the DCT output.

DCT, IDCT Discrete cosine transform or inverse discrete cosine transform.

DFT Discrete Fourier transform.

DHT Discrete Hadamard transform.

DMA Direct memory access.

DST Discrete sine transform.

DPCM Differential pulse code modulation.

dpi Dots per inch.

DSP Digital signal processing or digital signal processor.

DVD Digital versatile disk. A system for the delivery of video and audio information.

DVI Digital video interleave. Early video coding scheme from Intel.

EOB End of block.

FFT Fast Fourier transform.

fps Frames per second.

FSM Finite state machine.

GOB Group of blocks.

GOP Group of pictures.

GOPS Giga operations per second. Same as billion operations per second (BOPS).

HDTV High-definition TV. Several formats have been chosen for HDTV. For instance, MPEG-2 considers two HDTV formats, namely, 1440 pixels × 1152 lines and 1920 pixels × 1152 lines.

HVS Human visual system.

I/O Input-output.

IC Integrated circuit.

ISDN Integrated services digital network.

ISO International Organization for Standardization.

ISO/IEC The international Electrotechnical Commission of the ISO.

ITU International Telecommunications Union.

ITU-T The telecommunication standardization sector of ITU, formerly the CCITT.

JBIG Joint Binary Image Experts Group.

JFIF JPEG file interchange format.

JPEG Joint Photographic Experts Group.

KLT Karhunen-Loeve transform.

MAC Multiply-accumulate.

MAD Mean absolute difference. Same as MAE.

MAE Mean absolute error.

MAX MAX-1 and MAX-2 are small sets of multimedia acceleration extensions implemented in the PA-RISC processors from Hewlett-Packard.

MCU Minimum coded unit.

MMX A set of multimedia extensions for the Intel architecture.

MOPS Million operations per second.

MPEG Moving Pictures Expert Group.

mux Multiplexor.

NTSC A 525-line, 60-Hz television system developed by the National Television Systems Committee of the USA. NTSC is used in North America and Japan.

PAL A 625-line, 50-Hz television system developed in Germany. NTSC, PAL, and SECAM (from France) are the three main television transmission systems in the world.

PCM Pulse code modulation.

PHODS Parallel hierarchical one-dimensional search.

PLA Programmable logic array.

PNG Portable network graphics format. An image format recommended by W3C for storing images in the internet.

PSNR Peak signal-to-noise ratio.

P×64 Another name for the H.261 video teleconferencing standard.

QCIF Quarter CIF. A QCIF image is defined as 176 pixels × 144 lines.

QMF Quadrature mirror filter.

RAC ROM accumulator.

RISC Reduced instruction set computer.

RLC Run-length coder.

SIF Source input format. In a SIF image the luminance (Y) component is defined as 352 pixels × 240 lines at 30 Hz or 352 pixels × 288 lines at 25 Hz. Chroma components (Cb and Cr) are subsampled by a factor of two in both directions.

SIMD Single input multiple data.

SNR Signal-to-noise ratio.

SPIFF The still-picture interchange file format.

TI Texas Instruments.

TLB Translation look-aside buffer.

TSS Three-step-search.

TSVQ Tree-search vector quantization.

VIS The visual instruction set from Sun Microsystems.

VLC Variable length coder.

VLI Variable length integer.

VLIW Very long instruction word.

VLSI Very large scale integration.

VQ Vector quantization.

W3C The worldwide Web consortium. An industry consortium for the development of common standards for the evolution of the worldwide Web.

YCbCr A color coordinate system used for the transmission and storage of image and video signals.

YUV A color coordinate system used in the PAL TV system.

C

INTERNET DIRECTORY

The Internet is a rich source of materials on the image and video compression standards and their implementation. One can find standards documents, literature, and software implementations. Most of the discussions on data compression take place in the news groups *comp.compression* and *comp.compression.research*. A good start to a search for information is the monthly posted *FAQ* in these newsgroups. This monthly posting provides answers to the most frequently asked questions (FAQ). You can find this posting by searching for "comp.compression Frequently Asked Questions" in the subject lines. This posting is also available via anonymous ftp from rtfm.mit.edu in the directory usenet/news.answers/compression-faq or from other archive sites, such as ftp.uu.net. On the worldwide Web (WWW), it is available from http://www.cis.ohio-state.edu/-hypertext/faq/usenet/compression-faq/top.html. (For the remaining of this section, all WWW sites begin with http:// unless otherwise noted.)

In this appendix we provide internet addresses for some of the sites where literature and software implementations of the compression standards are currently available. Internet sites are constantly changing, so be sure to consult the latest "FAQ" postings or any of the WWW search engines for the most updated information.

- **Standards Documents and General Information.**

  The WWW site for ITU is www.itu.ch. The site includes information about activities within ITU and provides on-line access to most of the

ITU Recommendations. The ITU also provides most of the ITU Recommendations in postscript format in `gopher.itu.ch`. You can access the gopher ITU server using the command `telnet gopher.itu.ch 2000`. On the WWW, the site location is `gopher://gopher.itu.ch/`.

ISO also has a WWW site at `www.iso.ch` . The site allows you to search for ISO documents and provides useful information on the ISO activities and ISO members.

For videoconferencing, many of the relevant standards and other information is available from the WWW site of the International Multimedia Teleconferencing Consortium at `www.imtc.org/imtc/`. The documents for the ITU-T H.324 standard can also be obtained from `ftp.std.com` in `/vendors/PictureTel/h324/`.

The following sites

`www.mpeg.org` and

`/www-nt.e-technik.uni-erlangen.de/ hartung/links.html`

provide a number of links to MPEG-related sites, including software and hardware vendors.

If you do not have Internet access, the mailing addresses and telephone numbers for the international organizations are given below:

ITU
Palais des Nations
CH-1211
Geneva 10, Switzerland
Tel: (22) 730-5111,

ISO
1 Rue de Varembe
Case Postale 56
CH-1211
Geneva 20, Switzerland
Tel: (22) 749-0111.

In the United States, standards documents are also available from the American National Standards Institute (ANSI):

ANSI
11 West 42nd St.
New York, NY 10036
Tel: (212) 642-4900.

- **Sources for JPEG.**

  Source code for a JPEG coder from the Independent JPEG Group is available from `ftp.uu.net:/graphics/jpeg/`. For example, version 6a of the program is in the archive file jpegsrc.v6a.tar.gz. For additional information contact `jpeg-info@uunet.uu.net`.

  Other sources for JPEG are available from Stanford University at

  `havefun.stanford.edu:pub/jpeg`.

  This version also supports lossless JPEG. Another program for lossless JPEG is available from Cornell University at

  `ftp.cs.cornell.edu:/pub/multimed/ljpg.tar.Z`.

- **Sources for H.261 and H.263.**

  The IVS video conferencing system from the French research institute INRIA is available from `www.inria.fr/rodeo/ivs.html` or from

  `zenon.inria.fr:/rodeo/ivs/last_version/`.

  This is X-windows based and runs on most Unix workstations.

  Another H.261 coder is available from Stanford University at

  `havefun.stanford.edu:pub/p64/`.

  An implementation of the H.263 standard is available from Telenor Research at

  `www.nta.no/brukere/DVC/h263_software/`.

- **Sources for MPEG.**

  The official MPEG site is `www.cselt.stet.it/mpeg/`. For MPEG-1, sources are available from Stanford University at

  `havefun.stanford.edu:/pub/mpeg/`

  and from the University of California at Berkeley multimedia group at

  `mm-ftp.cs.berkeley.edu:/pub/multimedia/mpeg/`.

  The Berkeley and Stanford sites include implementations for MPEG-1 encoder and decoder.

For MPEG-2, software is available from U. C. Berkeley at

`mm-ftp.cs.berkeley.edu:/pub/multimedia/mpeg2/`

and the MPEG Software Simulation Group at

`www.mpeg.org/MSSG/` or `ftp.mpeg.org:/pub/mpeg/mssg/`.

This site also includes pointers to other locations related to MPEG.

REFERENCES

[1] *FlashPix Format Specification, Version 1.0*, 1996.

[2] PNG (portable network graphics) specification, version 1.0. http:-/www.w3.org/Graphics/PNG/, 1996.

[3] CCIR Recommendation 601-1. Encoding parameters of digital television for studios.

[4] A.N. Akansu and R.A. Haddad. *Multiresolution Signal Decomposition. Transforms, Subbands, and Wavelets*. Academic Press, San Diego, CA, 1992.

[5] T. Akiyama et al. MPEG2 video codec using image compression DSP. *IEEE Transactions on Consumer Electronics*, 3(40):466--472, Aug. 1994.

[6] J.D. Allen and S.M. Blonstein. The multiply-free Chen transform - a rational approach to JPEG. In *Picture Coding Symposium, Japan*, pages 237--240, 1991.

[7] S. Ammon and C. Erskine. Low-delay CELP speech compression and its implementation on a fixed-point DSP. *DSP & Multimedia Technology*, May 1994.

[8] P.H. Ang, P.A. Ruetz, and D. Auld. Video compression makes big gains. *IEEE Spectrum*, pages 16--19, Oct. 1991.

[9] M. Antonini, M. Barlaud, P. Mathieu, and I. Daubechies. Image coding using vector quantization in the wavelet transform domain. In *Proceedings IEEE International Conference on Acoustics, Speech, and Signal Processing (ICASSP)*, pages 2297--2300, 1990.

[10] K. Aono et al. A video digital signal processor with a vector-pipeline architecture. *IEEE Journal Solid-State Circuits*, 27(12):1886--1894, Dec. 1992.

[11] Y. Arai, T. Agui, and M. Nakajima. A fast DCT-SQ scheme for images. *Transactions of the IEICE*, E-71(11):1095--1097, Nov. 1988.

[12] R.B. Arps and T.K. Truong. Comparison of international standards for lossless still image compression. *Proceedings of the IEEE*, 82(6):889--899, June 1994.

[13] A. Artieri and O. Colavin. A chip set core for image compression. *IEEE Transactions on Consumer Electronics*, 36(3):395--402, August 1990.

[14] ATSC. United States Advanced Television Systems Committee Digital Audio Compression (AC-3) Standard, Doc. A/52/10, Nov. 1994.

[15] D. Auld. A flexible chip set for intra frame video compression. In *Digest of Papers, COMCON*, pages 330--332, 1991.

[16] C. Auyeung, J. Kosmach, M. Orchard, and T. Kalafatis. Overlapped block motion compensation. In *SPIE Visual communications and image processing*, pages 561--571. SPIE, Nov. 1992.

[17] D. Bailey et al. Programmable vision processor/controller for flexible implementation of current and future image compression standards. *IEEE Micro*, pages 33--39, Oct. 1992.

[18] A. Balkanski, S. Purcell, and J. Kirkpatrick. System for compression and decompression of video data using discrete cosine transform and coding techniques. U.S. Patent 5341318, Aug. 1994.

[19] M.A. Bayoumi and E.E. Swartzlander, Jr., editors. *VLSI Signal Processing Technology*. Kluwer Academic Publishers, Boston, 1994.

[20] T.C. Bell, J.G. Cleary, and I.H. Witten. *Text Compression*. Prentice Hall, Englewood Cliffs, New Jersey, 1990.

[21] G.B. Beretta et al. New Group 3 color facsimile standard: protocol testing and performance. In *SID 1995 Intern. Symposium Digest of Technical Papers*. SID, May 1995.

[22] T. Berger. *Rate Distortion Theory - A Mathematical Basis for Data Compression*. Prentice-Hall, Englewood Cliffs, New Jersey, 1971.

[23] V. Bhaskaran, K. Konstantinides, and R. Lee. Real-time MPEG-1 software decoding on HP workstations. In *Proceedings IS&T/SPIE Symp. on Electronic Imaging, Vol. 2419*, pages 466--473, Feb. 1995.

[24] V. Bhaskaran, K. Konstantinides, R.B. Lee, and J.P. Beck. Algorithmic and architectural enhancements for real-time MPEG-1 decoding on a general purpose RISC workstation. *IEEE Trans. on CSVT*, 5(5):380--386, Oct. 1995.

[25] H. Bheda and P. Srinivasan. A high-performance cross-platform MPEG decoder. In *Proceedings SPIE, vol. 2187*, pages 241--248, San Jose, CA, 1994.

[26] M. Bierling. Displacement estimation by hierarchical blockmatching. In *SPIE Vol. 1001 Visual Communications and Image Processing*, pages 942--951, May 1988.

[27] M. Bolton et al. A complete single-chip implementation of the JPEG image compression standard. In *Proceedings IEEE Custom Integrated Circuits Conference*, pages 12.2.1--12.2.4, 1991.

[28] M. Bosi et al. ISO/IEC MPEG-2 advanced audio coding. In *AES 101st Convention, Los Angeles, CA.*, preprint 4382, Nov. 8-11 1996.

[29] K. Brandenburg and M. Bosi. Overview of MPEG-Audio: Current and future standards for low bit-rate audio coding. In *AES 99th Convention, NewYork*, preprint 4130, Oct. 6-9 1995.

[30] D. Brinthaupt et al. A programmable audio/video processor for H.320, H.324, and MPEG. In *Proceedings IEEE International Solid-State Circuits Conference*, pages 244--245, 1996.

[31] D. Bursky. MPEG silicon puts quality video on PCs. Electronic Design, Penton Publishing Inc., Oct. 14 1994.

[32] M.D. Carr. New video coding standard for the 1990s. *Electronics and Communication Engineering Journal*, pages 119--124, June 1990.

[33] A. Chandrakasan, A. Burstein, and R.W. Brodersen. A low-power chipset for a portable multimedia I/O terminal. *IEEE Journal of Solid-State Circuits*, 29(12):1415--1428, Dec. 1994.

[34] S-F. Chang and D.G. Messerschmitt. Designing high-throughput VLC decoder, Part I-concurrent VLSI architectures. *IEEE Transactions on Circuits and Systems for Video Technology*, 2(2):187--196, June 1992.

[35] Chi-Fa Chen and K.K. Pang. Hybrid coders with motion compensation. *Multidimensional Systems and Signal Processing*, 3(12):241--266, Dec. 1992.

[36] C.T. Chen and A. Wong. A self-governing rate buffer control strategy for pseudoconstant bit rate video coding. *IEEE Transactions on Image Processing*, 2(1):50--59, Jan. 1993.

[37] L-G. Chen, W.T. Chen, Y.S. Jehng, and T.D. Chiueh. An efficient parallel motion estimation algorithm for digital image processing. *IEEE Transactions on Circuits and Systems for Video Technology*, 1(4):378--385, Dec. 1991.

[38] M-H Chen et al. VLSI implementation of single chip JPEG codec. In *Proceedings International Symp. on VLSI Technology, Systems, and Applications*, pages 189--193, 1993.

[39] W-H. Chen and C.H. Smith. Adaptive coding of monochrome and color images. *IEEE Transactions on Communications*, COM-25(11):1285--1292, Nov. 1977.

[40] X. Chen and A. Luthra. MPEG-2 multiview profile and its application in 3D TV. In *Proceedings SPIE - Multimedia Hardware Architectures, vol. 3021*, pages 212--223, 1997.

[41] L. Chiariglione. Development of multi-industry information technology standards. the MPEG case. In *Proceedings International workshop on HDTV '93*, Oct. 1993.

[42] L. Chiariglione. MPEG and multimedia communications. *IEEE Transactions on Circuits and Systems for Video Technology*, 7(1):5--18, Feb. 1997.

[43] N.I. Cho and S.U. Lee. Fast algorithm and implementation of 2-D discrete cosine transform. *IEEE Transactions on Circuits and Systems*, 38(3):297--305, Mar. 1991.

[44] O. Colavin, A. Artieri, J-F. Naviner, and R. Pacalet. A dedicated circuit for real-time motion estimation. In *Proceedings Euro ASIC '91, Paris, France*, pages 96--99, 1991.

[45] S. Cucchi and M. Fratti. A novel architecture for VLSI implementation of the 2-D DCT/IDCT. In *Proceedings ICASSP*, pages V--693--V--696. IEEE, 1992.

[46] R.L. de Queiroz and K.R. Rao. Human visual sensitivity-weighted progressive image transmission using the lapped orthogonal transform. *Journal of Electronic Imaging*, 1(3):328--338, July 1992.

[47] Y.F. Dehery, M. Lever, and P. Urcun. A MUSICAM source codec for digital audio broadcasting and storage. In *Proceedings IEEE International Conference on Acoustics, Speech, and Signal Processing (ICASSP)*, pages 3605--3608. IEEE, 1991.

[48] T. Demura et al. A single-chip MPEG2 video decoder LSI. In *Proceedings IEEE International Solid-State Circuits Conference*, pages 72--73, 1994.

[49] R. Dianysian and R.L. Baker. A VLSI chip set for real time vector quantization of image sequences. In *Proceedings IEEE International Symposium on Circuits and Systems (ISCAS)*, pages 221--224, 1987.

[50] W. Ding and B. Liu. Rate-quantization modelling for rate control of MPEG video coding and recording. In *Proceedings SPIE Vol. 2419*, pages 139--150, Feb. 1995.

[51] T. Dorcey. CU-SeeMe desktop videoconferencing software. *ConneXions*, 9(3), March 1995.

[52] S. Eckart. High performance software MPEG video player for PCs. In *Proceedings SPIE, vol. 2419*, pages 446--454, San Jose, CA, 1995.

[53] S. Eckart and C. Fogg. ISO/IEC MPEG-2 software video codec. In *Proceedings SPIE*, volume 2419, pages 100--109, Feb. 1995.

[54] D.F. Elliot and K.R. Rao. *Fast Transforms - Algorithms, Analyses, Applications*. Academic Press, New York, 1982.

[55] T. Enomoto and M. Yamashina. Video signal processor (VSP) ULSIs for video data coding. In *Proceedings VLSI Technology, Systems, and Applications*, pages 184--188, 1993.

[56] E. Feig and S. Winograd. Fast algorithms for the discrete cosine transform. *IEEE Transactions on Signal Processing*, 40(9):2174--2193, Sept. 1992.

[57] R.D. Fellman, R.T. Kaneshiro, and K. Konstantinides. Design and evaluation of an architecture for a digital signal processor for instrumentation applications. *IEEE Transactions on ASSP*, 38(3):537--546, March 1990.

[58] J. Feng, K-T Lo, H. Mehrpour, and A.E. Karbowiak. Adaptive block matching motion estimation algorithm using bit-plane matching. In *IEEE Proc. ICIP-95, Washington, DC.*, pages 496--499, Oct. 1995.

[59] J.A. Fisher. Replacing hardware that thinks (especially about parallelism) with a very smart compiler. In *Intern. specialist seminar on the design and application of parallel digital processors (Conf. Publ. No. 298)*, pages 153--159. IEEE, April 1988.

[60] P. Foley. The Mpact media processor redefines the multimedia PC. In *Proceedings of COMPCON '96*, pages 311--318. IEEE, 1996.

[61] E.D. Frimout, J. Biemond, and R.L. Lagendijk. Forward rate control for MPEG recording. In *SPIE Visual Commun. Image Processing*, pages 184--194, May 1993.

[62] H. Fujiwara et al. An all-ASIC implementation of a low bit-rate video codec. *IEEE Transactions on Circuits and Systems for Video Technology*, 2(2):123--134, June 1992.

[63] ITU-T Recommendation G.722. 7 kHz audio-coding within 64 kbit/s.

[64] ITU-T Recommendation G.723.1. Dual rate speech coder for multimedia communications transmitting at 5.3 and 6.3 kbit/s.

[65] CCITT Recommendation G.726. 40, 32, 24, 16 kbit/s adaptive differential pulse code modulation (ADPCM), 1990.

[66] CCITT Recommendation G.728. Coding of speech at 16 kbit/s using low-delay code excited linear prediction, Sept. 1992.

[67] D. Galbi et al. An MPEG-1 audio/video decoder with run-length compressed antialiased video overlays. In *Proceedings IEEE International Solid-State Circuits Conference*, pages 286--287, Feb. 1995.

[68] D. Le Gall. MPEG: A video compression standard for multimedia applications. *Communications of the ACM*, 34(4), April 1991.

[69] R. Gallager and D.V. Voorhis. Optimal source codes for geometrically distributed integer alphabets. *IEEE Trans. on Information Theory*, IT-21:228--230, March 1975.

[70] R.G. Gallager. *Information Theory and Reliable Communication*. John Wiley and Sons, NewYork, 1968.

[71] R.G. Gallager. Variations on a theme by Huffman. *IEEE Transactions on Information Theory*, 24(6):668--674, Nov. 1978.

[72] W. Gass. Architecture trends in MPEG decoders for set-top box. In *Proceedings IS&T/SPIE Symp. on Electronic Imaging, Vol. 3021*, pages 162--169, Feb. 1997.

[73] A. Gersho and R.M. Gray. *Vector Quantization and Signal Compression*. Kluwer Academic Publishers, Boston, 1992.

[74] M. Ghanbari. The cross-search algorithm for motion-estimation. *IEEE Transactions on Communications*, 38(7):950--953, July 1990.

[75] H. Gharavi and M. Mills. Blockmatching motion estimation algorithms - new results. *IEEE Transactions on Circuits and Systems*, 37(5):649--651, May 1990.

[76] B. Girod. The efficiency of motion-compensating prediction for hybrid coding of video sequences. *IEEE Journal on Selected Articles in Communications*, SAC-5(7):1140--1154, Aug. 1987.

[77] B. Girod. Motion-compensating prediction with fractional-pel accuracy. *IEEE Transactions on Communications*, 41(4):604--612, Apr. 1993.

[78] B. Girod, E. Steinbach, and N. Farber. Comparison of the H.263 and H.261 video compression standards. In *First International Symposium on Photonics Technologies and Systems for Voice, Video, and Data Communication*, pages 1--19, Oct. 1995.

[79] S.W. Golomb. Run-length encodings. *IEEE Trans. on Information Theory*, IT-12:399--401, July 1966.

[80] J. Golston. Single-chip H.324 videoconferencing. *IEEE Micro*, 16(4):21--33, Aug. 1996.

[81] C.A. Gonzales and E. Viscito. Motion video adaptive quantization in the transform domain. *IEEE Transactions on Circuits and Systems for Video Technology*, 1(4):374--378, Dec. 1991.

[82] B.M. Gordon and T.H.Y. Meng. A low power subband video decoder architecture. In *Proceedings IEEE International Conference on Acoustics, Speech, and Signal Processing (ICASSP)*, pages II--409--412, 1994.

[83] K. Guttag, R.J. Gove, and J.R. Van Aken. A single-chip multiprocessor for multimedia: the MVP. *IEEE Computer Graphics and Applications*, 12(6):53--64, November 1992.

[84] ITU-T Recommendation H.261. Line transmission of non-telephone signals. Video codec for audiovisual services at p x 64 kbits, March 1993.

[85] ITU-T Recommendation H.263. Video coding for low bitrate communication, 11/95.

[86] ITU-T Recommendation H.320. Line transmission of non-telephone signals. Narrow-band visual telephone systems and terminal equipment, 03/93.

[87] ITU-T Recommendation H.324. Terminal for low bitrate multimedia communication, 11/95.

[88] E. Hamilton. JPEG file interchange format, Version 1.02. C-Cube Microsystems, Literature Department, Sept. 1992.

[89] M.C. Hans and V. Bhaskaran. A fast integer-based, CPU scalable, MPEG-I Layer-2 decoder. In *AES 101st Convention, Los Angeles*, preprint 4359, November 8-11 1996.

[90] M. Harrand et al. A single chip videophone video encoder/decoder. In *Proceedings IEEE International Solid-State Circuits Conference*, pages 292--293, Feb. 1995.

[91] R. Hashemian. Design and hardware implementation of a memory efficient Huffman decoding. *IEEE Transactions on Consumer Electronics*, 40(3):345--352, Aug. 1994.

[92] B. Haskell, A. Puri, and A. Netravali. *Digital Video Compression Standard: An Introduction to MPEG2*. Chapman and Hall, 1996.

[93] J. Hennessy, N. Jouppi, F. Baskett, and J. Gill. MIPS: A VLSI processor architecture. In *Proceedings CMU Conference on VLSI Systems and Computations*, pages 337--346, Pittsburgh, PA, 1981.

[94] E. Hoffert et al. QuickTime: an extensible standard for digital multimedia. In *Digest of Papers COMPCON Spring 1992*, pages 15--20, Feb. 1992.

[95] E. Holmann, A. Yamada, T. Yoshida, and S. Uramoto. Real-time MPEG-2 software decoding with a dual-issue RISC processor. In W. Burleson, K. Konstantinides, and T. Meng, editors, *VLSI Signal Processing, IX*, pages 105--114. IEEE Press, 1996.

[96] R. Hopkins. Digital terrestrial HDTV for North America: the Grand Alliance HDTV system. *IEEE Transactions on Consumer Electronics*, 40(3):185--198, Aug. 1994.

[97] C. Horne, T. Naveen, A. Tabatabai, R. Eifrig, and A. Luthra. Study of the characteristics of the MPEG2 4:2:2 profile-Application of MPEG2 in studio environment. *IEEE Transactions on Circuits and Systems for Video Technology*, 6(3):251--272, June 1996.

[98] P.G. Howard and J.S. Vitter. Arithmetic coding for data compression. *Proceedings of the IEEE*, 82(6):857--865, June 1994.

[99] C-H. Hsieh and T-P. Lin. VLSI architecture for block-matching motion estimation algorithm. *IEEE Transactions on Circuits and Systems for Video Technology*, 2(2):169--175, June 1992.

[100] C-T. Hsieh and S.P. Kim. A concurrent memory-efficient VLC decoder for MPEG applications. *IEEE Transactions on Consumer Electronics*, 42(3):439--445, Aug. 1996.

[101] D.A. Huffman. A method for the construction of minimum-redundancy codes. *Proceedings Institute of Electrical and Radio Engineers*, 40(9):1098--1101, Sept. 1952.

[102] R. Hunter and A.H. Robinson. International digital facsimile coding standards. *Proceedings of the IEEE*, 68(7):854--867, Jul. 1980.

[103] G. Hutson, P. Shepherd, and J. Brice. *Colour Television. System Principles, Engineering Practice, and Applied Technology*. McGraw-Hill, London, second edition, 1990.

[104] T. Inoue et al. A 300 MHz 16b BiCMOS video signal processor. In *Proceedings IEEE International Solid-State Circuits Conference*, pages 36--37, 1993.

[105] International Telegraph and Telephone Consultative Committee. Standardization of Group 3 facsimile apparatus for document transmission, Recommendation T.4. ISO, 1980.

[106] International Telegraph and Telephone Consultative Committee. Facsimile coding schemes and coding control functions for Group 4 facsimile apparatus, Recommendation T.6. ISO, 1984.

[107] K. Ishihara et al. A half-pel precision MPEG2 motion-estimation processor with concurrent three-vector search. In *Proceedings IEEE International Solid-State Circuits Conference*, pages 288--289, Feb. 1995.

[108] ISO/IEC JTC1 CD 10918. Digital compression and coding of continuous-tone still images - part 1: Requirements and guidelines. ISO, 1993.

[109] ISO/IEC JTC1 CD 11172. Coding of moving pictures and associated audio for digital storage media up to 1.5 Mbits/s. International Organization for Standardization (ISO), 1992.

[110] ISO/IEC JTC1 CD 11544. Coded representation of picture and audio information - progressive bi-level image compression. ISO, 1993.

[111] ISO/IEC JTC1 CD 13818. Generic coding of moving pictures and associated audio. International Organization for Standardization (ISO), 1994.

[112] ISO/IEC JTC1 CD 13818-2 Amendment 2. Generic coding of moving pictures and associated audio: video. International Organization for Standardization (ISO), Jan. 1996.

[113] ISO/IEC JTC1 CD 13818-2 Amendment 3. Generic coding of moving pictures and associated audio: video. International Organization for Standardization (ISO), Sept. 1996.

[114] ISO/IEC JTC1/SC29/WG11/N0400. Test model 5: version 2. ISO, 1993.

[115] ITU (Q.5/8 Rapporteur). Amendments to ITU-T Rec. T.30 for enabling continuous-tone colour and grayscale mode for Group 3, COM 8-43-E, June 1994.

[116] ITU (Q.5/8 Rapporteur). Amendments to ITU-T Rec. T.4 to enable continuous-tone colour and grayscale mode for Group 3, COM 8-44-E, June 1994.

[117] A.K. Jain et al. $1.38cm^2$ 550 MHz microprocessor with multimedia extensions. In *Proceedings IEEE International Solid-State Circuits Conference*, pages 174--175, Feb. 1997.

[118] J.R. Jain and A.K. Jain. Displacement measurement and its application in interframe coding. *IEEE Transactions on Comunications*, 29(12):1799--1808, Dec. 1981.

[119] Y. Jang et al. A 0.8μ 100-MHz 2-D DCT core processor. *IEEE Transactions on Consumer Electronics*, 40(3):703--710, August 1994.

[120] N. Jayant. Signal compression: Technology targets and research directions. *IEEE Journal on Selected Articles in Communications*, 10(5):796--818, June 1992.

[121] N.S. Jayant and P. Noll. *Digital Coding of Waveforms, Principles and Applications to Speech and Video*. Prentice-Hall, Englewood Cliffs, New Jersey, 1984.

[122] Y-S. Jehng, L-G. Chen, and T-D. Chiueh. A motion estimator for low bit-rate codec. *IEEE Transactions on Consumer Electronics*, 38(2):60--69, May 1992.

[123] J. Jeong and W. Ahn. Subpixel accuracy motion estimation using a model for motion-compensated errors. In *Picture coding symposium, Lausanne, Switzerland*, pages 13.4--13.6, 1993.

[124] R.A. Vander Kam, P.W. Wong, and R.M. Gray. JPEG compression for a grayscale printing pipeline. In *Proceedings SPIE Vol. 2418*, pages 229--240, Feb. 1995.

[125] F.A. Kamangar and K.R. Rao. Fast algorithms for the 2-D discrete cosine transform. *IEEE Transactions On Computers*, C-31(9):899--906, Sept. 1982.

[126] S. Kappagantula and K.R. Rao. Motion compensated interframe image prediction. *IEEE Transactions on Communications*, 33(9):1011--1015, Sept. 1985.

[127] G. Keesman, I. Shah, and R.K. Gunnewiek. Bit-rate control for MPEG encoders. *Signal Processing: Image Communication*, 6(1):545--560, March 1995.

[128] J.C. Kieffer. Strong converses in source coding relative to a fidelity criterion. *IEEE Transactions on Information Theory*, IT-37(2):697--707, 1991.

[129] D. Kim et al. A real-time MPEG encoder using a programmable processor. *IEEE Transactions on Consumer Electronics*, 40(2):161--170, May 1994.

[130] P. Knebel et al. HP's PA7100LC: A low-cost superscalar PA-RISC processor. In *Digest of Papers COMPCON Spring 1993*, pages 441--447. IEEE, Feb. 1993.

[131] U. Ko et al. DSP core with parallel module testability. In *IEEE ASIC '89: 2nd Annual Seminar*, pages P6--3.1 -- P6--3.5, 1989.

[132] L. Kohn et al. The visual instruction set (VIS) in UltraSPARC. In *Digest of Papers COMPCON Spring 1995*, pages 462--469. IEEE, March 1995.

[133] R.K. Kolagotla, S-S Yu, and J.F. JáJá. VLSI implementation of a tree searched vector quantizer. *IEEE Transactions on Signal Processing*, 41(2):901--905, Feb. 1993.

[134] T. Komarek and P. Pirsch. Array architectures for block matching algorithms. *IEEE Transactions on Circuits and Systems*, 36(10):1301--1308, October 1989.

[135] K. Konstantinides. Fast subband filtering in MPEG audio coding. *IEEE Signal Processing Letters*, 1(2):26--28, Feb. 1994.

[136] Chung-Wei Ku, L. Chen, C. Chen, J. Jiu, and C. Huang. Investigation of a visual telephone prototyping on personal computers. *IEEE Transactions on Consumer Electronics*, 42(3):750--758, Aug. 1996.

[137] Jan L.P. De Lameillieure. A heuristic algorithm for the construction of a code with limited word length. *IEEE Transactions on Information Theory*, 33(5):438--443, July 1988.

[138] Jan L.P. De Lameillieure and I. Bruyland. Comment on algorithm for construction of variable length code with limited maximum word length. *IEEE Transactions on Communications*, 34(12):893--894, Dec. 1986.

[139] L.L. Larmore and D.S. Hirschberg. A fast algorithm for optimal length-limited Huffman codes. *Journal of the Association for Computing Machinery JACM*, 37(3):464--473, July 1990.

[140] B.G. Lee. A new algorithm to compute the discrete cosine transform. *IEEE Transactions on ASSP*, ASSP-32(6):1243--45, Dec. 1984.

[141] D.T. Lee. JPEG: New enhancements and future prospects. In *Proceedings IS&T 48-th Annual Conference*, pages 58--62, May 1995.

[142] E.A. Lee. Programmable DSP architectures: Part 1. *IEEE ASSP Magazine*, pages 4--19, Oct. 1988.

[143] J.C. Lee, E. Cheval, and J. Gergen. The Motorola 16-bit DSP ASIC core. In *IEEE Proceedings of ICASSP*, pages 973--976, 1990.

[144] J.H. Lee, M.K. Doh, and C.W. Lee. A VLSI chip for motion estimation of HDTV signals. *IEEE Transactions on Consumer Electronics*, 40(2):154--160, May 1994.

[145] L-W. Lee, J.F. Wang, J.Y. Lee, and J.D. Shie. Dynamic search window adjustment and interlaced search for block-matching algorithm. *IEEE Transactions on Circuits and Systems*, 3(1):85--87, Feb. 1993.

[146] R.B. Lee. Realtime MPEG video via software decompression on a PA-RISC processor. In *Digest of Papers COMPCON Spring 1995*, pages 186--192, March 1995.

[147] R.B. Lee. Subword parallelism with MAX-2. *IEEE Micro*, 16(4):51--59, Aug. 1996.

[148] S. Lee, J.M. Kim, and S-I. Chae. New motion estimation using low-resolution quantization for MPEG2 video encoding. In W. Burleson, K. Konstantinides, and T. Meng, editors, *VLSI Signal Processing, IX*, pages 428--437. IEEE Press, 1996.

[149] W. Lee and Y. Kim. MPEG-2 video decoding on programmable processors: computational and architectural requirements. In *Proceedings SPIE, vol. CR60*, pages 265--287, 1995.

[150] S-M. Lei and M-T. Sun. An entropy coding system for digital HDTV applications. *IEEE Transactions on Circuits and Systems for Video Technology*, 1(1):147--155, March 1991.

[151] M. Levy and A. Coyle. EDN's DSP-chip directory. *EDN*, 41(5):40--103, March 1 1996.

[152] A.S. Lewis and G. Knowles. VLSI architecture for 2-D Daubechies wavelet transform without multipliers. *Electronic Letters*, 27(2):171--173, Jan. 1991.

[153] A.S. Lewis and G. Knowles. Image compression using the 2-D wavelet transform. *IEEE Transactions on Image Processing*, 1(2):244--250, April 1992.

[154] D.B. Lidsky and J.M. Rabaey. Low power design of memory intensive functions. case study: vector quantization. In J. Rabaey, P.M. Chau, and J. Eldon, editors, *VLSI Signal Processing VII*, pages 378--387. IEEE, New York, 1994.

[155] H-D. Lin, A. Anesko, and B. Petryna. A 14-Gops programmable motion estimator for H.26X video coding. *IEEE Journal of Solid-State Circuits*, 31(11):1742--1750, Nov. 1996.

[156] E. Linzer and E. Feig. New DCT and scaled-DCT algorithms for fused multiply/add architectures. In *Proceedings IEEE International Conference on Acoustics, Speech, and Signal Processing (ICASSP)*, pages 80--91, 1991.

[157] B. Liu and A. Zaccarin. New fast algorithms for the estimation of block motion vectors. *IEEE Transactions on Circuits and Systems*, 3(2):148--157, Apr. 1993.

[158] H. Lohscheller. A subjectively adapted image communication system. *IEEE Transactions on Communications*, COM-32(12):1316--1322, Dec. 1984.

[159] H. Lohscheller and U. Franke. Colour picture coding - algorithm optimization and technical realization. *Frequenz*, 41:291--299, 11/12 1987.

[160] C.H. Lu. Comment on algorithm for construction of variable length code with limited maximum word length. *IEEE Transactions on Communications*, 34(2):373--375, Mar. 1988.

[161] M-I. Lu and C-F. Chen. A Huffman-type code generator with order-N complexity. *IEEE Transactions on Acoustics, Speech, and Signal Processing*, 38(9):1619--1626, Sept. 1990.

[162] D-S. Ma, J-F. Yang, and J-Y. Lee. Programmable and parallel variable-length decoder for video systems. *IEEE Transactions on Consumer Electronics*, 39(3):448--454, Aug. 1993.

[163] A.G. MacInnis. MPEG systems committee draft, ISO/IEC JTC1/SC2/WG11. In *Digest of Papers COMPCON, Spring '91*, pages 338--339. IEEE, 1991.

[164] A. Madisetti, R. Jain, R.L. Baker, and R. Dianysian. Architecture and integrated circuits for real time vector quantization of images. In *Proceedings IEEE International Conference on Acoustics, Speech, and Signal Processing (ICASSP)*, pages V--677--680, 1992.

[165] X. Maitre. 7 kHz audio coding with 64 kbit/s. *IEEE Journal on Selected Areas in Communications*, 6(2):283--298, Feb. 1988.

[166] S.G. Mallat. A theory for multiresolution signal decomposition: the wavelet representation. *IEEE Transactions on Pattern Analysis and Machine Intelligence*, 11(7):674--693, July 1989.

[167] G. Maturi. Single chip MPEG audio decoder. *IEEE Transactions on Consumer Electronics*, 38(3):348--356, Aug. 1992.

[168] J.C. Michalina. Application specific DSP at SGS-Thomson. In *IEEE Proceedings of International Symposium on Circuits and Systems (IS-CAS)*, pages 627--630, 1989.

[169] J.L Mitchell, D. Le Gall, and C. Fogg. *MPEG Video Compression Standard*. Chapman and Hall, 1996.

[170] J.L. Mitchell and W.B. Pennebaker. Optimal hardware and software arithmetic coding procedures for the Q-coder. *IBM Journal of Research and Development*, 32(6):727--736, Nov. 1988.

[171] M. M. Mizuki, U. Y. Desai, I. Masaki, and A. Chandrakasan. Block matching architecture with reduced power consumption and silicon area requirements. In *Proceedings IEEE International Conference on Acoustics, Speech, and Signal Processing (ICASSP)*, pages 3248--3251, 1996.

[172] M. Mizuno et al. A 1.5 W single-chip MPEG2 MP@ML encoder with low-power motion estimation and clocking. In *Proceedings IEEE International Solid-State Circuits Conference*, pages 256--257, Feb. 1997.

[173] J.W. Modestino and D.G. Daut. Combined source channel coding of images. *IEEE Transactions on Communications*, COM-27:1644--1659, Nov. 1979.

[174] J.W. Modestino, D.G. Daut, and A.L. Vickers. Combined source channel coding of images using the block cosine transform. *IEEE Transactions on Communications*, COM-29(9):1261--1274, Sept. 1981.

[175] G. Morrison. Video coding standards for multimedia: JPEG, H.261, MPEG. In *IEE Colloquium on Technology Support of Multimedia (Digest No. 088)*, pages 2.1--2.4, April 1992.

[176] A. Mukherjee, N. Ranganathan, J.W. Flieder, and T. Acharya. MAR-VLE: a VLSI chip for data compression using tree-based codes. *IEEE Transactions on Very Large Scale Integration (VLSI) Systems*, 1(2):203--214, June 1993.

[177] H. Murakami, S. Matsumoto, and H. Yamamoto. Algorithm for construction of variable length code with limited maximum word length. *IEEE Transactions on Communications*, 32(10):1157--1159, Oct. 1984.

[178] T. Murakami et al. A DSP architecture for 64Kbs motion video codec. In *Proceedings International Conference on Circuits and Systems (ISCAS)*, pages 227--230. IEEE, 1988.

[179] H.G. Musmann. The ISO audio coding standard. In *Proceedings GLOBECOM '90*, pages 511--517. IEEE, 1990.

[180] ISO/IEC JTC1/SC29/WG11 N1430. MPEG-2 DIS 13818-7 (MPEG-2 advanced audio coding, AAC). International Organization for Standardization (ISO), Nov. 1996.

[181] K. Nadehara et al. Real-time software MPEG-1 video decoder design for low-cost, low-power applications. In W. Burleson, K. Konstantinides, and T. Meng, editors, *VLSI Signal Processing, IX*, pages 438--447. IEEE Press, 1996.

[182] M. Nakagawa et al. DCT-based still image compression ICs with bit-rate control. *IEEE Transactions on Consumer Electronics*, 38(3):711--717, Aug. 1992.

[183] B. Natarajan, V. Bhaskaran, and K. Konstantinides. Low-complexity algorithm and architecture for block-based motion estimation via one-bit transforms. In *Proceedings IEEE International Conference on*

Acoustics, Speech, and Signal Processing (ICASSP), pages 3244--3247, 1996.

[184] J.B.O. Neal, Jr. and T.R. Natarajan. Coding isotropic images. *IEEE Transactions on Information Theory*, IT-23:697--707, 1977.

[185] T. Nguyen et al. Multimedia signal processor (MSP) summary. In *Hotchips Symposium, Stanford, California*, 18-20 Aug. 1996.

[186] K. Ogawa et al. A single chip compression/decompression LSI based on JPEG. *IEEE Transactions on Consumer Electronics*, 38(3):703--710, Aug. 1992.

[187] E. Ogura et al. A cost efficient motion estimation processor LSI using a simple and efficient algorithm. *IEEE Transactions on Consumer Electronics*, 41(3):690--698, Aug. 1995.

[188] N. Ohta. *Packet Video - Modeling and Signal Processing*. Artech House, Norwood, Massachussetts, 1994.

[189] S. Okuba. Reference model methodology - a tool for the collaborative creation of video coding standards. *Proceedings of the IEEE*, 83(2):139--150, Feb. 1995.

[190] F. Ono, S. Kino, M. Yoshida, and T. Kimura. Bi-level image coding with MELCODE - comparison of block type code and arithmetic type code. In *Proc. of IEEE GLOBECOM '89*, pages 255--260, November 1989.

[191] Y. Ooi, A. Taniguchi, and S. Demura. A 162 Mbit/s variable length decoding circuit using an adaptive tree search technique. In *IEEE 1994 Custom Integrated Circuits Conference*, pages 6.5.1--6.5.4, 1994.

[192] M.T. Orchard and G.J. Sullivan. Overlapped block motion compensation: an estimation-theoretic approach. *IEEE Transactions on Image Processing*, 3(5):693--699, Sept. 1994.

[193] D. Pan. An overview of the MPEG/Audio compression algorithm. In *Proceedings SPIE, Vol. 2187, San Jose, CA*, pages 260--273. SPIE, Feb. 1994.

[194] D. Pan. Digital audio compression. *Digital Technical Journal*, 5(2):28--40, Spring 1993.

[195] D. Pan. A tutorial on MPEG/Audio compression. *IEEE Multimedia*, 2(2):60--74, Summer 1995.

[196] K.K. Pang and T.K. Tan. Optimum loop filter in hybrid coders. *IEEE Circuits and Systems for Video Technology*, 4(2):158, April 1994.

[197] P. Papamichalis. MPEG audio compression: algorithms and implementation. In *DSP95, Intern. Conference on Digital Signal Processing*, pages 72--77 (Vol. 1), Limassol, Cyprus, June 26-28 1995. Univ. of Cyprus.

[198] H. Park and V.K. Prasanna. Area efficient VLSI architectures for Huffman coding. *IEEE Transactions on Circuits and Systems-II: Analog and Digital Signal Processing*, 40(9):568--575, Sept. 1993.

[199] S-H Park et al. Fast algorithm on MPEG/Audio subband filtering. In *AES 99th Convention, NewYork*, preprint 4090, Oct. 6-9 1995.

[200] P. Patel and D. Douglass. Architectural features of the i860 - microprocessor RISC core and on-chip caches. In *IEEE Proceedings of ICCD*, pages 385--390, 1989.

[201] D.A. Patterson and J.L. Hennessy. *Computer Architecture: A Quantitative Approach*. Morgan Kaufmann, San Mateo, California, 1990.

[202] A. Peled and B. Liu. A new hardware realization of digital filters. *IEEE Transactions on ASSP*, ASSP-22(6):456--462, December 1974.

[203] A. Peleg and U. Weise. MMX technology extension to the Intel architecture. *IEEE Micro*, 16(4):42--50, Aug. 1996.

[204] W.B. Pennebaker and J.L. Mitchell. *JPEG Still Image Data Compression Standard*. Van Nostrand Reinhold, New York, 1993.

[205] H.A. Peterson, H. Peng, J.H. Morgan, and W.B. Pennebaker. Quantization of color image components in the DCT domain. In *SPIE Vol. 1453 Human Vision, Visual Processing, and Digital Display II*, pages 210--222, Feb. 1991.

[206] P. Pirsch, N. Demassieux, and W. Gehrke. VLSI architectures for video compression - a survey. *Proceedings of the IEEE*, 83(2):220--246, Feb. 1995.

[207] A. Pirson et al. A programmable motion estimation processor for full search block matching. In *Proceedings IEEE International Conference on Acoustics, Speech, and Signal Processing (ICASSP)*, pages 3283--3286, 1995.

[208] S. Purcell. Mpact 2 media processor, balanced 2X performance. In *Proceedings IS&T/SPIE Symp. on Electronic Imaging, Vol. 3021*, pages 102--108, Feb. 1997.

[209] S.C. Purcell. The C-Cube CL550 JPEG image compression processor. In *Digest of Papers COMCON*, pages 318--323, 1991.

[210] A. Puri and R. Aravind. Motion-compensated video coding with adaptive perceptual quantization. *IEEE Transactions on Circuits and Systems for Video Technology*, 1(4):351--361, Dec. 1991.

[211] A. Puri, H.M. Hang, and D.L. Schilling. An efficient block-matching algorithm for motion compensated coding. In *Proceedings IEEE International Conference on Acoustics, Speech, and Signal Processing (ICASSP)*, pages 25.4.1--25.4.4, 1987.

[212] F. Pereira R. Koenen and L. Chiariglione. MPEG-4: Context and objectives. *EURASIP: Image Communications - special issue on MPEG-4*, 9(1):2--18, Feb. 1997.

[213] G. Radin. The 801 minicomputer. In *Proceedings Symp. on Architectural Support for Prog. Lang. and Oper. Systems*, pages 39--47, Palo Alto, CA, 1982.

[214] R. Rajagopalan, L. Liu, W. Ding, and E. Feig. Fast motion estimation algorithms for overlapped block motion compensation. In *Proceedings SPIE, vol. 2668*, San Jose, CA, 1996.

[215] K.R. Rao and J.J. Hwang. *Techniques and Standards for Image, Video, and Audio Coding*. Prentice Hall, 1996.

[216] K.R. Rao and P. Yip. *Discrete Cosine Transform - Algorithms, Advantages, Applications*. Academic Press, San Diego, California, 1990.

[217] S. Rathnam and G. Slavenburg. An architectural overview of the programmable multimedia processor, TM-1. In *Proceedings of COMPCON '96*, pages 319--326. IEEE, 1996.

[218] V. Ratnakar, E. Feig, E. Viscito, and S. Kalluri. Runlength encoding of quantized DCT coefficients. In *Proceedings SPIE*, volume 2419, pages 398--406, Feb. 1995.

[219] V. Ratnakar and M. Livny. RD-OPT: an efficient algorithm for optimizing DCT quantization tables. In *Data Compression Conference, Snowbird, Utah*, pages 333--339, March 1995.

[220] A. Razavi et al. VLSI implementation of an image compression algorithm with a new bit rate control capability. In *Proceedings IEEE International Conference on Acoustics, Speech, and Signal Processing (ICASSP)*, pages V:669--672, 1992.

[221] R.F. Rice. Some practical universal noiseless coding techniques. Technical Report JPL-79-22, Jet Propulsion Laboratory, Pasadena, CA, March 1979.

[222] O. Rioul and M. Vetterli. Wavelets and signal processing. *IEEE Signal Processing Magazine*, 8:14--38, Oct. 1991.

[223] P.A. Ruetz et al. A high-performance full-motion video compression chip set. *IEEE Transactions on Circuits and Systems for Video Technology*, 2(2):111--121, June 1992.

[224] C.P. Sandbank, editor. *Digital Television*. John Wiley and Sons, Chichester, England, 1990.

[225] K. Sayood. *Introduction to data compression*. Morgan Kaufmann, San Mateo, California, 1995.

[226] R.A. Schaphorst. Status of H.324 videoconferencing standard for the public switched telephone network and mobile radio. *Journal of Optical Engineering*, 35(1):109--112, Jan. 1996.

[227] V. Seferidis and M. Ghanbari. Generalized block matching motion estimation. In *Proceedings of SPIE: Visual communications and Image Processing*, pages 80--91, 1992.

[228] C.H. Sequin and D.A. Patterson. Design and implementation of RISC-I. Technical Report UCB/CSD 82/106, Computer Science Division (EECS), Univ. of California, Berkeley, October 1982.

[229] A. Shacham et al. Architectural considerations for SF-core based microprocessor. In *IEEE Proceedings of ICCD*, pages 21--24, 1991.

[230] Y. Shishikui. A study on modeling of the motion compensation prediction error signal. *IEICE Transactions on Communications, Japan,* E75-B(5):368--376, May 1992.

[231] S. Shlien. Guide to MPEG-1 audio standard. *IEEE Transactions on Broadcasting,* 40(4):206--218, Dec. 1994.

[232] T. Sikora. The MPEG-4 video standard verification model. *IEEE Transactions on Circuits and Systems for Video Technology,* 7(1):19--31, Feb. 1997.

[233] A.N. Skodras. Fast discrete cosine transform pruning. *IEEE Trans. on Signal Processing,* 42(7):1833--1837, July 1994.

[234] A.N. Skodras and C.A. Christopoulos. Split-radix fast cosine transform algorithm. *Intern. Journal of Electronics,* 74(4):513--522, 1993.

[235] M. Slater, editor. *A Guide to RISC Microprocessors.* Academic Press, San Diego, California, 1992.

[236] Gerrit Slavenburg, Selliah Rathnam, and Henk Dijkstra. The Trimedia TM-1 PCI VLIW media processor. In *Hotchips Symposium, Stanford, California,* 18-20 Aug. 1996.

[237] M.M. Stojancic and C. Ngai. Architecture and VLSI implementation of the MPEG-2:MP,ML video decoding process. *SMPTE Journal,* pages 62--72, Feb. 1995.

[238] K. Suguri et al. A real-time motion estimation and compensation LSI with wide search range for MPEG2 video encoding. *IEEE Journal of Solid-State Circuits,* 31(11):1733--1741, Nov. 1996.

[239] G.J. Sullivan and M.T. Orchard. Methods of reduced-complexity overlapped block motion compensation. In *IEEE Proc. ICIP-94, Austin, Texas,* pages 957--961, Oct. 1994.

[240] M.T. Sun, L. Wu, and M.L. Liou. A concurrent architecture for VLSI implementation of discrete cosine transform. *IEEE Transactions on Circuits and Systems,* CAS-34(8):992--994, August 1987.

[241] I. Tamitani et al. An encoder/decoder chip set for the MPEG video standard. In *Proceedings IEEE International Conference on Acoustics, Speech, and Signal Processing (ICASSP),* pages V--661--664, 1992.

[242] Texas Instruments. *TMS320 DSP, Product Overview*, 1994.

[243] C.C. Todd et al. AC-3: Flexible perceptual coding for audio transmission and storage. In *Proceedings AES 96th Convention, Amsterdam*, preprint 3796, Feb. 26-March 1 1994.

[244] S. Todd, G.G. Langdon, Jr. , and J. Rissanen. Parameter reduction and context selection for compression of gray-scale images. *IBM J. Research and Development*, 29(2):188--193, March 1985.

[245] M. Toyokura et al. A video DSP with macroblock-level-pipeline and a SIMD type vector-pipeline architecture for MPEG2 codec. In *Proceedings IEEE International Solid-State Circuits Conference*, pages 74--75, 1994.

[246] M. Tremblay, D. Greenley, and K. Normoyle. The design of the microarchitecture of UltraSPARC-I. *IEEE Proceedings*, 83(12):1653--1663, Dec. 1995.

[247] M. Tremblay, J.M. O'Connor, V. Narayanan, and L. He. VIS speeds new media processing. *IEEE Micro*, 16(4):10--20, Aug. 1996.

[248] T. Turletti. The INRIA videoconferencing system. *ConneXions*, VIII(10), Oct. 1994.

[249] G. Tziritas and C. Labit. *Motion Analysis for Image Sequence Coding*. Elsevier Science b. V., Amsterdam, The Netherlands, 1994.

[250] S. Undy et al. A low-cost graphics and multimedia workstation chip-set. *IEEE Micro*, 14(2):10--22, April 1994.

[251] S. Uramoto et al. A 100-MHz 2-D discrete cosine transform core processor. *IEEE Journal of Solid-State Circuits*, 27(4):492--499, April 1992.

[252] A. van der Werf et al. I.McIC: A single-chip MPEG2 video encoder for storage. In *Proceedings IEEE International Solid-State Circuits Conference*, pages 254--255, Feb. 1997.

[253] S. Vernon. Design and implementation of AC-3 coders. *IEEE Transactions on Consumer Electronics*, 41(3):754--759, Aug. 1995.

[254] D.C. Van Voorhis. Constructing codes with bounded codeword lengths. *IEEE Transactions Information Theory*, 20(2):288--290, Feb. 1974.

[255] G.K. Wallace. The JPEG still picture compression standard. *IEEE Transactions on Consumer Electronics*, 38(1):xviii--xxxiv, Feb. 1992.

[256] B.W.Y. Wei and T.H. Meng. A parallel decoder of programmable Huffman codes. *IEEE Transactions on Circuits and Systems for Video Technology*, 5(2):175--178, April 1995.

[257] M.J. Weinberger, G. Seroussi, and G. Sapiro. LOCO-I: A low complexity, context-based, lossless image compression algorithm. In *Proceedings Data Compression Conference*, pages 140--149. IEEE Press, April 1996.

[258] N.D. Wells. MPEG: Status of digital coding standardization. In *IEE Colloquium on Digital satellite technology and electronic newsgathering, Digest No. 006*, 1993.

[259] P.T. Whitcomb and H.M. Ahmed. A VLSI architecture for the CCITT G.722 codec. In *Proceedings IEEE GLOBECOM '89*, pages 1262--1266. IEEE, 1989.

[260] S.A. White. Applications of distributed arithmetic to digital signal processing: A tutorial review. *IEEE ASSP Magazine*, 6(3):4--19, July 1989.

[261] I.H. Witten, R.M. Neal, and J.G. Cleary. Arithmetic coding for data compression. *Communications of the Association for Computing Machinery ACM*, 30(6):520--540, June 1987.

[262] S. Wolter, D. Birreck, and R. Laur. Classification for 2D-DCTs and a new architecture with distributed arithmetic. In *Proceedings IEEE International Symposium on Circuits and Systems*, pages 2204--2207, June 1991.

[263] J.W. Woods and S.D. O'Neil. Subband coding of images. *IEEE Transactions on Acoustics, Speech, and Signal Processing*, 34(5):1278--1288, Oct. 1986.

[264] K. Xie, L.V. Eycken, and A. Oosterlinck. A new block-based motion estimation algorithm. *Signal Processing: Image Communication*, 4:507--517, May 1992.

[265] K-M. Yang, H. Fujiwara, T. Sakaguchi, and A. Shimazu. VLSI architecture of a versatile variable length decoding chip for real-time

video codecs. In *Proceedings IEEE Region 10 Conference on Computer and Communication Systems (TENCON '90), Hong Kong*, pages 551--554, Sept. 1990.

[266] K-M. Yang, M-T. Sun, and L. Wu. A family of VLSI designs for the motion compensation block-matching algorithm. *IEEE Transactions on Circuits and Systems*, 36(10):1317--1325, October 1989.

[267] H. Yeo and Y.H. Hu. A novel modular systolic array architecture for full-search block matching motion estimation. *IEEE Transactions on Circuits and Systems for Video Technology*, 5(5):407--416, Oct. 1995.

[268] H. Yeo and Y.H. Hu. A novel architecture and processor-level design based on a new matching criterion for video compression. In W. Burleson, K. Konstantinides, and T. Meng, editors, *VLSI Signal Processing, IX*, pages 448--457. IEEE Press, 1996.

[269] C-G. Zhou et al. MPEG video decoding with the UltraSPARC visual instruction set. In *Digest of Papers COMPCON Spring 1995*, pages 470--475. IEEE, March 1995.

INDEX